Navigation: Science and Technology

Volume 14

This series *Navigation: Science and Technology (NST)* presents new developments and advances in various aspects of navigation - from land navigation, marine navigation, aeronautic navigation to space navigation; and from basic theories, mechanisms, to modern techniques. It publishes monographs, edited volumes, lecture notes and professional books on topics relevant to navigation - quickly, up to date and with a high quality. A special focus of the series is the technologies of the Global Navigation Satellite Systems (GNSSs), as well as the latest progress made in the existing systems (GPS, BDS, Galileo, GLONASS, etc.). To help readers keep abreast of the latest advances in the field, the key topics in NST include but are not limited to:

- Satellite Navigation Signal Systems
- GNSS Navigation Applications
- Position Determination
- Navigational instrument
- Atomic Clock Technique and Time-Frequency System
- X-ray pulsar-based navigation and timing
- Test and Evaluation
- User Terminal Technology
- Navigation in Space
- New theories and technologies of navigation
- Policies and Standards

This book series is indexed in **SCOPUS** and **EI Compendex** databases.

Kegen Yu
Editor

Positioning and Navigation Using Machine Learning Methods

Springer

Editor
Kegen Yu
School of Environment Science and Spatial Informatics
China University of Mining and Technology
Xuzhou, Jiangsu, China

ISSN 2522-0454 ISSN 2522-0462 (electronic)
Navigation: Science and Technology
ISBN 978-981-97-6198-2 ISBN 978-981-97-6199-9 (eBook)
https://doi.org/10.1007/978-981-97-6199-9

This Springer imprint is published by the registered company Springer Nature Singapore Pte Ltd.
The registered company address is: 152 Beach Road, #21-01/04 Gateway East, Singapore 189721, Singapore

Preface

Over the past few decades, global navigation satellite systems (GNSS) have made great advances especially due to governments' strong support and funding in the modernization and updating of the systems. GNSS has been dominantly used for positioning, navigation, and timing (PNT) and associated location-based services (LBS) in outdoor environments. To enable PNT in GNSS-denied environments such as indoor environments and underground space, a range of local positioning systems has been developed, which makes use of different types of signals such as radio signals, acoustic signals, infrared, visible light, and magnetic field. However, it is still a challenge to meet the increasing stricter requirements on PNT in areas such as reliability, accuracy, integrability, and safety to enable better LBS especially in complex environments.

Machine learning has been applied to a variety of fields with remarkable success. It has also been leveraged by researchers and engineers to handle complex and challenging problems in PNT to enhance performance and related LBS. Machine learning can usually provide a better solution to PNT than physics-based approaches in the case where the application/service is demanding, the environment is complex, or both. This book completely focuses on positioning and navigation using machine learning methods. Specifically, five chapters are related to GNSS positioning and navigation, while nine chapters deal with local positioning and navigation.

Specifically, Chap. 1 focuses on the use of machine learning to purify pseudorange measurements for GNSS positioning. The Gradient Boosting Decision Trees method is used to correct pseudorange errors in static scenario, while a random forest (RF)-based pseudorange correction method is used in the dynamic scenario. Chapter 2 deals with the performance enhancement of inertial navigation system (INS) especially during GNSS outage. A deep learning network is proposed to assist the INS. The network extracts the spatial features from the inertial measurement unit signals and track their temporal characteristics to mitigate the measurement errors. Chapter 3 analyzes the demand for integrity monitoring in high-precision positioning, overviews the developments of integrity monitoring in civil aviation applications, discusses the challenges on the calculation procedure of a few parameters for high-precision positioning, and provides preliminary studies on the generalization of the integrity monitoring algorithms. Chapter 4 presents a tropospheric model for real-time precise tropospheric corrections so as to enhance GNSS positioning especially over China. With the aid of machine learning, the model combines numerical weather prediction forecast and GNSS observations to make use of their complementary advantages. Chapter 5 deals with GNSS time series prediction in the presence of color noise. It presents a dual variational modal decomposition long short-term memory (LSTM) model to effectively handle the noise in GNSS time series prediction, where LSTM is a deep learning model. Chapter 6 presents an efficient real-time autonomous solution that enables the unmanned aerial vehicle (UAV) to navigate through a dynamic urban or suburban environment. Particularly, deep supervised learning is exploited to determine the optimal path for UAV to reach the destination, based on a cellular network instead of GNSS. Chapter 7 investigates the impact of a range of factors such as devices, testers, materials, and dates on the stability of the magnetic field. It studies compensation methods suited for different types of magnetic features, which are used to develop learning-based optimization strategies for online localization. Chapter 8 focuses on positioning in the GNSS-denied indoor environments, using portable smart terminal platforms. It presents an integrated positioning system based on near-ultrasonic and inertial navigation technologies. Both theoretical and practical requirements are considered to achieve privacy-secure solutions. Chapter 9 presents a smartphone-based scalable floor identification method, without relying on site survey, and some other assumptions. It consists of ground floor detection, semi-supervised learning-based floor association, and deep learning-based floor identification. The first two are conducted in the training phase, while the last one is done in the identification phase. Chapter 10 presents an infrastructure-independent multi-floor indoor localization method. It utilizes the strong feature extraction capability of the sequence-to-sequence deep learning model to predict real-time step action and extract vertical movement information. By configuring calibration nodes, localization is extended to three-dimensional applications. Chapter 11 presents a Wi-Fi localization algorithm, which transforms the received signal strength indicator data by translation and scaling. The back-propagation neural network is used to construct the ranging model. The initial values of weights and bias of the network are optimized by a genetic algorithm. Sequential quadratic programming is used for position determination. Chapter 12 studies a number of indoor positioning technologies which make use of artificial intelligence, including both traditional machine learning and deep learning methods, and sensor fusion approaches. The aim is at solving the problems of inadequate accuracy and efficiency in indoor positioning based on wireless signals. Chapter 13 focuses on high-precision positioning in mm Wave multi-input multi-output systems. It presents a two-dimensional adaptive grid refinement method to improve the sparse Bayesian learning framework, which considers the joint sparsity of the angle domain and time delay domain. A low-complexity grid evolution algorithm is also introduced to handle the grid mismatch problem. Chapter 14 focuses on non-line-of-sight (NLOS) identification and mitigation for UWB localization. One-dimensional wavelet packet analysis and convolutional neural network are combined for NLOS identification. Two different error models are developed to reduce ranging errors in LOS/NLOS environments, respectively. An improved Chan-Kalman localization algorithm is also presented.

I sincerely thank all the chapter authors for their precious contributions. Thanks also go to my colleagues for useful discussions, who are Yunjia Wang, Guoliang Chen, Zengke Li, Nanchan Zheng, Kefei Zhang, Jingxiang Gao, Qianxin Wang, Qiuzhao Zhang, Dongsheng Zhao, Shubi Zhang, Guobin Chang, Hongdong Fan, Zhongyuan Wang, Zhiping Liu, Hefang Bian, and Yongbo Wang. I would also like to thank the staff members of Springer Nature for their constant support over the entire process from proposal to final publication. Finally, I acknowledge the support by the Priority Academic Program Development of Jiangsu Higher Education Institutions, and the National Natural Science Foundation of China under Grant 42174022.

Xuzhou, China Kegen Yu

Contents

Chapter 1
GNSS Pseudorange Correction Using Machine Learning in Urban Areas

Qi Cheng and Rui Sun

Abstract GNSS signals are easily blocked and reflected by high buildings in urban areas, causing non-line-of-sight (NLOS) and multipath errors. These errors deteriorate the accuracy of position. In this chapter, machine learning based correction method is proposed to mitigate the NLOS/multipath errors in pseudorange. The results of a static and a dynamic experiments demonstrate the effectiveness of the proposed method. In the static experiment, the improvements of positioning accuracy in horizontal were 75.6 and 75.6%, and in 3D were 71.4 and 70.9%, compared with two conventional positioning methods. In the dynamic experiment, the two variations of pseudorange error correction model (PBC and GBC) are used to improve positioning accuracy in urban environments. PBC model achieved positional accuracy improvements in horizontal of 42.9 and 41.1%, and in 3D accuracy of 60.1 and 45.7% compared with comparative methods 1 and 2. GBC achieved improvements in horizontal of 40.8 and 38.9%, and in 3D 63.3 and 50.0%, compared with comparative methods, respectively.

1.1 Introduction

Global navigation satellite systems (GNSS) are widely used to provide positioning services in urban areas. Although GNSS work well in open environments, in complex urban environments, the GNSS signal is easily to be blocked, diffracted, or reflected by high-rise buildings, resulting in non-line-of-sight (NLOS) reception and multipath interference, which reduces the accuracy of the positioning obtained through GNSS. In this chapter, the NLOS and multipath are defined as follows. NLOS means at least one reflected signal is received, and the direct signal is not available. Multipath

Q. Cheng
Department of Land Surveying and Geo-informatics, The Hong Kong Polytechnic University, Hong Kong, China
e-mail: qi001.cheng@connect.polyu.hk

R. Sun (✉)
College of Civil Aviation, Nanjing University of Aeronautics and Astronautics, Nanjing, China
e-mail: rui.sun@nuaa.edu.cn

K. Yu (ed.), *Positioning and Navigation Using Machine Learning Methods*, Navigation: Science and Technology 14, https://doi.org/10.1007/978-981-97-6199-9_1

consists of a direct signal and at least one reflected signal. Both NLOS and multipath can deteriorate the positioning accuracy [1]. Multipath error is about several meters, being positive or negative. NLOS error is always positive, reaching tens, or even hundreds of meters in harsh urban canon scenarios. Unlike other GNSS errors, including orbit and clock errors of satellites, and atmospheric errors, which can be mitigated or eliminated by models or differencing, NLOS/multipath errors are hard to be modelled (at least currently), and cannot be differenced, especially for dynamic applications in challenging environments. This is because NLOS/multipath errors significantly depend on the surrounding environments and change rapidly in different location. Therefore, the mitigation of NLOS/multipath is essential to provide more accurate positioning services for urban applications. In this chapter, machine learning is investigated to correct the NLOS/multipath errors of pseudorange, then to improve the accuracy of positioning of both static and dynamic receiver. Since no extra sensor is needed, the proposed algorithm is suitable for low-cost applications, and easily be accepted by public market.

1.2 Literature Review

Various technologies have been investigated to mitigate the range errors caused by NLOS/multipath effects and to improve the positioning accuracy. These methods mainly include hardware design, vision aided and measurement-based modelling.

Hardware design methods can be divided into two types, antenna design and receiver design. Several kinds of antennas can often be used for NLOS/multipath mitigation. A typical example is the choke-ring antenna. The special designed bottom of this antenna can block reflected signals from the ground. Therefore, it is useful to mitigate the reflected signals of satellites with a low elevation angle [2–6]. However, the reflected signals in urban areas are mainly from high buildings, which disable the ground-based choke-ring antenna. Another example is the dual-polarisation antenna, which captures both right-hand circularly polarised (RHCP) and left-hand circularly polarised (LHCP) signals. In general, direct signals are with RHCP, and reflected signals are with LHCP in a high possibility, depending on the satellite elevation and the reflection surface roughness. By analysing the GNSS data from the dual-polarisation antenna, NLOS/multipath effect can be mitigated [7–11]. However, reflected signals are sometimes also with RHCP, which deteriorates this mitigation, since these reflected signals may be mistakenly considered as line-of-sight (LOS) signals. An antenna array, which is with more than one antenna, may be able to estimate the direction of incoming signals. Therefore, it can be used for multipath mitigation [12–16]. A rotating antenna can also be used for NLOS identification, by analysing the Doppler shift [17]. Nevertheless, the performance and bulky size limit their wide application.

Signal processing design in a receiver has also been investigated by many researchers. Narrow correlator was proposed to mitigate the influence of code multipath [18]. It performs well in mitigating long-delay multipath signals but is less effective in short-delay multipath mitigation. Double delta correlators, which are formed by two early and two late correlators [19], such as high resolution correlator [19], Ashtech's strobe correlator [20], and pulse aperture correlator [21], can be used for multipath mitigation, due to smaller code multipath error envelope. Multipath estimating delay lock loop (MEDLL) can also be used to mitigate both code and phase multipath errors, by estimating the parameters of multipath signals, using the maximum likelihood estimation theory [22]. These signal processing methods are designed to mitigate only certain types of multipath effect. However, no single method exists that accounts for mitigation of both NLOS and multipath effects.

The vision aided methods use vision sensors, or other sensors and techniques playing a similar role as vision, to mitigate NLOS/multipath errors. The basic principle of these methods is to check whether there is an obstacle between the antenna and a satellite. If there is, this satellite is considered as NLOS. Omnidirectional infrared cameras [23], fisheye cameras [24–27] and 3D lidar [28–30] have been investigated as ways to detect NLOS signals in urban environments. The corresponding satellite elevation and azimuth are firstly calculated, seen from the antenna's view. These visual sensors are installed close to the antenna, pointing to the sky. The borderline between the sky and the surrounding obstacles can be determined by these visual sensors. This borderline is compared with the tracking satellites to determine visibility of each satellite. Invisible but tracked satellites are determined as NLOS and excluded to improve the accuracy of position. 3D map has also attracted significant attention to mitigate NLOS signals [31–33]. Instead of using a visual sensor to draw the outline of buildings, the digital information of these buildings can be stored beforehand. The coordinates of the corners of these buildings nearest to the road are stored. An initial position of a receiver is required to determine the satellite visibility, with the help of known satellites' positions. Those signals with an obstacle on the propagation path are excluded. The vision aided methods, however, cannot identify multipath signals.

GNSS raw measurements, including carrier phase, pseudorange, carrier-to-noise-ratio (C/N_0), Doppler shift and their derivatives, have also been investigated for NLOS/multipath mitigation in decades. C/N_0 is the ratio of received carrier power to noise density. It is a measure of the strength of GNSS signal relative to the level of noise and interference in the received signal. A signal with higher C/N_0 often has less carrier and code tracking loop jitter, i.e., less noisy range measurements [34]. Due to the energy loss of GNSS signals after reflection, NLOS tends to have a lower C/N_0. A predetermined C/N_0 threshold was used to detect NLOS [1, 35]. Another term, signal-to-noise ratio (SNR), is sometimes used interchangeably with C/N_0. SNR is the ratio between the power of a signal or carrier and the power of noise in a given specific bandwidth, which is related to the bandwidth of a receiver. In contrast, C/N_0 is independent from the receiver bandwidth. This method is less effective if a reflector has a high reflection coefficient, such as metal or glass, which is common in urban areas. A developed version based on C/N_0 was using C/N_0 time series to identify and exclude NLOS [36], which can partially avoid the C/N_0 fluctuation of NLOS. The C/N_0 of multipath is more complex than that of NLOS. It becomes stronger or weaker with the superposition of direct and indirect signals. Since this superposition is relatively independent in different frequencies, the C/N_0 or SNR of dual-frequency and three-frequency can be compared to detect multipath signals [37, 38].

Elevation angle is another important factor often used to exclude poor GNSS signals. GNSS signals with a low elevation angle suggest a high possibility of NLOS or multipath. Increasing mask angle is a common strategy to mitigate multipath influence on the accuracy of position or velocity [39–41]. Elevation angle based weighted least squares method (WLSM) is another example, which is also used for atmospheric error mitigation.

Receiver autonomous integrity monitoring (RAIM) has also played an important role in NLOS/multipath mitigation. RAIM was originally developed for use in aviation, to detect errors in the position calculated by the receiver. These errors could result in the aircraft being off-course or even in a collision. If RAIM detects an error, the pilot is alerted immediately and then takes action to ensure the safety of the aircraft. The basic principle of RAIM is measurements consistency check [42]. Assuming that most GNSS measurements are normal, consistency check can detect and exclude the faulty measurements with sufficient redundancy. This concept has been borrowed to improve the accuracy of positioning in urban areas [43–49]. However, this kind of method faces two challenges in urban areas: (1) insufficient satellites, and (2) multiple errors. These make it difficult to identify and separate the contaminated signals.

More researchers have noticed that a single factor is not reliable for contaminated GNSS signals identification. Multi-feature based machine learning algorithms have thus been proposed to address the unreliability of using only C/N_0, elevation angle or pesudorange residual [50–54]. These machine learning based methods tried to classify GNSS signals into different types. The basic principle of classification is using machine learning to train the rules between multi-feature and the types of GNSS signals. The trained data need to be labelled as LOS, multipath and NLOS in advance, based on different techniques, such as ray-tracing [55, 56]. The calculated errors of pseudorange by known positions of receiver and satellites can also be used to label the types of GNSS signals, according to the value of calculated errors [52]. Part or all of the features, including C/N_0, elevation, azimuth, and residuals of pseudorange, are used as inputs in a typical machine learning method, such as decision trees (DT), gradient-boosting decision trees (GBDT) and support vector machine (SVM). The labelled GNSS signals are the corresponding outputs. A well-trained rule can be obtained through abundant real data training. The test GNSS signals can be classified as different types before positioning. Strategies can then be divided into different types based on the classification. The first one is to exclude the detected NLOS signals [57]. Although this can avoid large pseudorange errors, the geometrical configuration of satellites may be deteriorated. Reducing the weights of these detected NLOS/multipath signals is another method to improve the accuracy of positioning [58]. The results of classification can also be combined with other methods, such as shadow matching [53, 59].

Almost all the current methods using machine learning are for GNSS signal reception classification. However, problems still exist. The first one is how to accurately label GNSS signals, since it determines the training result. Another problem is how to make the best use of these classified signals. Exclusion may deteriorate the geometry of satellites and then reduce positioning accuracy. The suitable weights of each signal are difficult to be estimated in weighting least squares method. To avoid these problems, another important function of machine learning, regression, has attracted increasing attentions on NLOS/multipath mitigation. In the regression, the detailed pseudorange error is predicted using machine learning, instead of the type of GNSS signal. The corrected pseudorange can be used for positioning directly, without deteriorating geometry configuration of satellites. The key issue affecting the positioning performance, however, is whether the pseudorange is appropriately corrected. If we could design a robust algorithm to obtain the pseudorange error accurately from each observed satellite, it would potentially be possible to achieve a high accuracy positioning solution based on the pseudorange error correction method.

1.3 Methodology

The NLOS/multipath errors of GNSS will repeat for a receiver in the same environments due to the periodicity of satellites' orbits [60]. For example, a NLOS error repeats when the positions of receiver, satellite and reflector do not change. This characteristic has been used to build model between NLOS/multipath errors and GNSS measurements in specific environments. Firstly, NLOS/multipath errors in pseudorange are calculated with the known coordinates of the receiver and satellites. Then the pesudorange error prediction and correction model is trained using machine learning with these errors (outputs) and corresponding GNSS measurements (inputs). When a user comes to the same routes (with the a priori data), the pseudorange error corrections can be provided in real-time using the trained model.

The framework of the proposed algorithm is shown in Fig. 1.1, which describes a machine learning based pseudorange error correction algorithm that can be used to improve the accuracy of positioning in urban areas. In this algorithm, machine learning is used to train the rules during the offline periods in advance. In the training process, firstly, the pseudorange errors are calculated using the known position of both receiver and satellite. Then, several related inputs from GNSS measurements are determined. Lastly, machine learning is used to train the data to generate the rules between these inputs and corresponding labelled pseudorange errors. During the online period, these trained rules can be used to predict the pseudorange errors of new coming GNSS signals. After the correction of these pseudoranges, more accurate positioning results can be calculated.

1.3.1 Several Common Inputs

The received GNSS raw measurements and their derivatives can be used to evaluate and estimate the pseudorange error. In general, using more types of inputs

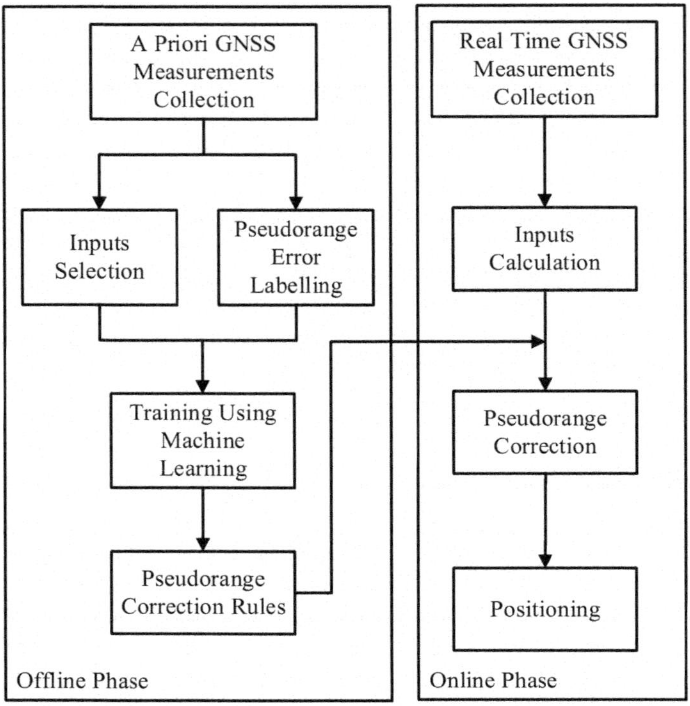

Fig. 1.1 Framework of proposed pseudoenage error correction algorithm

can improve the performance of the training and prediction in machine learning, but increase the computational load. Here, some common inputs are introduced, including C/N_0, pseudorange residuals η, satellite elevation angle θ_e, and position of receiver. Part of them will be used in the following static and dynamic experiments.

(1) C/N_0: C/N_0 can reflect the strength of a signal. Under the same noise power, the C/N_0 of a reflected (i.e. NLOS) is often smaller than that of a direct signal (i.e. LOS). As for multipath signals, the value of C/N_0 may become larger or smaller than that of a LOS signal. C/N_0 is the most common and effective indicator of the pseudorage error estimation.
(2) Pseudorange residual, η: the pseudorange residual is the inconsistency between the measured and calculated pseudoranges, expressed as η. It is worth noting that pseudorange residual is a different concept from pseudorange error. Pseudorange error is the difference between measured pseudorange and ideal pseudorange (without any noise or errors), which means pseudorange error cannot be calculated in practice, due to the lack of true position of receiver. In contrast, the pseudorange residual can be calculated through the following equation:

$$\eta = \rho - G \cdot r \tag{1.1}$$

where ρ denotes the measured pseudorange; G is the design matrix; r is the receiver state, including three-dimensional position and the clock offset of the receiver. The calculation of r could be obtained based on the least squares estimation in (1.2).

$$r = (G^T G)^{(-1)} G^T \rho. \tag{1.2}$$

Pseudorange residual has been used to detect the NLOS/multipath signals with sufficient GNSS measurements [52]. Therefore, the pseudorange residual is selected as one of the potential inputs for the pseudorange error prediction.

(3) Satellite elevation angle, θ_e: GNSS signals are less likely to be blocked or reflected by the surrounding environment in higher elevation angles and therefore with less NLOS/multipath effects. Weighting the measurements based on the elevation angle to reduce the NLOS/multipath effect is widely used in the position calculation [61]. The satellite elevation angle is therefore also used for the pseudorange error prediction.

(4) Positional information: Since the NLOS/multipath errors are environment related, the initial positional information can help users to find corresponding surroundings, which can be obtained using the single point positioning initially.

1.3.2 Labelling Process

Accurate labelling of pseudorange errors is important before the training of machine learning. The range errors of contaminated signals (i.e. due to multipath or NLOS) are typically a few tens or hundreds of meters in urban canyons. Once the reference trajectory of the training dataset is known, the pseudorange errors can be calculated. Generally, the observed pseudorange ρ of a satellite can be expressed in (1.3) and (1.4) as follows:

$$\rho = R + c\left(\delta t^r - \delta t^{sv}\right) + I + T + M + \varepsilon \tag{1.3}$$

$$R = \sqrt{(x^{sv} - x^r)^2 + (y^{sv} - y^r)^2 + (z^{sv} - z^r)^2} \tag{1.4}$$

where, R is the geometric range between the observed satellite and the receiver; (x^{sv}, y^{sv}, z^{sv}) and (x^r, y^r, z^r) are the coordinates of the satellite and receiver in an earth centred earth fixed (ECEF) coordinate system; c is the velocity of light in a vacuum; δt^{sv} is the satellite clock offset time; δt^r is the receiver clock offset; I and T denotes the atmospheric delays; M is the NLOS/multipath error; ε represents receiver noise, which is relatively small and negligible.

After the corrections of related errors, the corrected pseudorange ρ^c is:

$$\rho^c = R + c(\Delta\delta t^r - \Delta\delta t^{sv}) + \Delta I + \Delta T + M + \varepsilon \tag{1.5}$$

where, the geometric range R can be calculated based on the known positions between the receiver and observed satellite (broadcast ephemeris); $\Delta\delta t^r$ is the residual of the receiver clock offset after correction, in which the calculated receiver clock error is from the pseudorange positioning equations with the known ground truth; $\Delta\delta t^{sv}$ is the residual of the satellite clock offset after correction (broadcast ephemeris); $\Delta I + \Delta T$ are the residuals of the ionospheric and tropospheric delays after the corrections from the Klobuchar and Saastamoinen models. The pseudorange error $\Delta\rho$, dominated by NLOS/multipath, can be calculated:

$$\Delta\rho = \rho^c - R = c(\Delta\delta t^r - \Delta\delta t^{sv}) + \Delta I + \Delta T + M + \varepsilon. \tag{1.6}$$

We can calculate the corresponding pseudorange error $\Delta\rho$ following above steps, for every set of inputs from the GNSS measurement, containing part of carrier-to-noise ratio C/N_0, pseudorange residuals η, satellite elevation θ_e, and positional information in the offline labelling phase. After obtaining the labelled data, machine learning can be used to train the pseudorange correction model. The details of dataset and training for the static and dynamic experiments are discussed in the next sections.

1.4 Static Experiments

1.4.1 Algorithm Framework

The Gradient Boosting Decision Trees (GBDT) method is used to correct pseudorange errors in this section. The detailed framework of the proposed GBDT based GNSS pseudorange prediction and correction algorithm is presented in Fig. 1.2.

In the offline phase, GNSS measurements are collected from known points in an urban canyon and a reference station in an open area to generate the training dataset. LOS, NLOS and multipath signals are all contained in this training dataset. The corresponding pseudorange errors are computed as described in the part of *labelling process*. C/N_0, θ_e and η are selected as the inputs and the corresponding pseudorange error is labelled as the output. During the offline phase, the GBDT algorithm is used to fit the calculated pseudorange error, thereby obtaining the rules, which can reflect the relationship between the inputs and the output. The details for the GBDT based training process are introduced further in the subsequent sections, *training process of GBDT*.

In the online phase, inputs (C/N_0, η and the θ_e) from new GNSS measurements in the urban canyon are used together with the rules extracted from the offline phase to predict the pseudorange errors. According to the predicted results, two variations of positioning algorithm are tested: (1) positioning solutions based on corrected

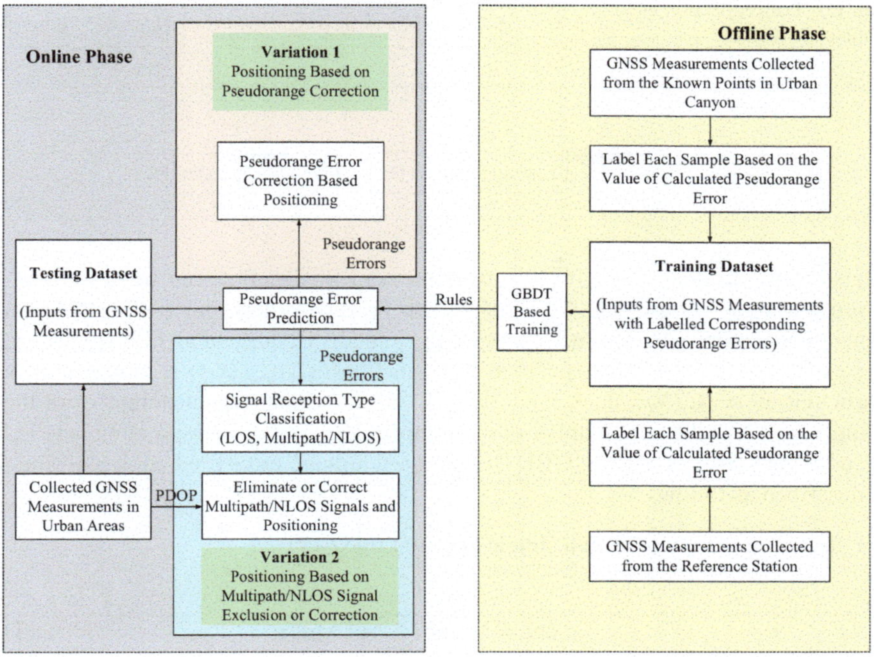

Fig. 1.2 Framework of the GBDT pseudorange correction based positioning algorithm

pseudoranges; and (2) positioning solutions based on exclusion or correction of the NLOS/multipath signals.

1.4.2 Data Collection

GNSS data were collected from several points to generate the training and testing dataset. To avoid a biased fitting, training dataset D1 contains GNSS measurements from two points in an urban canyon and one reference station. The data from urban areas mainly include multipath/NLOS signals, while the data from reference station mainly contain LOS signals. The testing dataset was also from the urban canyon. The details are in Fig. 1.3.

1.4.3 Training Process of GBDT

In this section, GBDT is used to train models for the pseudorange error prediction and correction. GBDT uses gradient boosting regression technique to minimise the

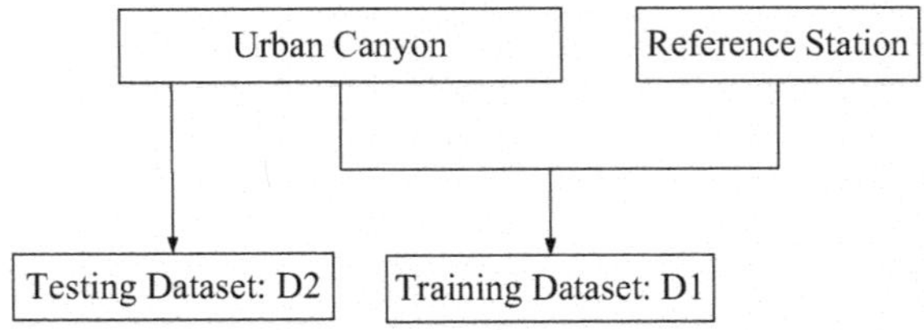

Fig. 1.3 Data collection in static experiments

decision tree training error [62]. The problem in this section can be defined as: given a training sample $\{\boldsymbol{x}_i, \Delta\rho_i\}_1^N$ of known $(\boldsymbol{x}, \Delta\rho)$ values, the goal is to find a function that maps $\boldsymbol{x}$ to $\Delta\rho$, such that over the joint distribution of all $(\boldsymbol{x}, \Delta\rho)$ values, the expected value of some specified loss function $L(\Delta\rho_i, f(\boldsymbol{x}_i))$ is minimised. In particular, $\boldsymbol{x}_i = (C/N_{0i}, \eta_i, \theta_{ei})$, $i = 1, 2, 3, \ldots, N$. i is the sequence number of the sample, and N is the total number of the samples. $\Delta\rho_i$ is the corresponding labelled pseudorange error of $\boldsymbol{x}_i$. The GBDT based pseudorange error prediction algorithm is described as follows.

(1) Initiate predictions with a simple decision tree $f_0(\boldsymbol{x})$:

$$f_0(\boldsymbol{x}) = \underset{\gamma}{\operatorname{argmin}} \sum_{i=1}^{N} L(\Delta\rho_i, \gamma) \tag{1.7}$$

where, $f_0(\boldsymbol{x})$ is a regression decision tree containing only one root node, and γ is a constant value which is the output of $f_0(\boldsymbol{x})$. In order to ensure that the loss function $L(\Delta\rho_i, f(\boldsymbol{x}_i))$ decreases in each iteration, the weak learner $h_m(\boldsymbol{x}_i; \boldsymbol{a}_m)$, $m = 1, \ldots, M$, is created in the direction of steepest descent (i.e., a negative gradient direction). m is the sequence number of iterations. $h_m(\boldsymbol{x}_i; \boldsymbol{a}_m)$ is a decision tree with the parameter $\boldsymbol{a}$, which determines the splitting variable, split locations, and terminal node of the individual tree.

(2) For $m = 1, \ldots, M$

(2.1) Compute the negative gradient according to:

$$\tilde{y}_i = -\left[\frac{\partial L(\Delta\rho_i, f(\boldsymbol{x}_i))}{\partial f(\boldsymbol{x}_i)}\right]_{f(\boldsymbol{x})=f_{m-1}(\boldsymbol{x})} \tag{1.8}$$

where, the loss function $L(\Delta\rho_i, f(\boldsymbol{x}_i))$ is the square loss function, $\frac{1}{2}(\Delta\rho_i - f(\boldsymbol{x}_i))^2$;

(2.2) Create a new dataset based on replacing $\Delta\rho_i$ in the training dataset by $\tilde{y}_i$. The new dataset is expressed as:

$$T_m = \{(\boldsymbol{x}_1, \tilde{y}_1), (\boldsymbol{x}_2, \tilde{y}_2), \ldots, (\boldsymbol{x}_i, \tilde{y}_i), \ldots, (\boldsymbol{x}_N, \tilde{y}_N)\}. \tag{1.9}$$

A new weak predictor $h_m(\boldsymbol{x}_i; a_m)$ is created in (1.10) by training the new dataset T_m to minimise the loss function.

$$a_m = \underset{a}{\operatorname{argmin}} \sum_{i=1}^{N} (\tilde{y}_i - h_m(x_i; a))^2. \tag{1.10}$$

(2.3) Update the original predictor with the new predictor multiplied by learning rate β to form a stronger predictor:

$$f_m(\boldsymbol{x}) = f_{m-1}(\boldsymbol{x}) + \beta h_m(\boldsymbol{x}; \boldsymbol{a}_m) \tag{1.11}$$

where β is usually chosen to be a value between 0 and 1 to prevent over-fitting. $h_m(\boldsymbol{x}; \boldsymbol{a}_m)$ is the weak predictor and $f_{m-1}(\boldsymbol{x})$ is the strong predictor from the previous iteration.

(3) Output $f_M(\boldsymbol{x})$ as the final predictor after the iteration termination:

$$f_M(\boldsymbol{x}) = f_0(\boldsymbol{x}) + \sum_{m=1}^{M} \beta h_m(\boldsymbol{x}; \boldsymbol{a}_m) \tag{1.12}$$

(4) Once the final predictor $f_M(\boldsymbol{x})$ (i.e. the rules of the GBDT method) is obtained, the corresponding pseudorange errors of the newly collected GNSS measurements can be predicted. The input $\boldsymbol{x} = (C/N_0, \eta, \theta_e)$ is used together with the rules to predict the pseudorange errors for each observed satellite.

1.4.4 Position Calculation with Two Variations

The detailed two variations of positioning method based on the predicted pseudorange errors are described below.

(1) Positioning based on pseudorange error correction.

The newly collected pseudoranges can be corrected by subtracting the predicted pseudorange errors according to:

$$\begin{cases} \rho_1^c = \rho_1 - \widetilde{\Delta\rho_1} \\ \rho_2^c = \rho_2 - \widetilde{\Delta\rho_2} \\ \vdots \\ \rho_i^c = \rho_i - \widetilde{\Delta\rho_i} \end{cases} \tag{1.13}$$

where ρ_i^c is the corrected pseudorange of the ith satellite, and $\widetilde{\Delta\rho_i}$ is the predicted pseudorange error of the ith satellite. With the corrected pseudoranges, the position coordinates are calculated based on the least squares method.

(2) Positioning based on NLOS/multipath signal exclusion or correction.

The framework of this algorithm is illustrated in Fig. 1.4. Instead of correcting pseudorange directly, the predicted pseudorange errors are used to determine the types of signals, by comparing their absolute values with a proposed threshold p. If a predicted absolute pseudorange errors $|\widetilde{\Delta\rho_i}|$ is less than the threshold p, the corresponding pseudorange can be consider a LOS signal. Otherwise, it will be considered a NLOS/multipath signal. Nevertheless, we do not remove all these NLOS/multipath signals since this could degrade the satellite geometry significantly. The value of the PDOP is calculated for all tracking satellites at each epoch as a reference. Then, the candidate PDOPs are calculated for each time excluding one satellite. If the exclusion of this satellite increases PDOP, the pseudorange of this satellite will be corrected. Alternatively, if the removal of this satellite does not cause the PDOP to increase, this satellite will be excluded before positioning.

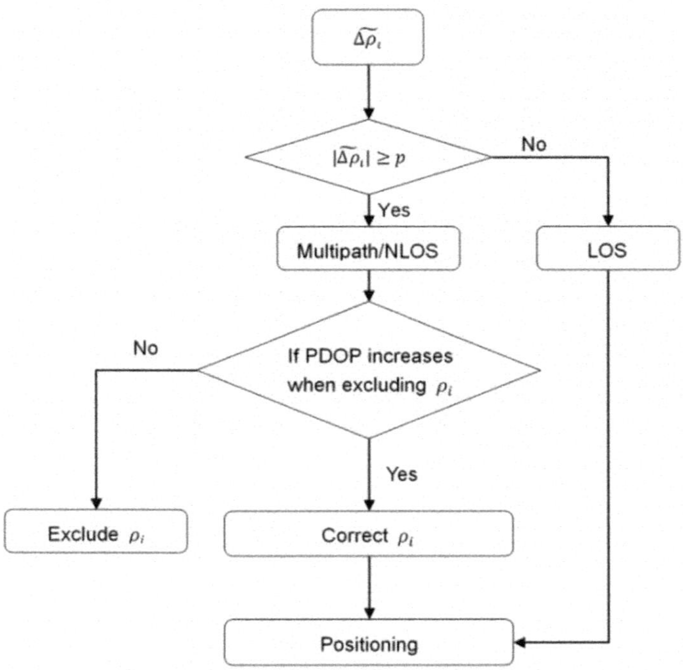

Fig. 1.4 Positioning based on NLOS/multipath signal exclusion or correction

1.4.5 Static Case with Wide Road and High-Rise Buildings on One Side

The static case was carried out at a street in Hong Kong, with high rise buildings on one side, covered by glass. The training dataset (D1) contained two parts, including urban data and open-sky data. The urban data were collected from two points (P1 and P2), for about 20 min with a sampling frequency of 5 Hz using a NovAtel OEM6 geodetic receiver, seen in Fig. 1.5. The open-sky data were from the SatRef HKSC station, for 4 h with an interval of five seconds using a LEICA GR50 geodetic receiver. The testing dataset (D2) was formed using the other 30 min' data collected in P1 at the sampling rate 5 Hz. Table 1.1 provides a summary of the datasets used in this case.

For better comparisons, four methods are tested in this section, including two conventional positioning methods (positioning with standard outlier detection and exclusion, and positioning with C/N0 and elevation angle-based NLOS/ multipath

Fig. 1.5 The locations of P1 and P2

Table 1.1 Summary of the training dataset D1 in the static case

	Training dataset (D1)					Testing dataset (D2)	
Collected location	Urban canyon				Reference station	Urban canyon (P1)	
	P1		P2				
Pseudorange error	Large	Small	Large	Small	Small	Small	Large
Sample size	2400	1600	2400	1600	1600	14971	4686
Total	9600					19657	

exclusion), pseudorange correction, and NLOS/multipath signal exclusion or correction (with a threshold $p = 50$ m). The results for the four methods are depicted in Fig. 1.6 and Table 1.2. The positioning accuracy of the pseudorange correction method is highest, with a 3D RMSE of 23.27 m. This value is about 60 m for the NLOS/multipath signal exclusion or correction method, and about 80 m for the two conventional positioning methods. This shows a significant improvement in the accuracy after the corrections of pseudorange.

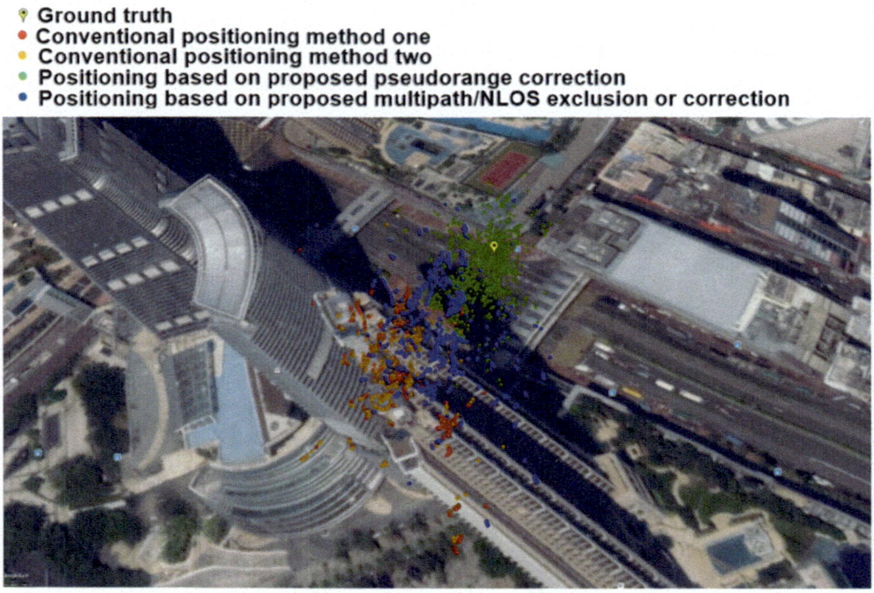

Fig. 1.6 Horizontal positioning results in the static case

Table 1.2 Positioning accuracy comparison in the static case

RMSE (m)	E	N	U	3D	Horizontal
Conventional positioning method one	35.70	56.27	40.51	81.26	66.64
Conventional positioning method two	35.80	56.06	46.24	80.01	66.52
Positioning based on pseudorange correction	8.77	13.69	16.65	23.27	16.26
Positioning based on NLOS/multipath signal exclusion or correction	23.94	40.06	38.94	60.78	46.67

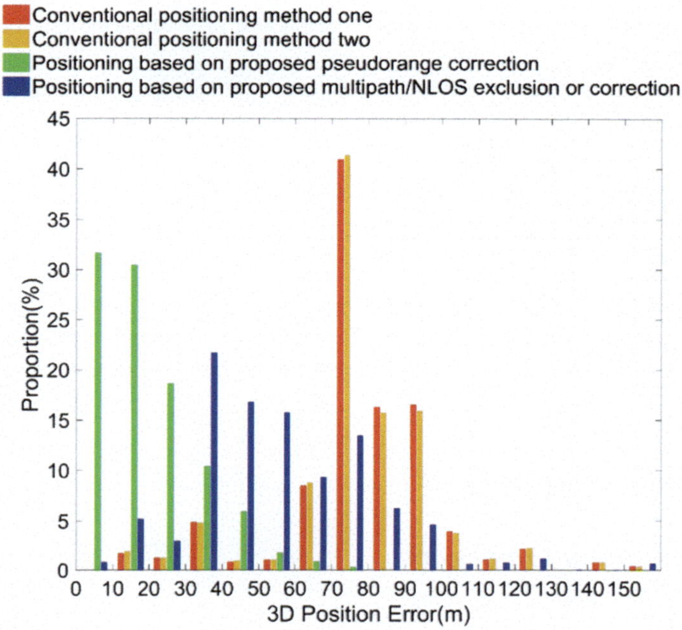

Fig. 1.7 3D Positioning accuracy histogram of the static case

The 3D positioning error distribution of the four methods are also depicted in Fig. 1.7. With the corrections of pseudorange, the epochs with a positioning accuracy within 30 m have increased significantly, while for conventional methods, most errors fall within 60–90 m. For the NLOS/multipath exclusion or correction based positioning, most errors are in 30–60 m, which is also worse than the pseudorange correction results.

The positioning performance is further analysed by comparing with conventional method one (which showed a similar performance to the conventional method two). In Table 1.3, the horizontal and 3D positioning results improved in 98% and 97% of the epochs, respectively, with the pseudorange correction, while only around 3% of the epochs got worse, due to the inaccurate predicted errors. The NLOS/multipath exclusion or correction method, meanwhile, improved the positioning accuracy of about 81% (3D) and 91% (horizontal) of the epochs, while the positioning accuracy for 9% (3D and horizontal) of the epochs did not change. The worse epochs for the horizontal positioning are only 0.4% but with 10% for 3D positioning, since the height is more easily affected by the reduction of satellite number in challenging urban areas.

In summary, the positioning results in the static case show that the pseudorange correction using machine learning can perform better than not only the two traditional

Table 1.3 Algorithm performance evaluation with proportion of epochs in the static case

Proportion of epochs (%)	3D			Horizontal		
	Better	Equal	Worse	Better	Equal	Worse
Positioning based on pseudorange corrections	96.50	0.00	3.50	97.80	0.00	2.20
Positioning based on NLOS/multipath signal exclusion or correction	81.13	8.77	10.10	90.53	9.11	0.36

methods, but also the NLOS/multipath exclusion or correction method. The pseudorange correction method does not reduce the satellite number but correct pseudorange NLOS/multipath errors, hence finally improving the positioning accuracy.

1.5 Dynamic Experiments

1.5.1 Algorithm Framework

In this section, a random forest (RF) based pseudorange correction method, with two variations, grid-based correction (GBC) and point-based correction (PBC), is used in the dynamic experiment. The related framework is in Fig. 1.8. The offline training and online testing parts are similar to the static case. One difference is that a priori data were collected at a specific area with the receiver on a vehicle, instead of several static points. Another difference is that the inputs are changed due to the moving receiver. For GBC, C/N_0 and elevation angle are selected as the inputs, while for PBC, C/N_0, elevation angle, and positional information are used. The calculated pseudorange error still serves as the output for the machine learning in both variations. The training dataset are divided according to the constellation type of the data in this section, including GPS and BDS.

RF is used to train the rules, which can map the relation between the selected inputs and labelled output. For the PBC, all the a priori data are used together to generate the rules for GPS and BDS respectively for the whole area. In contrast, for GBC, hexagonal grids are first designed to cover this area. Then, the data in each hexagonal grid are used to train the rules of this grid for GPS and BDS.

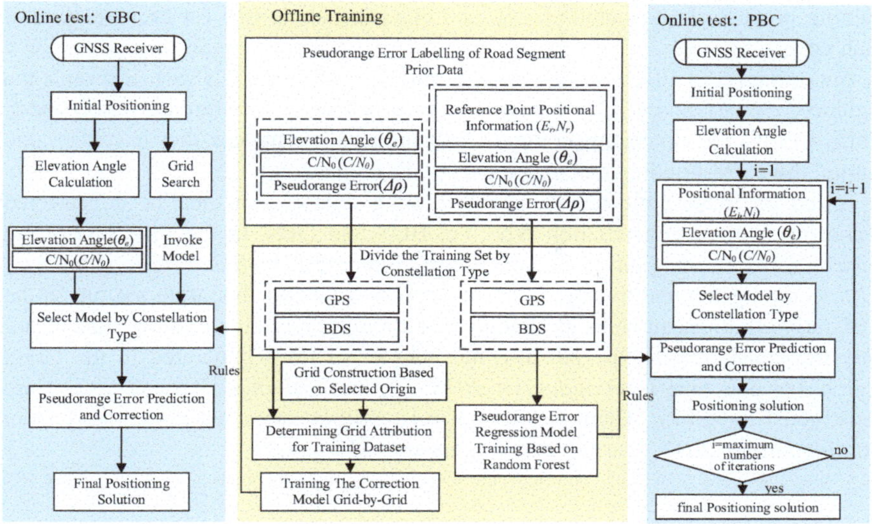

Fig. 1.8 Framework of the RF pseudorange correction based positioning algorithm

Based on the trained rules using RF in the offline phase, the pseudorange errors will be predicted and corrected in the online phase. For the PBC, an iteration process is needed, since the accurate positional information (one of the inputs) is lacked. The result of single point positioning (without NLOS/multipath correction) is used as the positional information at the first iteration. Then, the positioning result at current iteration will be one of the inputs for the next iteration, until the maximum number of iterations. For the GBC, the initial position is used to determine the hexagonal grid, in which the receiver is located. Then the corresponding rules of this grid will be used to predict and correct the pseudorange error. The position coordinates can then be calculated based on the corrected pseudoranges.

1.5.2 Training Data Generation

The inputs and output are determined as follows. Firstly, the a priori data were collected by high-grade GNSS/IMU integration systems mounted on the roof of a moving vehicle, along the routes in a target region between 14:00 and 18:00, Beijing time, every day for almost two months. Reference trajectories were estimated using the post-processing of the GNSS/INS integration in the tightly-coupled mode through Inertial Explorer (IE) software, whose accuracy can reach cm to dm level. After obtaining the reference trajectories of the receiver, the corresponding pseudorange errors can be calculated as described in the previous section *labelling process*. The raw GNSS measurements in this integration systems were also collected for the training process. They include C/N$_0$, and elevation angle (θ_e) for each satellite at each epoch.

Accordingly, in PBC, every set of inputs from the raw GNSS measurements, include C/N_0, θ_e, and (E_r, N_r) (positional information from reference trajectories), while in GBC, the inputs only contain C/N_0 and θ_e. Nevertheless, they have the same output, the corresponding calculated pseudorange error $\Delta\rho$.

Based on these collected data, the training dataset can then be constructed according to their constellation type, i.e. BDS and GPS. In the PBC algorithm, therefore, two training datasets, namely *trainingdataset_BDS* and *trainingdataset_GPS*, can be generated to extract their respective rules for the target region. In the GBC algorithm, meanwhile, the number of training datasets takes account of not only the constellation type but also the number of grids generated in the target region. For example, $trainingdataset\_BDS_{grid\_n}$ and $trainingdataset\_GPS_{grid\_n}$ are constructed for grid *n*. Then, two sets of rules will be extracted respectively from $trainingdataset\_BDS_{grid\_n}$ and $trainingdataset\_GPS_{grid\_n}$.

1.5.3 Random Forest (RF) Based Pseudorange Error Prediction

After the generation of the training datasets, the RF is used to train the rules for pseudorange correction. RF is an ensemble learning algorithm that combines bagging with the idea of a feature subspace. It can avoid the problems of insufficient precision and overfitting that may occur with a single decision tree [63]. Therefore, RF has been widely used to solve nonlinear problems [64, 65].

To clearly describe the training process, the input vector of the training dataset is expressed as $\boldsymbol{x}$ and the corresponding output as $\mathbf{y}$. The input vector $\boldsymbol{x}$ in PBC can be expressed as $\boldsymbol{x} = [(E, N), C/N_0, \theta_e]$, and in GBC as $\boldsymbol{x} = [C/N_0, \theta_e]$. When there are K samples in a training dataset $\boldsymbol{L}$, then it can be expressed as $\boldsymbol{L} = \{(\boldsymbol{x}_q, \boldsymbol{y}_q)\}_{q=1}^{K}$. Figure 1.9 illustrates the use of random forest for regression model training and prediction of the pseudorange error. The specific steps are as follows:

(1) Note that there are K samples in the training dataset $\boldsymbol{L}$. Assuming there are K_s ($K_s < K$) sample in a subset, a set of training subsets $\boldsymbol{L}_t$ is generated by randomly sampled with replacement from the $\boldsymbol{L}$, with equal probability $1/K$ for each sample:

$$\boldsymbol{L}_t = \{(\boldsymbol{x}_q, \boldsymbol{y}_q)\}_{q=1}^{K_s}, t = 1,2, \ldots, m \tag{1.14}$$

where m is the total number of the sample subsets. Each subset is trained using a single regression tree.

(2) s (s is no more than the total number of input features) features are randomly sampled from the input features and one of them is selected for node splitting in

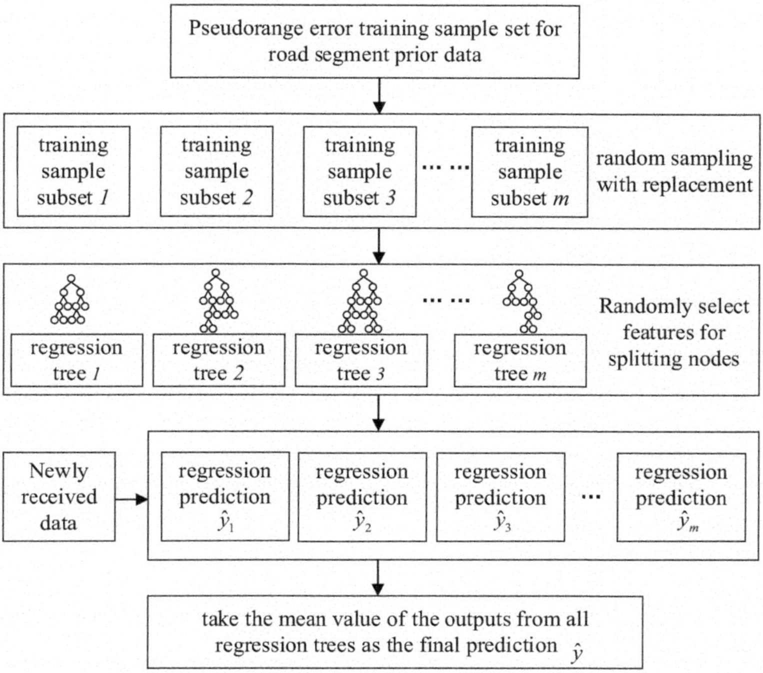

Fig. 1.9 Random forest regression model

order to generate a regression tree. At the node of the decision tree, the calculation principle governing the feature selected for node splitting is to minimize the mean square error; That is, for an arbitrary feature (represented by A), the dataset will be divided into two datasets D_1 and D_2 at an arbitrary divide point j corresponding to A. The aim here is to find the feature and the splitting point j, that minimizes the mean square error (MSE) of D_1 and D_2, respectively, and the sum of them:

$$\underset{(A,j)}{(\min)}\left[\min_{c_1}\sum\nolimits_{x_q \varepsilon D_1(A,j)}(y_q - c_2)^2 + \min_{c_2}\sum\nolimits_{x_q \varepsilon D_2(A,j)}(y_q - c_2)^2\right]. \quad (1.15)$$

In (1.15), x_q is a value of feature A; y_q is the corresponding output; c_1 is the mean value of the sample outputs in dataset D_1; c_2 is the mean value of the sample outputs in dataset D_2. After m decision trees are generated, the regression model of RF can be represented as $y = \frac{1}{m}\sum_1^m F_{RF}(\boldsymbol{x})$.

(3) The features of the newly received data $\boldsymbol{x}$ are inputs, and the mean value of the predicted outputs $\{\hat{y}_1, \hat{y}_2, \ldots, \hat{y}_m\}$ of all regression trees is then calculated to obtain the predicted pseudorange error value $\hat{y}$ of the RF output. Finally, the predicted pseudorange correction value $\Delta\hat{\rho}$ is:

$$\Delta\hat{\rho} = \hat{y} = \frac{1}{m}\sum_{t=1}^{m}\hat{y}_t. \quad (1.16)$$

1.5.4 Positioning with Two Variations

(1) Positioning with PBC

The basic steps of the positioning with PBC are as follows. Since the positional information is one of the inputs, a initial position (E_0, N_0, U_0) is obtained by the conventional single point positioning algorithm with outlier detection and exclusion (CSPP 1), in which an Efficient Leave One Block Out (ELOBO) method is used to exclude gross errors before positioning [66]. The initial position, C/N_0 and elevation angle are used together to predict pseudorange error $\Delta\widehat{\rho}_k$ for the satellite k. Assuming that the observed pseudorange is $\widetilde{\rho}_k$, the corrected pseudorange ρ_k' can be obtained as:

$$\rho_k' = \tilde{\rho}_k - \Delta\hat{\rho}_k. \quad (1.17)$$

After obtaining all the corrected pseudoranges at an epoch, the positioning solution (E_1, N_1, U_1) can be calculated. The new position is used to update the inputs to correct these pseudoranges, until obtaining $\left(E_{i_M}, N_{i_M}, U_{i_M}\right)$, where i_M represents the maximum number of iterations. Therefore, the final positioning solution is output as $\left(E_{i_M}, N_{i_M}, U_{i_M}\right)$.

(2) Positioning with GBC

Since GNSS data quality is susceptible to surrounding environment, the characteristics of pseudorange error will be similar in a small area. Considering this issue, the size of the hexagonal grids for the multipath correction map needs to be careful adjusted. In contrast, the details of road, such as road sections, lanes, and directions, are not distinguished.

After determining these hexagonal grids, the training dataset needs to be divided according to the hexagonal grid size, as will be described in detail below. At each epoch, the training sample belongs to a hexagonal grid where the receiver is located in, as shown in Fig. 1.10.

Assuming that there is a hexagonal grid with a center point o, and the receiver is at a point B. To easier describe the steps, a rectangular plane coordinate system is established, with a x-axis and y-axis, shown in Fig. 1.10. Let the radius of this hexagon be a and the coordinates of B be (w, q). The determination of whether point B is located inside the hexagon with the center point o is described as follows.

(1) Determine whether point B belongs to the inside of the outer rectangle of the hexagon. If $q \geq a$ or $w \geq \frac{\sqrt{3}}{2}a$, point B does not belong to the hexagon.

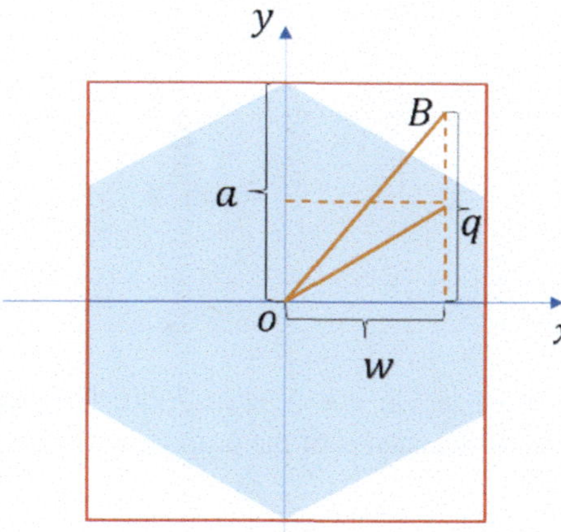

Fig. 1.10 Point location determination in the grid

Otherwise, point B is located inside the outer rectangle (represented by the red rectangle in Fig. 1.10) of the hexagon.

(2) Determine whether $a - q \geq \frac{w}{\sqrt{3}}$. If this condition is false, point B does not belong to the hexagon or is on the boundary of the hexagonal grid. If the condition is true, point B is located inside the hexagon. After the division of the training set, the pseudorange error correction models are trained grid-by-grid according to the above division results.

The process of GBC based positioning is then as follows. An initial position is also calculated to determine which hexagonal grid the user is located in. Then, the correction model of this corresponding hexagonal grid is used to correct received pseudoranges. Finally, the more accurate position can be obtained using these corrected pseudoranges.

1.5.5 Field Test and Analysis of Results

To verify the proposed algorithms, a 20-min dynamic experiment was conducted to obtain the testing data in the same route as the training data, with a sampling rate 1 Hz, using the vehicle-grade receiver (Allystar EVK-2024 module), and a single-band active patch antenna (Allystar AGR 6301). GNSS measurements from both GPS and BDS were collected for testing. The reference trajectory was obtained using the post-processing with a tightly-coupled integration mode, in Inertial Explorer software, with data from a tactical INS (NovAtel SPAN®-LCI) and antenna (NovAtel GPS-GGG-703-HV). The positioning accuracy of the integrated navigation system can reach 2 cm in horizontal and 5 cm in vertical, according to the NovAtel SPAN®-LCI product performance specification. During the testing period, the average velocity was 26.63 km/h, with the maximum velocity about 54 km/h. The parameters of

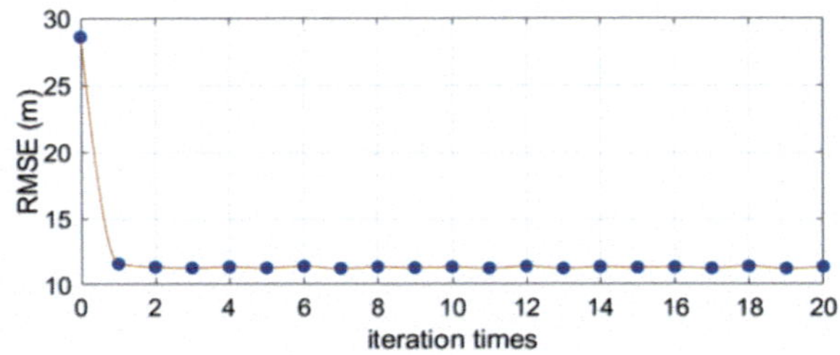

Fig. 1.11 RMSE with iteration times

RF were set as: $s = 2$, $m = 2000$. To achieve better positioning performance, the sensitivity analysis of the proposed correction models was carried out.

1.5.6 PBC Analysis

The iteration times of PBC have an influence on both the efficiency and accuracy. Therefore, it is essential to analyse the performance of different iteration times. The root mean square errors (RMSEs) of positioning under different number of iterations are shown in Fig. 1.11. The RMSE is about 28 m for the initial positioning, using CSPP 1 (without pseudorange error correction). It is reduced to about 12 m after one iteration. This means that the iteration is effective to improve the positioning accuracy. As the number of iterations increases, however, the effect of accuracy improvement gradually becomes weaker and tends to converge. Too many iteration times will increase the computational burden. The suitable number should be determined for the compromise of efficiency and accuracy.

This chapter determine the iteration threshold using the proportion of convergent epochs. If the RMSE of the jth iteration is equal to that of the $(j + 1)$th and $(j + 2)$th iterations, this epoch is considered convergence at the jth iteration. The proportion of convergent epochs under different number of iterations is shown in Table 1.4 and Fig. 1.12. It is indicated that after the fifth iteration, more than 80% epochs converged. After the tenth iteration, this proportion was stable at 83.2%, so the number of iterations was set as ten in this section.

1.5.7 GBC Analysis

For the GBC algorithm, the spatial resolution of GBC is set to 25 m, according to Smolyakov et al. [67]. The heatmap in Fig. 1.13 is used to show the RMSEs of the pseudoranges in each hexagonal grid. Since BDS has better visibility than GPS in the observed area in China, the RMSE level of the pseudorange of BDS is generally lower than that of GPS.

Table 1.4 The proportion of convergent epochs under different iteration times

Times	Proportion (%)	Times	Proportion (%)
1	0.4	10	83.2
2	20.3	11	83.2
3	62.5	12	83.2
4	75.8	13	83.2
5	80.9	14	83.2
6	82.0	15	83.2
7	82.5	16	83.2
8	82.9	17	83.2
9	83.0	18	83.2

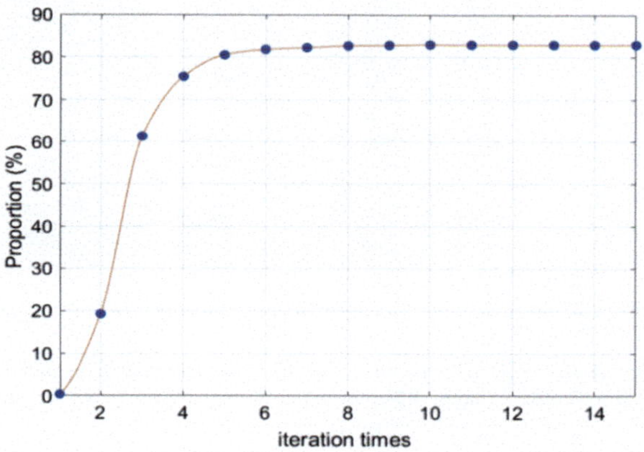

Fig. 1.12 The proportion of convergent epochs under different iteration times

1.5.8 Positioning Performance Evaluation

For better comparison, CSPP 1 (introduced in *Positioning with PBC*) and CSPP 2 are used as the comparative methods. CSPP 2 is based on CSPP 1, with an extra elevation angle based weighting strategy and a C/N_0 threshold of 25 dB-Hz. The weighting matrix is:

$$W = diag(\frac{1}{\sin(\theta_{e1})^2} \cdots \frac{1}{\sin(\theta_{en})^2}) \tag{1.18}$$

where, θ_{en} means the elevation angle of satellite n; $diag()$ means diagonal matrix.

The positioning results of the four algorithms are shown in Fig. 1.14, with a reference trajectory in black dots. The red and blue dots are the positioning results of CSPPs 1 and 2, while the green and yellow dots are the positioning results using the proposed PBC and GBC algorithms. The whole experiment can be divided into three parts according to the environments, with narrow street, obstacles such as tall buildings and overpasses. Specially, the details of each part are zoomed in, for

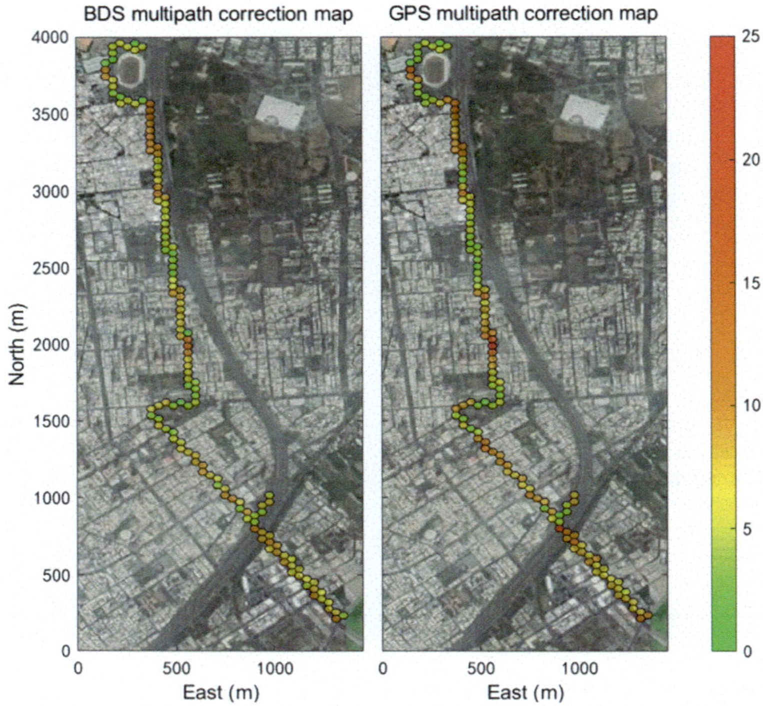

Fig. 1.13 BDS/GPS heatmap in the target urban region of the satellite pseudorange, where the different colors mean different value of the root mean square error (m)

clearly comparison. It is indicated that the proposed algorithms deliver more accurate positioning results in the NLOS/multipath contaminated areas.

Table 1.5 shows the positioning accuracy of the different methods, in terms of RMSE, and the corresponding positioning error time series are shown in Figs. 1.15 and 1.16. It can be found that the performance of PBC and GBC is comparable, with RMSEs of 5.6 and 5.8 m in horizontal, and 11.4 and 10.5 m in 3D. Compared with the results of CSPPs 1 and 2, PBC can achieve improvements in accuracy by 37.3 and 36.5% in east, by 49.2 and 46.7% in north and by 63.2 and 47.3% in the up direction. For GBC, the improvements are 34.7% and 33.8%, 50.8% and 48.3%, and 67.3% and 53.2%, respectively. This can be validated by the results in Figs. 1.15 and 1.16, where the errors of CSPPs 1 and 2 fluctuate violently. In contrast, both the proposed algorithms can effectively reduce this fluctuation, although the performance of GBC is slightly more stable. This is caused by their different dataset construction in the training process. For PBC, the training set contains the prior data of the entire road segment, while for GBC, the dataset is divided according to the hexagonal grids. This means that the small individual grids in GBC reflect relatively more accurate environments than the entire road segments that make up the PBC data.

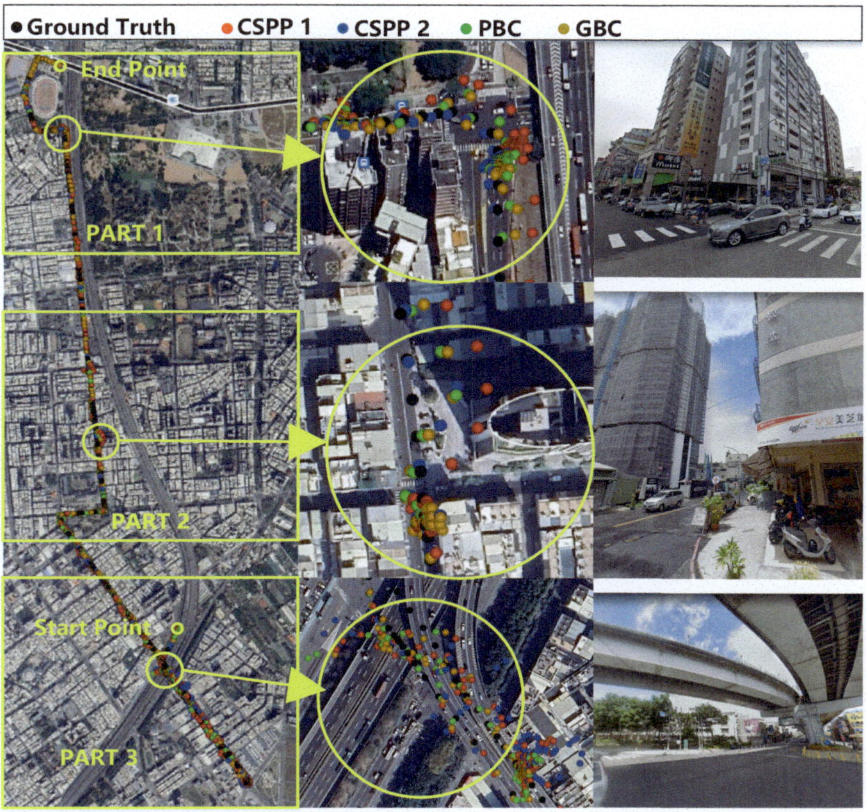

Fig. 1.14 Experimental results and environments

Table 1.5 Comparison of positioning accuracy in dynamic experiments

RMSE (m)	One-dimensional			Horizontal	3D
	East	North	Up		
CSPP 1	7.5	6.3	26.9	9.8	28.6
CSPP 2	7.4	6.0	18.8	9.5	21.0
PBC	4.7	3.2	9.9	5.6	11.4
Improvements (%)	37.3	49.2	63.2	42.9	60.1
	36.5	46.7	47.3	41.1	45.7
GBC	4.9	3.1	8.8	5.8	10.5
Improvements (%)	34.7	50.8	67.3	40.8	63.3
	33.8	48.3	53.2	38.9	50.0

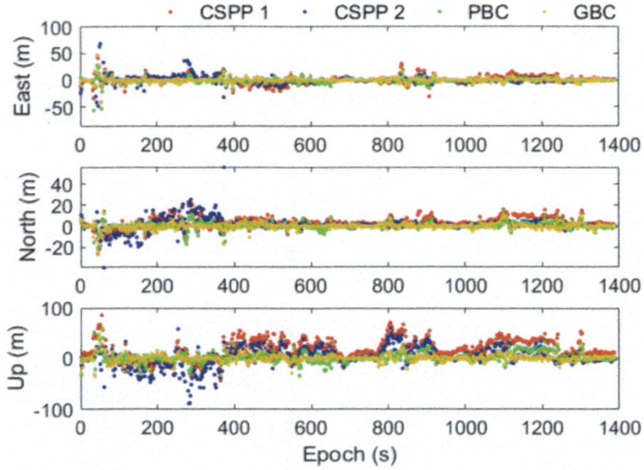

Fig. 1.15 Positioning error in local coordinate system

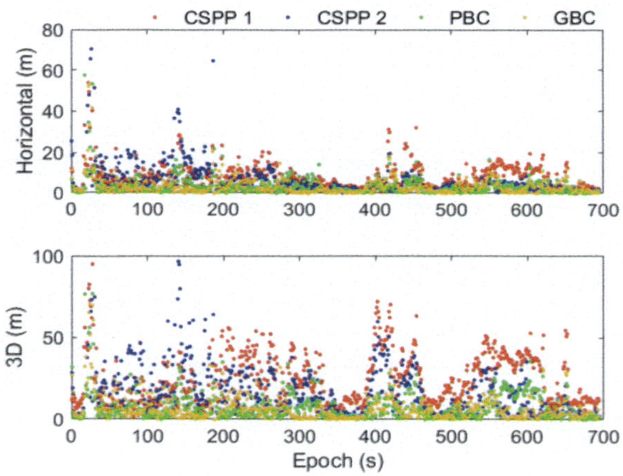

Fig. 1.16 Horizontal and 3D positioning error

In Table 1.5, compared with CSPPs 1 and 2, the error reductions in the up direction are 17.0 and 8.9 m for PBC, and 18.1 and 10.0 m for GBC. They are about 3 m in east and north direction. This means that NLOS/multipath errors can cause larger up direction error than the horizontal error.

The error in east is larger than that in north direction for all algorithms. This is because that the overall orientation of the driving route in the test case is north–south, while the east–west direction is cross-street at most epochs. Given that the satellite visibility along the street is better than that in the cross-street direction, there are large outliers in the eastward error in Fig. 1.15 compared to north direction, which considerably increases the RMSE value. Even, after correcting NLOS/multipath errors, the accuracy in east direction is still higher than that in north direction.

The horizontal and 3D error distributions of the four algorithms are shown in Fig. 1.17. The shape of the error distribution after pseudorange correction changes

Fig. 1.17 Horizontal and 3D error distribution

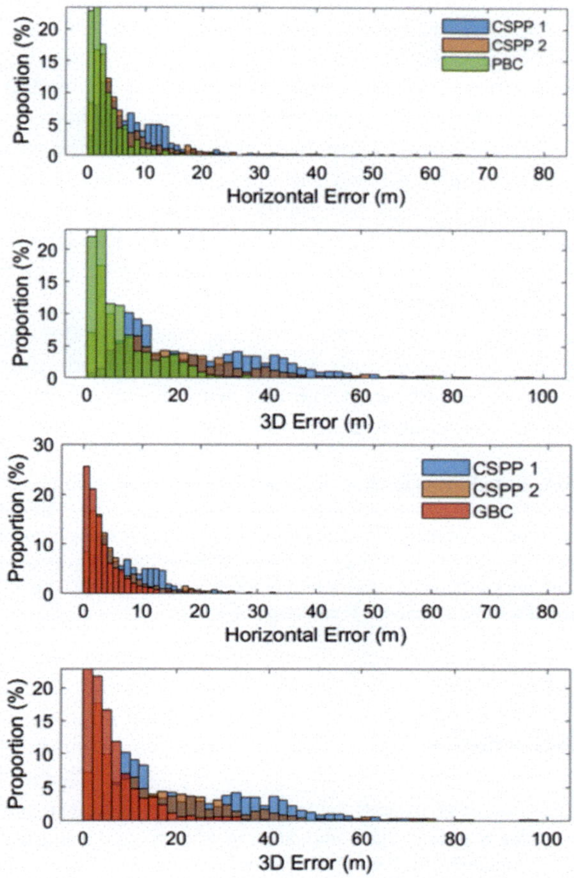

significantly. In horizontal, most errors of CSPPs 1 and 2 are between 0 to 20 m, while the counterparts of PBC and GBC are within 10 m. Meanwhile, most 3D errors of CSPPs 1 and 2 are between 0 to 60 m, which is much larger than that in horizontal, due to the large error component in the up direction. After the corrections of pseudorange, the 3D errors fall within the range of 0 to 20 m, which are obviously better than the results of CSPPs 1 and 2. In summary, the proposed PBC and GBC models can improve the positioning accuracy remarkably.

1.6 Conclusion

In this chapter, we have proposed a machine learning based GNSS pseudorange correction method to improve the accuracy of positioning in urban areas.

In the static experiments, GBDT was used to predict and correct the pseudorange errors. Based on the corrected pseudoranges, the improvements of positioning accuracy in horizontal were 75.6% and 75.6%, and in 3D were 71.4% and 70.9%, compared to two conventional positioning methods, respectively.

In the dynamic experiment, RF is used for the pseudorange correction in urban environments with two variations, PBC and GBC. PBC model achieved improvements in horizontal positioning accuracy of 42.9% and 41.1%, and in 3D accuracy of 60.1% and 45.7% compared with CSPP 1 and 2. GBC achieved improvements in horizontal of 40.8% and 38.9%, and in 3D 63.3% and 50.0%, compared with CSPPs 1 and 2, respectively.

The proposed machine learning based pseudorange correction methods do not require additional sensors and can be used in real-time. Therefore, it is suitable for the users with low-cost receivers in urban areas, such as smartphone. It can potentially benefit the public widely.

Acknowledgements Authors thanks Guanyu Wang and Linxia Fu for data processing. The authors would also like to thank these who collected the data. This work was jointly supported by the sponsorship of the University Grants Committee of Hong Kong under the scheme Research Impact Fund (Grant No. R5009-21), the Research Institute of Land and System, Hong Kong Polytechnic University, the National Natural Science Foundation of China (Grant No. 41974033, 42174025), and the Natural Science Foundation of Jiangsu Province (Grant No. BK20211569).

References

1. Groves PD, Jiang Z (2013) Height aiding, C/N0 weighting and consistency checking for GNSS NLOS and multipath mitigation in urban areas. J Navig 66(5):653–669
2. Tranquilla JM, Carr JP, Al-Rizzo HM (1994) Analysis of a choke ring groundplane for multipath control in global positioning system (GPS) applications. IEEE Trans Antenn Propag 42(7):905–911
3. Blum R, Bischof R, Sauter UH, Foeller J (2016) Tests of reception of the combination of GPS and GLONASS signals under and above forest canopy in the Black Forest, Germany, using choke ring antennas. Int J Forest Eng 27(1):2–14
4. Taghdisi E, Ghaffarian S, Mirzavand R, Mousavi P (2020) Compact substrate integrated choke ring ground structure for high-precision GNSS applications. In: IEEE international symposium Antenn Propaga North American Radio Science meeting, pp 1705–1706
5. Lin D, Wang E, Wang J (2022) New choke ring design for eliminating multipath effects in the GNSS system. Int J Antenn Propaga 2022:1–6
6. Rykała Ł, Rubiec A, Przybysz M, Krogul P, Cieślik K, Muszyński T, Rykała M (2023) Research on the positioning performance of GNSS with a low-cost choke ring antenna. Appl Sci 13(2):1007
7. Jiang Z, Groves PD (2014) NLOS GPS signal detection using a dual-polarisation antenna. GPS Solut 18:15–26
8. Palamartchouk K, Clarke PJ, Edwards SJ, Tiwari R (2015) Dual-polarization GNSS observations for multipath mitigation and better high-precision positioning. In: Proceedings of the 28th international technical meeting of the Satellite division of The Institute of Navigation, pp 2772–2779

9. Guermah B, Sadiki T, El Ghazi H (2017) Fuzzy logic approach for GNSS signal classification using RHCP and LHCP antennas. In: IEEE 8th annual ubiquitous computing, electronics and mobile communication conference (UEMCON), pp 203–208
10. Egea-Roca D, Tripiana-Caballero A, López-Salcedo JA, Seco-Granados G, De Wilde W, Bougard B, ..., Popugaev A (2018) GNSS measurement exclusion and weighting with a dual polarized antenna: the FANTASTIC project. In: 8th international conference on the localization and GNSS, pp 1–6
11. Ge X, Liu X, Sun R, Fu L, Qiu M, Zhang Z (2023) A weighted GPS positioning algorithm for urban canyons using dual-polarised antennae. J Locat Based Serv 17(3):185–206
12. Seco Granados G (2000) Antenna arrays for multipath and interference mitigation in GNSS receivers. Universitat Politècnica de Catalunya
13. Closas P, Fernández-Prades C (2011) A statistical multipath detector for antenna array based GNSS receivers. IEEE Trans Wirel Commun 10(3):916–929
14. Daneshmand S, Broumandan A, Sokhandan N, Lachapelle G (2013) GNSS multipath mitigation with a moving antenna array. IEEE Trans Aerosp Electron Syst 49(1):693–698
15. Vagle N, Broumandan A, Jafarnia-Jahromi A, Lachapelle G (2016) Performance analysis of GNSS multipath mitigation using antenna arrays. J Glob Position Syst 14(1):1–15
16. Razgūnas M, Rudys S, Aleksiejūnas R (2023) GNSS 2×2 antenna array with beamforming for multipath detection. Adv Space Res 71(10):4142–4154
17. Suzuki T, Matsuo K, Amano Y (2020) Rotating GNSS antennas: simultaneous LOS and NLOS multipath mitigation. GPS Solut 24:1–13
18. Van Dierendonck AJ, Fenton P, Ford T (1992) Theory and performance of narrow correlator spacing in a GPS receiver. Navig 39(3):265–283
19. McGraw GA, Braasch MS (1999) GNSS multipath mitigation using gated and high-resolution correlator concepts. In: Proceedings of the 1999 national technical meeting of the Institute of Navigation, pp 333–342
20. Garin L, van Diggelen F, Rousseau JM (1996) Strobe and edge correlator multipath mitigation for code. In: Proceedings of the 9th international technical meeting of the satellite division of the Institute of Navigation, pp 657–664
21. Irsigler M, Hein GW, Eissfeller B (2004) Multipath performance analysis for future GNSS signals. In: Proceedings of the 2004 national technical meeting of the Institute of Navigation, pp 225–238
22. Van Nee RD, Siereveld J, Fenton PC, Townsend BR (1994) The multipath estimating delay lock loop: Approaching theoretical accuracy limits. In: Proceedings of the 1994 IEEE position, location and navigation, pp 246–251
23. Meguro J, Murata T, Takiguchi J, Amano Y, Hashizume T (2009) GPS multipath mitigation for urban area using omnidirectional infrared camera. IEEE Trans Intell Transp Syst 10(1):22–30
24. Shytermeja E, Garcia-Pena A, Julien O (2014) Proposed architecture for integrity monitoring of a GNSS/MEMS system with a fisheye camera in urban environment. In: International conferences localization GNSS 2014, Helsinki, Finland, 24–26 June 2014, pp 1–6
25. Tokura H, Kubo N (2016) Effective satellite selection methods for RTK-GNSS NLOS exclusion in dense urban environments. In: Proceedings of the ION GNSS + 2016, Portland, Oregon, September, pp 304–312
26. Moreau J, Ambellouis S, Ruichek Y (2017) Fisheye-based method for GPS localization improvement in unknown semi-obstructed areas. Sensors 17(1):119
27. Horide K, Yoshida A, Hirata R, Kubo Y, Koya Y (2019) NLOS satellite detection using fish-eye camera and semantic segmentation for improving GNSS positioning accuracy in urban area. Proc ISCIE Int Symp Stochastic Syst Theory Appl 2019:212–217
28. Wen W, Zhang G, Hsu LT (2019a) GNSS NLOS exclusion based on dynamic object detection using LiDAR point cloud. IEEE Trans Intell Transp Syst 22(2):853–862
29. Wen W, Zhang G, Hsu LT (2019b) Correcting NLOS by 3D LiDAR and building height to improve GNSS single point positioning. Navig 66(4):705–718
30. Hassan T, Fath-Allah T, Elhabiby M, Awad A, El-Tokhey M (2022) Detection of GNSS no-line of sight signals using LiDAR sensors for intelligent transportation systems. Surv Rev 54(385):301–309

31. Pinana-Diaz C, Toledo-Moreo R, Betaille D, Gomez-Skarmeta AF (2011) GPS multipath detection and exclusion with elevation-enhanced maps. In: 14th international IEEE conference on intelligent transportation systems, pp 19–24
32. Peyraud S, Bétaille D, Renault S, Ortiz M, Mougel F, Meizel D, Peyret F (2013) About non-line-of-sight satellite detection and exclusion in a 3D map-aided localization algorithm. Sensors 13(1):829–847
33. Peyret F, Bétaille D, Carolina P, Toledo-Moreo R, Gómez-Skarmeta AF, Ortiz M (2014) GNSS autonomous localization: NLOS satellite detection based on 3-D maps. IEEE Robot Autom Mag 21(1):57–63
34. Petovello M (2009) Carrier-to-noise density and AI for INS/GPS integration. Inside GNSS 4(5):20–29
35. Wen H, Pan S, Gao W, Zhao Q, Wang Y (2020) Real-time single-frequency GPS/BDS code multipath mitigation method based on C/N0 normalization. Measurement 164:108075
36. Kubo N, Kobayashi K, Furukawa R (2020) GNSS multipath detection using continuous time-series C/N0. Sensors 20(14):4059
37. Strode PR, Groves PD (2016) GNSS multipath detection using three-frequency signal-to-noise measurements. GPS Solut 20:399–412
38. Zhang Z, Li B, Gao Y, Shen Y (2019) Real-time carrier phase multipath detection based on dual-frequency C/N0 data. GPS Solut 23:1–13
39. Heng L, Walter T, Enge P, Gao GX (2014) GNSS multipath and jamming mitigation using high-mask-angle antennas and multiple constellations. IEEE Trans Intell Transp Syst 16(2):741–750
40. Dyukov A (2016) Mask angle effects on GNSS speed validity in multipath and tree foliage environments. Asian J Appl Sci 4(2):309–321
41. Malicorne M, Macabiau C, Calmettes V, Bousquet M (2002) Effects of masking angle and multipath on Galileo performances in different environments. INS 2001. 8th Saint Petersburg Conf Integr Navig Syst 36(1):96–107
42. Hewitson S, Wang J (2006) GNSS receiver autonomous integrity monitoring (RAIM) performance analysis. GPS Solut 10:155–170
43. Jiang Z, Groves PD, Ochieng WY, Feng S, Milner CD, Mattos PG (2011) Multi-constellation GNSS multipath mitigation using consistency checking. In: Proceedings of the 24th international technical meeting of the satellite division of the Institute of Navigation, pp 3889–3902
44. Salos Andrés CD (2012) Integrity monitoring applied to the reception of GNSS signals in urban environments (Doctoral dissertation)
45. Castaldo G, Angrisano A, Gaglione S, Troisi S (2014) P-RANSAC: an integrity monitoring approach for GNSS signal degraded scenario. Int J Navig 2014:173818
46. Bijjahalli S, Gardi A, Sabatini R (2018) GNSS performance modelling for positioning and navigation in Urban environments. In: 2018 5th IEEE international workshop metrology AeroSpace, pp 521–526
47. Angrisano A, Gaglione S, Crocetto N, Vultaggio M (2020) PANG-NAV: a tool for processing GNSS measurements in SPP, including RAIM functionality. GPS Solut 24(1):19
48. Liu J, Rizos C, Cai BG (2020) A hybrid integrity monitoring method using vehicular wireless communication in difficult environments for GNSS. Veh Commun 23:100229
49. Sun R, Qiu M, Liu F, Wang Z, Ochieng WY (2022) A Dual w-test based quality control algorithm for integrated IMU/GNSS navigation in urban areas. Remote Sens 14(9):2132
50. Yozevitch R, Moshe BB, Weissman A (2016) A robust GNSS LOS/NLOS signal classifier. Navig 63(4):427–440
51. Hsu L T (2017) GNSS multipath detection using a machine learning approach. In: 2017 IEEE 20th the IEEE international conference on intelligent transportation systems Yokohama, Japan, 16–19 Oct 2017, pp 1–6
52. Sun R, Wang G, Zhang W, Hsu LT, Ochieng WY (2020) A gradient boosting decision tree based GPS signal reception classification algorithm. Appl Soft Comput 86:105942
53. Xu H, Angrisano A, Gaglione S, Hsu LT (2020) Machine learning based LOS/NLOS classifier and robust estimator for GNSS shadow matching. Satell Navig 1(1):1–12

54. Suzuki T, Amano Y (2021) NLOS multipath classification of GNSS signal correlation output using machine learning. Sensors 21(7):2503
55. Lau L, Cross P (2007) Development and testing of a new ray-tracing approach to GNSS carrier-phase multipath modelling. J Geod 81:713–732
56. Nicolas ML, Jacob M, Smyrnaios M, Schæn S, Kürner T (2011) Basic concepts for the modelling and correction of GNSS multipath effects using ray tracing and software receivers. In: 2011 IEEE-APS topical conference on antennas and propagation in wireless communications, pp 890–893
57. Zhu F, Ba T, Zhang Y, Gao X, Wang J (2020) Terminal location method with NLOS exclusion based on unsupervised learning in 5G-LEO satellite communication systems. Int J Satell Commun Netw 38(5):425–436
58. Li L, Elhajj M, Feng Y, Ochieng WY (2023) Machine learning based GNSS signal classification and weighting scheme design in the built environment: a comparative experiment. Satell Navig 4(1):1–23
59. Ng HF, Zhang G, Hsu LT (2021) Robust GNSS shadow matching for smartphones in urban canyons. IEEE Sens J 21(16):18307–18317
60. Sun R, Fu L, Wang G, Cheng Q, Hsu LT, Ochieng WY (2021) Using dual-polarization GPS antenna with optimized adaptive neuro-fuzzy inference system to improve single point positioning accuracy in urban canyons. Navig 68(1):41–60
61. Eueler HJ, Goad CC (1991) On optimal filtering of GPS dual frequency observations without using orbit information. Bull Géodésique 65:130–143
62. Friedman JH (2001) Greedy function approximation: a gradient boosting machine. Ann Stat 1189–1232
63. Breiman L (2001) Random Forests. Mach Learn 45:5–32
64. Ham J, Chen Y, Crawford MM, Ghosh J (2005) Investigation of the random forest framework for classification of hyperspectral data. IEEE Trans Geosci Remote Sens 43(3):492–501
65. Stumpf A, Kerle N (2011) Object-oriented mapping of landslides using Random Forests. Remote Sens Environ 115(10):2564–2577
66. Biagi L, Caldera S (2013) An efficient leave one block out approach to identify outliers. J Appl Geodesy 7(1):11–19
67. Smolyakov I, Rezaee M, Langley RB (2020) Resilient multipath prediction and detection architecture for low-cost navigation in challenging urban areas. Navig 67(2):397–409

Chapter 2
Deep Learning-Enabled Fusion to Bridge GPS Outages for INS/GPS Integrated Navigation

Yimin Zhou, Yaohua Liu, and Jin Hu

Abstract The low-cost inertial navigation system (INS) suffers from bias and measurement noise, which would result in poor navigation accuracy during the global positioning system (GPS) outages. Aiming to bridge the GPS outages duration and enhance the navigation performance, a deep learning network architecture named GPS/INS neural network (GI-NN) is proposed to assist the INS. The GI-NN combines a convolutional neural network and a gated recurrent unit neural network to extract the spatial features from the inertial measurement unit (IMU) signals and track their temporal characteristics. The relationship among the attitude, specific force, angular rate and the GPS position increment is modelled, while the current and previous IMU data are used to estimate the dynamics of the vehicle via the proposed GI-NN. Numerical simulations, real field tests and public data tests are performed to evaluate the effectiveness of the proposed algorithm. Compared with the traditional machine learning algorithms, the results illustrate that the proposed method can provide more accurate and reliable navigation solution in the GPS denied environments.

2.1 Introduction

The integrated navigation system based on the Inertial Navigation Systems (INS) and Global Positioning System (GPS) is a high-precision position and navigation solution for most unmanned ground vehicles or unmanned aerial vehicles (UAVs) when GPS signals are available [1–3]. However, when the UAVs equipped with

Y. Zhou (✉)
Shenzhen Institute of Advanced Technology, Chinese Academy of Sciences, Xueyuan Avenue, Xili Univerrsity Town, Shenzhen 1068, China
e-mail: ym.zhou@siat.ac.cn

Y. Liu
Artificial Intelligence and Digital Economy Guangdong Provincial Laboratory, Shenzhen, China

J. Hu
School of Information and Electrical Engineering, Hunan University of Science and Technology, Xiangtan, China

K. Yu (ed.), *Positioning and Navigation Using Machine Learning Methods*, Navigation: Science and Technology 14, https://doi.org/10.1007/978-981-97-6199-9_2

GPS pass through environments with weak GPS signals, such as dense skyscraper or tunnels, the INS/GPS integrated navigation system would turn to only INS mode with the presence of the inertial measurement unit (IMU) bias drift and scale factor instability [4], so the navigation accuracy would sharply decreases. Therefore, it is necessary to explore more robust integrated navigation systems to adapt to various complex environments.

To address the issues of GPS interruption on the navigation performance, this chapter proposes a deep learning network structure-the GPS/INS Neural Network (GI-NN) to assist in the INS position navigation. The GI-NN combines the Convolutional Neural Network (CNN) and Gated Recurrent Unit Neural Network (GRUNN) to extract spatial features from the IMU signals and track their temporal information. It establishes a relationship model among attitude, specific force, angular rate, and GPS position increments, so as to dynamically estimate the vehicle motion state via the current and past IMU data. Furthermore, a hybrid fusion strategy is designed to fit the nonlinear relationship between the sensor measurements and GPS position increments: when GPS is available, INS data is fused with GPS data using a Kalman filter to obtain more reliable position and velocity information. Meanwhile, the GPS and INS data are stored in an onboard computer for the GI-NN model training. When GPS signals are unavailable, the trained GI-NN model is used to predict the GPS position increments and generate virtual GPS position values to continue to fuse with INS data, thereby maintaining high navigation accuracy. Finally, the effectiveness of the algorithm is validated through simulation, actual experiments and public datasets. Comparative experimental results with the traditional machine learning algorithms demonstrate that the proposed method can provide a more accurate and reliable navigation solution in GPS interrupted environments.

2.2 INS/GPS Integrated Navigation System

INS and GPS each has distinct advantages and notable shortcomings. INS is an autonomous navigation system that does not rely on external electromagnetic information, featuring characteristics like good concealment, strong resistance to the interference, and short-term high precisions. However, the long-term errors of the INS diverge without bounds over time due to the errors accumulating in the inertial devices. GPS has long-term stable positioning accuracy, and errors do not accumulate over time, but its output data rate is relatively low. Therefore, a combined system can be constructed to overcome their individual shortcomings and a robust position navigation solution can provide with higher accuracy than a single navigation system, by fully leveraging the complementary characteristics of the INS/GPS.

2.2.1 *Inertial Navigation Coordinate Systems*

When describing the motion of a mobile robot, it is essential to establish both the robot coordinate system and the reference coordinate system for the relative motion beforehand. The common coordinate systems in the inertial navigation mainly include: the Inertial Coordinate System (ICS), the Earth-Centered Earth-Fixed (ECEF) Coordinate System, Local Horizontal Coordinate System and the Body Coordinate System.

1. *Inertial Coordinate System*

The ICS remains stationary or moves at a constant velocity (without acceleration) in the space. All measurements of the inertial devices are the results under the inertial frame. As illustrated in Fig. 2.1a, the center of the Earth serves as the origin of the ICS. The Z-axis aligns with the Earth's rotation axis, pointing toward the North Pole, while the X-axis lies within the equatorial plane, directed towards the vernal equinox. The Y-axis, forming a right-handed coordinate system with the X and Z axes, completes the orientation.

2. *Earth-Centered Earth-Fixed Coordinate System*

The ECEF Coordinate System shares the same origin and Z-axis definition as the ICS, but it rotates synchronously with the Earth. As shown in Fig. 2.1b, the origin coincides with the center of the Earth, the Z-axis points to the North Pole, the X-axis points to the intersection of the equator and the prime meridian, and the Y-axis in the equatorial plane forms a right-handed Cartesian coordinate system with the X and Z axes.

3. *Local Horizontal Coordinate System*

The Local Horizontal Coordinate System is commonly used to describe the attitude and velocity of a vehicle in near-Earth motion, also referred to as the navigation coordinate system. As shown in Fig. 2.2a, the origin is the center of the coordinate system of the inertial device, with the X-axis pointing to local geographic north, the Y-axis

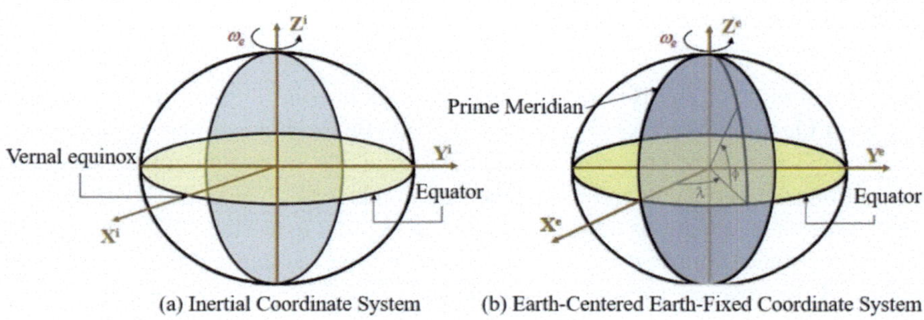

Fig. 2.1 ICS and ECEF coordinate system

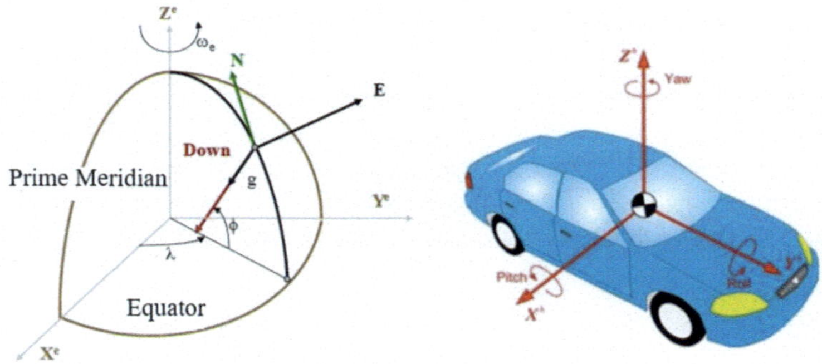

(a) Local Horizontal Coordinate System (b) Body Coordinate System

Fig. 2.2 Local horizontal coordinate system and body coordinate system

pointing to local geographic east, and the Z-axis forming a right-handed Cartesian coordinate system with the X and Y axes, perpendicular to the ellipsoidal surface of the Earth. Since the Z-axis can be oriented upward or downward perpendicular to the Earth's ellipsoidal surface, there are two types of the Local Horizontal Coordinate System: one is the North-East-Down (NED), and the other is the East-North-Up (ENU). The NED coordinate system is adopted in this chapter.

4. *Body Coordinate System*

In practical applications, the measurement axes of the accelerometer and gyroscope in the MEMS-IMU are determined by the axes of the motion platform on which the device is mounted, forming the Body Coordinate System. As shown in Fig. 2.2b, the origin is set at the mass center of the vehicle, the Y-axis points forward along the vehicle, the X-axis is perpendicular to the Y-axis and points sideways along the vehicle, and the Z-axis forms a right-handed Cartesian coordinate system with the X and Y axes, pointing vertically along the vehicle, following the right-hand rule.

2.2.2 *Kalman Filter Model*

The Kalman Filter (KF) [5] is an efficient algorithm in control theory, also known as the Linear Quadratic Estimator (LQE), which can be used to estimate the unknown variables via a series of observations and measurements, providing greater accuracy than estimates based on the individual measurements [6, 7]. The corresponding state and measurement equations must be established in order to use the KF for the system state estimation. Assuming that the actual state at time k is derived from the state at time $k-1$, the discrete form of the system state equation can be written as,

$$x_k = F_k x_{k-1} + B_k u_k + \omega_k \tag{2.1}$$

where x_k is the state vector at the kth moment; F_k is the state transition model applied to the previous state x_{k-1}; B_k is the coefficient vector related to the control vector u_k; ω_k is the process noise assumed as a zero mean with normal distribution and covariance matrix Q_k.

The measurement equation can be defined as,

$$z_k = H_k x_k + v_k \tag{2.2}$$

where z_k is the measurement vector; H_k is the observation model which maps the actual state space into the observed space; v_k is the observation noise assumed to be zero mean Gaussian white noise with covariance R_k.

If the estimated state vector x_k and measurement vector z_k can be written in the form of Eqs. (2.1) and (2.2), and the noise ω_k and v_k follow the zero mean Gaussian white noise distribution, the process of KF algorithm can be divided into two parts,

(1) Time Update:

$$\hat{x}_k^- = F_K \hat{x}_{k-1} + B_k u_k \tag{2.3}$$

$$P_k^- = F_K P_{k-1} F_k^T + Q \tag{2.4}$$

(2) Measurement Update:

$$K_k = P_k^- H_k^T (H_k P_k^- H_k^T + R)^{-1} \tag{2.5}$$

$$\hat{x}_k = \hat{x}_k^- + K_k \left(z_k - H_k \hat{x}_k^- \right) \tag{2.6}$$

$$P_k = (I - K_k H_k) P_k^- \tag{2.7}$$

In Eqs. (2.3) and (2.4), $\hat{x}_k^-$ is the a priori estimate of the system state, *and* P_k^- is the a priori error covariance. P_k denotes the predicted state covariance matrix, and K_k is the Kalman filter gain matrix. The essence of the Kalman filter lies in the iterative process of time update and measurement update, as depicted in Fig. 2.3. First, the filter undergoes initialization, which requires the initial values for the state estimate x_0 and the estimated mean square error P_0. P_0 is derived based on x_0 and is typically set to a relatively high value. Besides, the initial estimates of the system noise covariance matrix Q and the measurement noise covariance matrix R are required, which are based on prior state information about the system for the optimal state estimation. In the first step of the time update process, the system state is recursively updated from time k-1 to time k, denoted as $\hat{x}_k^-$. The second step of the time update involves calculating the covariance P_k^- of $\hat{x}_k^-$. This process utilizes all the available information at time $k - 1$ to obtain the expected value of the state error variance at time k.

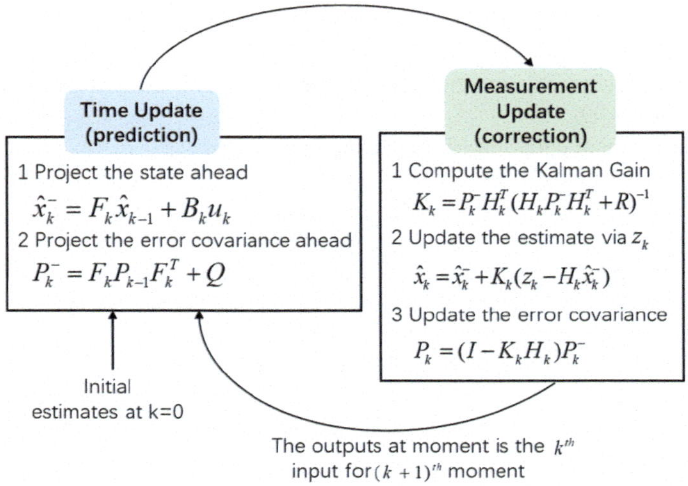

Fig. 2.3 The flowchart of the KF Algorithm

After updating the estimated state of KF, whenever a measurement is obtained from an external source, the state estimate is adjusted accordingly. Specifically, the Kalman gain K is calculated based on the measurement covariance R and the a priori error P_k^- to minimize the estimated mean square error. If the measurement noise increases or the process noise decreases, K will decrease accordingly, conversely K will increase. In the INS/GPS integrated navigation, when GPS is relatively accurate and less affected by the noise interference, the value of K will increase correspondingly. After receiving the new measurement z_k at time k, the KF compares z_k with $H_k\hat{x}_k^-$ based on the a priori state estimate. Simultaneously, K is determined by weighting their difference, updating the system state to the optimal estimate, thus completing the correction process. Upon the state estimate correction, KF updates the posteriori covariance matrix P_k based on the priori error covariance matrix P_k^-. Then the KF prepares for the next iteration by treating the posterior estimate as the new a priori estimate for the upcoming KF time update and measurement update cycle.

2.3 The INS/GPS Fusion Algorithm Based on Kalman Filter

The architecture of the INS/GPS integrated navigation can be classified into loosely coupled, tightly coupled and ultra-tightly coupled types. Loosely coupled structure has the advantages such as simplicity and good robustness, making it widely used in the navigation of UAVs. Therefore, this section takes loosely coupled as the basic framework, and utilize the KF to fuse the INS/GPS navigation information, aiming to achieve integrated navigation positioning.

2.3.1 INS Position Estimation Algorithm

The process of the INS pose estimation is illustrated in Fig. 2.4. The raw measurements of the gyroscopes are used to calculate the attitude matrix of the carrier. Through this attitude matrix, the specific force information measured by the accelerometer along the axes of the carrier coordinate system is transformed into a specific coordinate system (such as the navigation coordinate system), then the navigation computation is performed.

The INS position estimation algorithm first requires to associate the measurements of the IMU with the navigation information. Since the navigation position information is obtained by integrating the IMU measurements, the most direct method to associate the two is to establish differential equations between them, namely attitude differential equations, velocity differential equations and position differential equations. By solving the corresponding differential equations, the related pose information can be obtained. In order to establish a comprehensive and general INS differential equation, taking medium-to-high precision inertial navigation as an example, based on reference [8], the attitude differential equation with the navigation coordinate system (N-frame) as the reference frame is written as,

$$\dot{C}_b^n = C_b^n(\omega_{nb}^b \times) \tag{2.8}$$

where $\times$ is the antisymmetric matrix formed by the three-dimensional angular velocity vector, and the matrix C_b^n is the direction cosine transformation matrix from the body coordinate system (b-frame) to the navigation frame (n-frame). Since the output of the gyroscope is the angular velocity ω_{ib}^b of the b-frame relative to the inertial frame (i-frame), and ω_{nb}^b represents the projection of the angular velocity of the b-frame relative to the n-frame in the b-frame, it is necessary to transform Eq. (2.8) as,

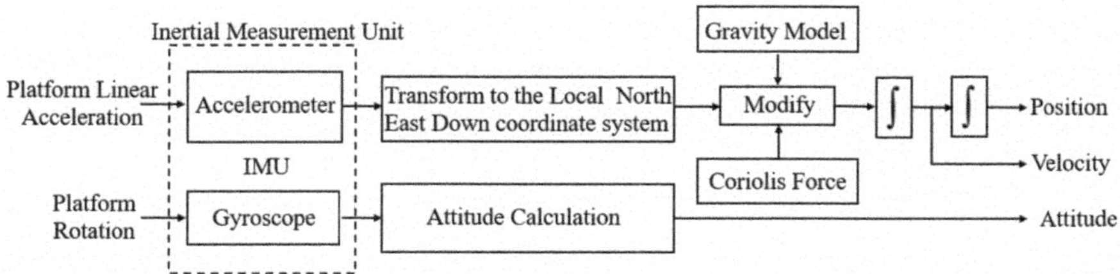

Fig. 2.4 The flowchart of the INS position and attitude calculation

$$\dot{C}_b^n = C_b^n[\omega_{ib}^b\times] - [\omega_{in}^n\times]C_b^n \tag{2.9}$$

where ω_{in}^n is the rotation from the n-frame to the i-frame, including two components: the rotation of the navigation coordinate frame caused by the Earth rotation and the rotation of the n-frame due to the motion of the INS near the Earth surface caused by the curvature of the Earth, namely $\omega_{in}^n = \omega_{ie}^n + \omega_{en}^n$,

$$\omega_{ie}^n = \left[0\ \omega_{ie}\cos L\ \omega_{ie}\sin L\right]^T \tag{2.10}$$

$$\omega_{en}^n = \left[-\frac{v_N}{R_M+h}\ \frac{v_E}{R_N+h}\ \frac{v_E}{R_N+h}\tan L\right]^T \tag{2.11}$$

where ω_{ie} is the Earth rotation angular velocity; L and R are the geographical latitude and height, respectively; v_N and v_E are the velocities in the northward and eastward directions of the vehicle, respectively; RN and R_M are the radii of the Earth meridian and equatorial circles, respectively; h.

Correspondingly, the velocity differential equation and the position differential equation are demonstrated in Eqs. (2.12) and (2.13), respectively, where $\omega_{ie}^n \times v^n$ represents the acceleration caused by the vehicle moving on the rotating surface of the Earth, and $\omega_{en}^n \times v^n$ is the centripetal acceleration caused by the vehicle movement on the Earth surface,

$$\dot{v}^n = C_b^n f_{ib}^b - \left(2\omega_{ie}^n + \omega_{en}^n\right) \times v^n + g^n \tag{2.12}$$

$$\dot{r}^n = v^n - \omega_{en}^n \times r^n. \tag{2.13}$$

2.3.2 *Differential Equations of INS Errors*

The accuracy of the INS is affected by various factors, including initial alignment errors, inertial sensor errors, and limitations of the processed algorithms [9]. To fully understand the impact of these errors on the output parameters of the INS (position, velocity, attitude), it is necessary to establish differential equations of the INS errors. Hence, the KF can be utilized to estimate and compensate for these errors. The INS attitude error can be derived as,

$$\dot{\phi} = \phi \times \omega_{in}^n + \delta\omega_{in}^n - \varepsilon^n \tag{2.14}$$

where Φ is the attitude angle error; ω_{in}^n and $\delta\omega_{in}^n$ are the angular velocity of the rotation of the navigation coordinate frame related to the inertial coordinate frame and its error; ε^n is the gyroscope drift vector of the navigation coordinate frame.

The INS velocity error equation is described as,

$$\delta \dot{V}^n = -\phi^n \times f^n + \delta V^n \times \left(2\omega_{\text{ie}}^n + \omega_{en}^n\right) + V^n \times \left(2\delta\omega_{\text{ie}}^n + \delta\omega_{en}^n\right) + \nabla^n \tag{2.15}$$

where δV^n and V^n are the velocity error and velocity in the east, north and upward, respectively; f^n is the specific force; ω_{ie}^n and ω_{en}^n are the earth self-rotation rate and the angle rate relative to the earth in the navigation coordinate system respectively; ∇^n is the accelerometers bias of the navigation frame.

The INS position error equation is written as,

$$\begin{cases} \delta\dot{L} = \frac{\delta V_N}{R_M+h} - \frac{\delta h V_N}{(R_M+h)^2} \\ \delta\dot{\lambda} = \frac{\delta V_E secL}{R_N+h} + \frac{\delta L V_E tanL secL}{R_N+h} - \frac{\delta h V_E secL}{(R_N+h)^2} \\ \delta\dot{h} = \delta V_U \end{cases} \tag{2.16}$$

where δL, $\delta\lambda$ and δh are the errors of the latitude, longitude and height; δV_N, δV_U and δV_E denote the velocity errors in the north, east and upward directions respectively; R_M and R_N are the radiuses of the curvature in the meridian and prime vertical.

2.3.3 *Design of Loosely Coupled INS/GPS Integrated Navigation System*

As depicted in Fig. 2.5, the GPS and INS are first decoupled and operated independently to provide the navigation outputs separately. To improve the output performance, the outputs of both GPS and INS are subtracted and fed back to the KF. The errors of the INS are then estimated based upon the error differential equations of the INS. After the error correction and compensation, the outputs of the INS are realized in the form of position, velocity, and attitude for the integrated navigation output.

According to Sect. 2.2.2 the KF model in the state equation of INS can be established as,

$$\delta\dot{x} = F\delta x + G\omega \tag{2.17}$$

The state vector δx is selected as a 15-dimensional column vector, including position, velocity, attitude, accelerometer biases, and gyroscope drift error components,

$$\delta x = [\,\delta r^T\ \delta v^T\ \delta\phi^T\ \delta\omega^T \delta f^T\,]^T \tag{2.18}$$

where $\delta r = [\delta L, \delta\lambda, \delta h]^T$ is the position error vector in the longitude, latitude, and altitude directions; $\delta v = [\delta v_n, \delta v_e, \delta v_d]^T$ is the velocity error vector in the north, east, and down directions; $\delta\phi = [\delta\omega_x, \delta\omega_y, \delta\omega_z]^T$ denotes the error vector for

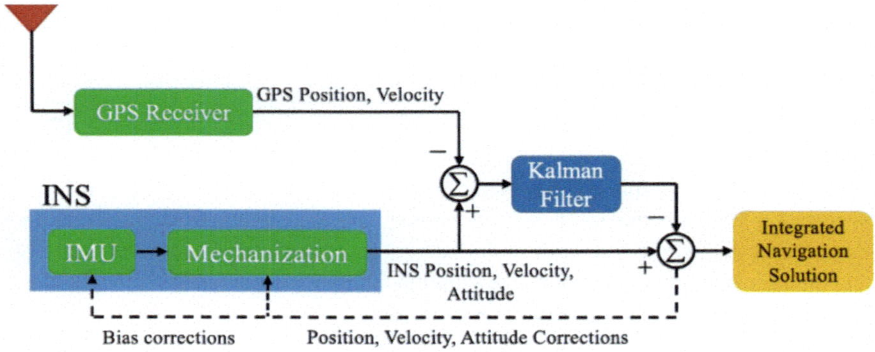

Fig. 2.5 A block diagram of a loosely coupled INS/GPS integration

pitch, roll, and yaw angles; $\delta\omega = \left[\delta\omega x, \delta\omega y, \delta\omega z\right]^T$ is the gyroscope error vector; $\delta f = [\delta f_x, \delta f_y, \delta f_z]^T$ is the accelerometer error vector; ω denotes the Gaussian white noise with unit variance.

In Eq. (2.17), G represents the noise distribution vector, including the variance of the state vector,

$$G = [\sigma_{r,1\times3}\sigma_{v,1\times3}\sigma_{\phi,1\times3}\sigma_{\omega,1\times3}\sigma_{f,1\times3}]^T \tag{2.19}$$

and F in Eq. (2.17) is the dynamic coefficient matrix composed of the error models of the INS position, velocity, attitude and inertial devices. Its specific form can be derived based on Eqs. (2.14)–(2.16),

$$\mathrm{F} = \begin{bmatrix} 0_{3\times3} & F_r & 0_{3\times3} & 0_{3\times3} & 0_{3\times3} \\ 0_{3\times3} & 0_{3\times3} & F_v & 0_{3\times3} & R_b \\ 0_{3\times3} & F_\phi & 0_{3\times3} & R_b & 0_{3\times3} \\ 0_{3\times3} & 0_{3\times3} & 0_{3\times3} & F_\omega & 0_{3\times3} \\ 0_{3\times3} & 0_{3\times3} & 0_{3\times3} & 0_{3\times3} & F_f \end{bmatrix} \tag{2.20}$$

Therefore, the system equations for the loosely coupled INS/GPS integrated navigation can be written as,

$$\begin{bmatrix} \delta\dot{r} \\ \delta\dot{v} \\ \delta\dot{\phi} \\ \delta\dot{\omega} \\ \delta\dot{f} \end{bmatrix} = \begin{bmatrix} 0_{3\times3} & F_r & 0_{3\times3} & 0_{3\times3} & 0_{3\times3} \\ 0_{3\times3} & 0_{3\times3} & F_v & 0_{3\times3} & R_b \\ 0_{3\times3} & F_\phi & 0_{3\times3} & R_b & 0_{3\times3} \\ 0_{3\times3} & 0_{3\times3} & 0_{3\times3} & F_\omega & 0_{3\times3} \\ 0_{3\times3} & 0_{3\times3} & 0_{3\times3} & 0_{3\times3} & F_f \end{bmatrix} \begin{bmatrix} \delta r \\ \delta v \\ \delta\phi \\ \delta\omega \\ \delta f \end{bmatrix} + \begin{bmatrix} \sigma_r \\ \sigma_v \\ \sigma_\phi \\ \sigma_\omega \\ \sigma_f \end{bmatrix} \omega \tag{2.21}$$

Similarly, the measurement equations for the INS/GPS integrated navigation can be written as,

$$\delta z_k = H_k \delta x_k + \eta_k \tag{2.22}$$

According to the loosely coupled integrated navigation framework, the measurement vector δz_k should include the difference between the predicted values from the INS and the corresponding GPS measurements, as follows,

$$\delta z_k = \begin{bmatrix} r_{INS} - r_{GPS} \\ v_{INS} - v_{GPS} \end{bmatrix} = \begin{bmatrix} L_{INS} - L_{GPS} \\ \lambda_{INS} - \lambda_{GPS} \\ h_{INS} - h_{GPS} \\ v_{n,INS} - h_{n,GPS} \\ v_{e,INS} - h_{e,GPS} \\ v_{d,INS} - h_{d,GPS} \end{bmatrix} \tag{2.23}$$

In Eq. (2.22), η_k is a zero-mean, variance R_k measurement noise vector, and H_k is the measurement matrix at time t, representing the linear combination relationship of the state vector without noise. In the loosely coupled integrated navigation framework, since the measurements directly come from the position and velocity error states, H_k can be written as,

$$H_k = \begin{bmatrix} I_{6\times6} \; 0_{6\times9} \end{bmatrix} \tag{2.24}$$

where I represents the identity matrix, thus the complete measurement equation for loosely coupled INS/GPS can be written as,

$$\begin{bmatrix} r_{INS} - r_{GPS} \\ v_{INS} - v_{GPS} \end{bmatrix} = \begin{bmatrix} I_{3\times3} & 0_{3\times3} & 0_{3\times9} \\ 0_{3\times3} & I_{3\times3} & 0_{3\times9} \end{bmatrix} \delta x_k + \begin{bmatrix} \eta_r \\ \eta_v \end{bmatrix} \tag{2.25}$$

So far, the derivation of the system state equation and observation equation for the loosely coupled INS/GPS integrated navigation can be obtained. When the GPS signal is available, a KF is constructed based on the derived system state and observation equations, performing the KF time update and measurement update processes iteratively to estimate the INS error compensation after the INS/GPS fusion. In turn, the INS navigation results can be corrected to improve the accuracy of the INS.

2.4 The INS/GPS Integrated Navigation Algorithm in GPS Denied Environments

From the previous analysis, it is evident that when the GPS signal is available, the KF can effectively fuse the INS and GPS measurements, thereby correcting the INS errors. However, in the scenarios where the GPS signal is interrupted, such as when a mobile robot enters a tunnel or encounters dense tall buildings, the KF may be unable to obtain new GPS measurements for an extended period, leading to the inability to complete the measurement update process and consequently failing to correct the INS errors. This would result in a rapid degradation of the accuracy of the INS. Therefore, considering the cost and size constraints of mobile robots, we propose a navigation algorithm assisted by the deep learning. In the GPS signal denied environments, a pre-trained deep learning model is utilized to assist in correcting INS errors, thereby mitigating the INS error drift.

2.4.1 GRU Neural Network

Here, the INS and GPS data are assumed as a kind of time series data so that a deep learning model is designed based on the Gated Recurrent Unit (GRU) Neural Network to assist in the INS navigation. Unlike the feedforward Neural Network, the connections between RNN nodes can form a directed graph along the sequence, allowing for more effective processing of the input time-series data, via their internal states [10–12]. However, it could face the vanishing gradient problem, which makes it unable to find appropriate gradients for long-term memory. To solve this issue, Kyunghyun Cho et al. [13] have developed the GRU Neural Network, similar to (Long Short-Term Memory) LSTM, with a forget gate but fewer parameters. Therefore, we use the GRU neural network to predict the GPS position increments when GPS fails.

A GRU memory unit is illustrated in Fig. 2.6, where h_{t-1} is the hidden state of the previous moment, h_t is the current output of the hidden state, and x_t is the input data in the current moment. There are two gates in the GRU structure, i.e., the update gate and the reset gate. The update gate is responsible for determining how much of the previous hidden state is to be retained and which portion of the new proposed hidden state (derived from the reset gate) is to be added on the final hidden state. The reset gate is responsible to decide which portions of the previous hidden state are to be combined with the current input to form a new hidden state.

The forward passing equations of the GRU can be written as,

$$r_t = \sigma(W_{xr}x_t + W_{hr}h_{t-1} + b_r) \tag{2.26}$$

$$z_t = \sigma(W_{xz}x_t + W_{hz}h_{t-1} + b_z) \tag{2.27}$$

$$\tilde{h}_t = \tanh(W_{xh}x_t + W_{hh}(r_t \odot h_{t-1}) + b_h) \tag{2.28}$$

$$h_t = z_t h_{t-1} + (1 - z_t) \odot \tilde{h}_t \tag{2.29}$$

where W_{xr}, W_{hr},W_{xz},W_{hz}, W_{xh} and W_{hh} are the weight matrices between the input layer, update gate, reset gate and hidden state;b_r,b_z and b_h are the bias vectors of the update gate, reset gate and hidden state respectively. σ and *tanh* are the activation

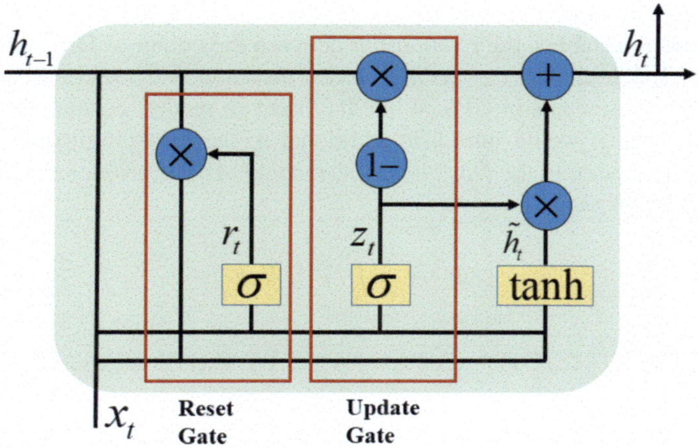

Fig. 2.6 Illustration of the GRU structure

functions to transform the hidden states to the nonlinear states. ⊙ represents the Hadamard product operation.

2.4.2 A Deep Learning-Assisted INS/GPS Integrated Navigation Structure

The main idea of the INS/GPS integrated navigation framework assisted by the deep learning models proposed in this Chapter is to utilize the deep learning to build a model between the navigation information and INS outputs (velocity, attitude, specific force and angular rate), while maintaining high navigation accuracy when GPS fails. Navigation information generally includes position error or position increments between the INS and GPS. The main motivation for using the deep learning methods is that the INS/GPS system is a nonlinear and complex system, making it difficult to establish an accurate mathematical model. The accuracy of the traditional filtering-based methods largely depends on the quality of the mathematical model. When some sensors fail, such as due to the GPS signal interruption, filtering-based methods also become ineffective. Deep learning-based methods are data-driven and do not require a precise mathematical model of the system, making them suitable for handling nonlinear systems. Compared to the filters, deep learning methods have stronger learning capabilities, while the filter parameters are fixed with limited adaptability to different scenarios.

Currently, many ML-aided models have been proposed to describe the relationship between the navigation information and the INS outputs, almost all of which can be divided into 3 classes in terms of the outputs, i.e., $O_{INS} - \delta P_{INS}$, $O_{INS} - X_k$ and $O_{INS} - \Delta P_{GPS}$. The $O_{INS} - \delta P_{INS}$ model can be designed to find the relationship between the INS information and the position error of GPS & INS. The $O_{INS} - X_k$ model intends to establish the relationship between the output of INS and the state vector X_k of KF. In the $O_{INS} - \Delta P_{GPS}$ model, the input is the INS and the output is the position increments of GPS. While the first two models contain the INS and GPS information, it would introduce additional mixed errors compared with the $O_{INS} - \Delta P_{GPS}$ model. In the $O_{INS} - \Delta P_{GPS}$ model [14], the position increments of the GPS can be denoted as,

$$\Delta P_{GPS} = \iint \dot{V}_n(t)dtdt \tag{2.30}$$

$$= \iint (C_b^n f_{ib}^b(t) - \left(2\omega_{ie}^n(t) + \omega_{en}^n(t)\right) \times V_n(t) + G_n)dtdt \tag{2.31}$$

where V_n is the velocity of the vehicle in the navigation coordinate system, and G_n is the gravity vector. Therefore, this chapter selects the ($O_{INS} - \Delta P_{GPS}$) model to construct the deep learning-assisted INS/GPS integrated navigation structure.

In Eq. (2.30), ω_{ie}^n and G_n are affected by the longitude and latitude, while ω_{en}^n is related to V_n. Since the motion range of the mobile robots in actual scenarios is small, the variations in longitude and latitude are minimal. Therefore, C_b^n, f_{ib} and V_n are the main factors affecting ΔP_{GPS}, where C_b^n can be expressed as follows,

$$\begin{bmatrix} \cos\theta\cos\psi & -\cos\gamma\sin\psi + \sin\gamma\sin\theta\cos\psi & \sin\gamma\sin\psi + \cos\gamma\sin\theta\cos\psi \\ \cos\theta\sin\psi & \cos\gamma\cos\psi + \sin\gamma\sin\theta\sin\psi & -\sin\gamma\cos\psi + \cos\gamma\sin\theta\sin\psi \\ -\sin\theta & \sin\gamma\cos\theta & \cos\gamma\cos\theta \end{bmatrix} \tag{2.32}$$

where $\{\theta,\gamma,\psi\}$ represent pitch, roll and yaw angles. The attitude angles are mostly obtained by integrating the angular velocity ω_{ib}^b measured by the gyroscope. In summary, ΔP_{GPS} is mainly determined by f_{ib}, ω_{ib}^b, θ, γ and ψ. Therefore, these variables are selected as inputs to the deep learning model, which learns to fit the mathematical relationship between the position increments and the INS/GPS integrated navigation system when GPS signal is normal.

The trained deep learning model (INS/GPS Neural Network, GI-NN) is deployed on the mobile robot initially. After the robot starts to move, the deep learning-enabled INS/GPS integrated navigation can be divided into two modes. As shown in Fig. 2.7, when GPS is available, the navigation system can operate in the online training mode. The INS would provide the velocity V_{INS}, position P_{INS}, and attitude A_{INS}, which are fused with the position P_{GPS} provided by the GPS using a KF. Simultaneously, the estimated velocity error δV, position error δP and attitude error δA are fed back to the inertial navigation system to reduce its position drift. Furthermore, the INS/GPS data is stored in the onboard computer for training the GI-NN deep learning model. Typically, the GI-NN deep learning model is primarily in a training state as GPS interruptions only represent a small portion of the mobile robot motion duration. During the long-term training process, IMU data is continuously sent to the GI-NN deep learning model for training, ensuring that GI-NN is adequately trained with

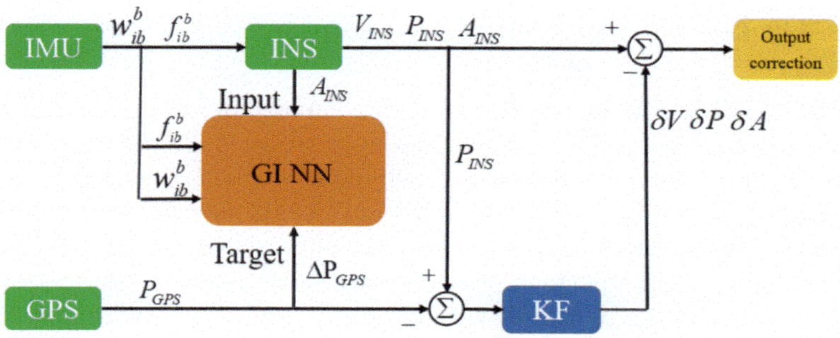

Fig. 2.7 The INS/GPS integrated navigation system based on GI-NN in the training mode

abundant data. Through this training process, the weights between the hidden layer neurons can be adjusted to better map the input–output relationship.

Once the GPS signal becomes unavailable, the deep learning-enabled INS/GPS integrated navigation system enters a prediction mode. During the GPS outage, the KF cannot acquire new GPS observations, resulting in the inability to update KF estimations. Consequently, the INS/GPS integrated navigation system can be effectively transformed into an independent inertial navigation system, resulting in more accumulated errors over time. The block diagram of the prediction mode is shown in Fig. 2.8. The trained GI-NN deep learning model predicts the GPS position increments ΔP_{GPS}. By summing all ΔP_{GPS} increments according to Eq. (2.33), a virtual GPS position can be obtained,

$$P_{GPS}(k) = P_{GPS_0} + \sum_{i=0}^{k} \Delta P_{GPS_i} \tag{2.33}$$

where P_{GPs0} is the initial position when GPS fails at the kth moment. This virtual $P_{GPS}(k)$ is then substituted for the interrupted GPS position increments and fused with INS by KF. The hybrid structure of the INS/GPS will maintain the navigation continuously when GPS signals are lost.

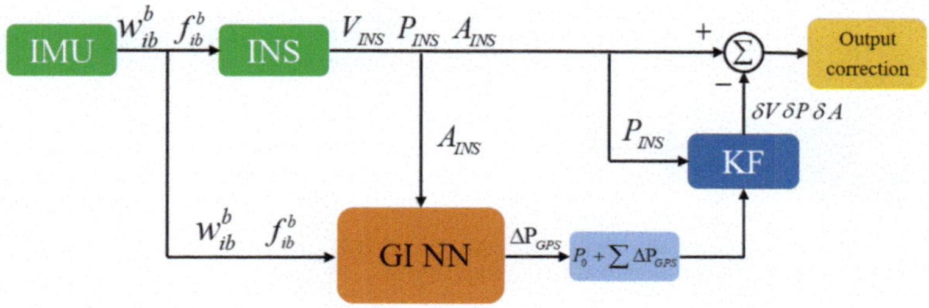

Fig. 2.8 The INS/GPS integrated navigation system based on GI-NN in the prediction mode

2.4.3 The GI-NN Architecture

As illustrated in Fig. 2.9, the GI-NN is mainly composed of multiple convolutional layers and recurrent layers. The convolutional layers are acted as the feature extractors to learn the sparse feature representation and model implicit dependencies among various sensors, so as to provide abstract representations of the input sensor data in the feature maps. The recurrent layers can model the temporal dynamics of the activation of the feature maps. The input to the network can be expressed as a time series vector $S_t = (S_1, S_2, \ldots, S_i, \ldots, S_k)$, while each element S_i represents 9 measurements of the accelerometer, gyroscope and attitude sensor at each time step, and k represents the time window size. The predicted output is the GPS position increments ΔP_{GPS}^{k+1} in the time step $k+1$ based on the previous k IMU measurements, hence, the prediction problem can be formulated as,

$$\Delta P_{GPS}^{k+1} = GI - NN(S_1, S_2, \ldots, S_k) \tag{2.34}$$

One-dimensional convolution (1DCNN) is used to deal with each element in S_t, and the convolution kernel filter is applied to acquire the convolution information of the local perceptual domain by the sliding filter. To model the implicit dependencies and extract the global features among multi sensors, four 1D convolutional layers of 512 features with a kernel size of 1 are stacked to embed the raw data with higher dimension. The formulation of each convolutional layer l is,

$$F_t^{(l)} = RELU\left(W^{(l)} * F_t^{l-1}\right) + b^{(l)} \tag{2.35}$$

where $W^{(l)}$ and $b^{(l)}$ are the learned parameters, and rectified linear unit (RELU) is adopted as the activation function. After stacking four convolutional layers, the acquired feature map can contain the correlated features from different sensors data.

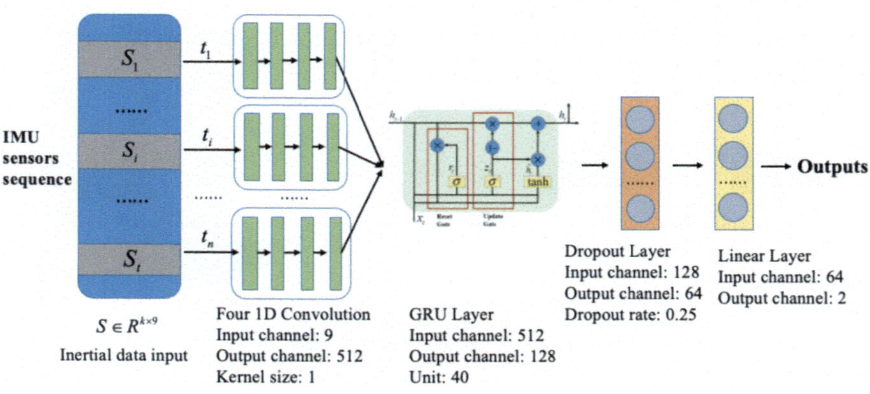

Fig. 2.9 The architecture of the GI-NN

The GRU network is used to capture the temporal sequential dependency, which is proposed to address the exploding and vanishing gradient issue of the traditional Recurrent Neural Network (RNN), described as,

$$h_t = GRU(F_t, h_{i,t-1}) \tag{2.36}$$

where h_t is the output of the GRU layer at time step t. Finally, a dropout layer with a probability of 0.25 is also applied to avoid the overfitting, and a linear layer is added to transform the high dimensional data as the output data dimension to generate the navigation solutions.

2.5 Experimental Results and Analysis

To fully test and validate the performance of the proposed algorithm, simulation experiments, physical experiments, and experiments using public datasets are conducted, and comparisons are implemented with different algorithms: (1) INS; (2) Multi-layer Perceptron (MLP) model; (3) Long Short-Term Memory (LSTM) neural network; (4) GI-NN model.

2.5.1 Simulation Experiment

In the simulation tests, the NaveGo [15] toolbox, developed via MATLAB, is utilized to generate the UAV flight trajectories. After generating the motion trajectories, the related INS and GPS output data are obtained. The specific parameters of the sensors during the simulation process are presented in Table 2.1. The deep learning framework PyTorch 1.8 is employed, and the experimental setup consists of an Intel Core i7-6700 processor running at 3.4 GHz with 16 GB RAM. Initially, a UAV flight trajectory of 430 s is generated using NaveGo, including IMU, GPS and navigation data, namely attitude angles, velocity and position. The simulated trajectory, as shown in Fig. 2.10, includes UAV climbing, straight flight and turning, covering the basic motion states.

To train the network model more effectively, 80% of the trajectory data from 0 to 250 s are used for the model training, with the remaining 20% reserved for the performance evaluation and parameter tuning. To enhance the testing efficiency, two typical scenarios, the straight-line motion phase (260–290 s) and the turning phase (370–400 s), are selected for testing. Regarding RNN, the sequence length is utilized to determine how much contextual information is sent to the model on each occasion. Considering the high sampling rate of the IMU sensor, a length of 40 and a time window of 200 ms are applied to the IMU sequence data, dividing it into multiple non-overlapping blocks as model inputs.

Table 2.1 The specifications of the navigation sensors

Sensors	Results	Accuracy
IMU	Gyroscope static biases	3°/s
	Gyroscope angular random walk	$2°/\sqrt{\text{hr}}$
	Accelermeter static biases	50 mg
	Accelermeter velocity random walk	$0.2\ \text{m/s}/\sqrt{hr}$
	Sampling frequency	200
GPS	Position	1 m
	Frequency	5 Hz

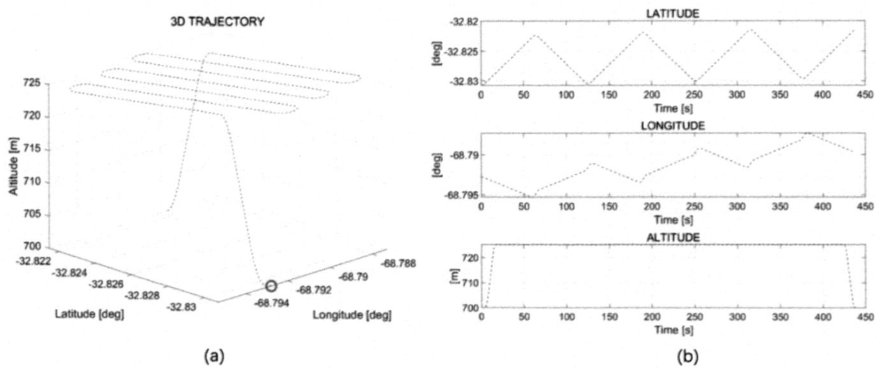

Fig. 2.10 The simulation trajectory. **a** The 3D simulation trajectory. **b** The position of the simulation trajectory

To mitigate the risk of overfitting and expedite training, the Adam optimizer [16] with a cosine annealing restart scheduler [17] is employed for training, with a learning rate initialized to 0.0001, a batch size set to 64, and 200 training epochs executed.

Based on the GI-NN-assisted integrated navigation system, it initially operates in the training mode, during which a 30-s GPS outage period is designed, occurring between 260 and 290 s. As depicted in Fig. 2.11, the north and east position errors are compared for the INS, MLP, LSTM, and GI-NN methods, represented by black, blue, orange and red curves, respectively. For simplicity of the observation, a logarithmic transformation is applied to the final results. Due to the lack of true GPS position information, the position error gradually increases over time. In the case of INS, MLP, LSTM, and GI-NN, the maximum east position errors are 166.8 m, 11.9 m, 7.3 m, and 5.4 m, respectively, while the maximum north position errors are 142.9 m, 7.4 m, 4.5 m and 3.4 m, respectively. Hence, the results indicate that the proposed GI-NN algorithm outperforms INS, MLP, and LSTM, thereby validating that GI-NN can more accurately predict the position increments.

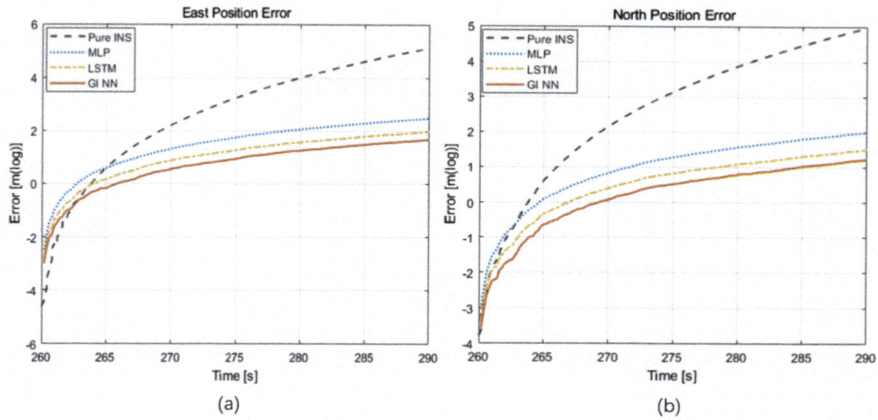

Fig. 2.11 The position errors among different algorithms during the period of 260–290 s in the simulation test. **a** The east position error. **b** The north position error

Another GPS outage scenario is simulated under more complex conditions, namely, a turning and heading change scenario. Figure 2.12 illustrates the navigation results during the period from 370 to 400 s. The maximum positioning errors in the north direction for MLP, LSTM, and GI-NN are reduced by 62.2%, 72.9%, and 86.5%, respectively, while in the east direction, they are reduced by 12.1%, 4.7%, and 44.5%, respectively. In Fig. 2.12b, although LSTM exhibits better northward position error compensation performance in the short term, its error increases faster than GI-NN with an increase in prediction time interval, thus indicating the overall superior performance of GI-NN over LSTM. Compared to Fig. 2.11, the prediction error variation of GI-NN is larger, indicating that the position error is impacted by different scenarios. Furthermore, the error curve of the GI-NN is more stable than that of the LSTM and MLP, demonstrating that GI-NN possesses stronger robustness. Therefore, even under severe maneuvering conditions, GI-NN can still maintain satisfactory navigation accuracy.

2.5.2 Field Tests

To further validate the proposed method, actual UAV flight tests are conducted. A quadcopter platform has been established, as depicted in Fig. 2.13a, for collecting flight data containing various real-world noises. The autopilot of the quadcopter is PIXHAWK, a high-performance autopilot suitable for various robot platforms such as fixed-wing aircraft, multi-rotor aircraft, helicopters, cars and boats. The primary IMU sensors in PIXHAWK are Invensense MPU 6000 and ST Micro LSM 303D, with gyroscope bias of 5 °/s and accelerometer bias of 60 mg [18]. To acquire position reference data, two GPS receivers with different accuracies are simultaneously

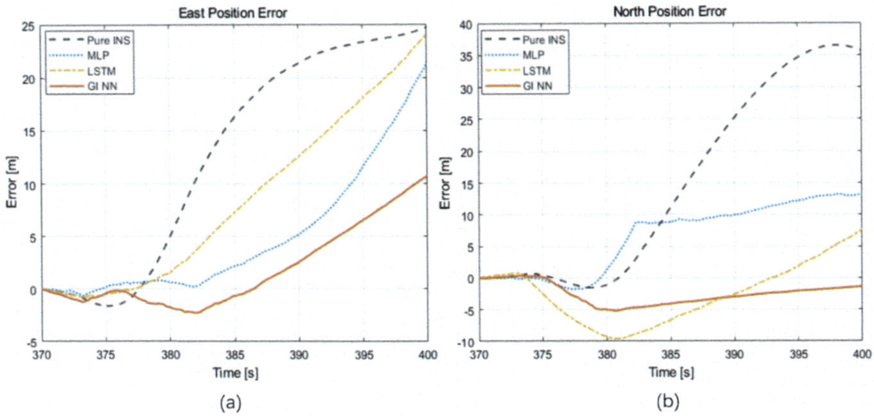

Fig. 2.12 The position errors among different algorithms during the period of 370–400 s in the simulation test. **a** The east position error. **b** The north position error

installed on the quadcopter: the ublox-neo-m8n GPS receiver and the ublox-neo-6 m GPS receiver. The positioning accuracy of the ublox-neo-m8n GPS receiver exceeds that of the ublox-neo-6 m. Therefore, data from the ublox-neo-m8n GPS receiver are used as the position references. Further, by employing Kalman Filtering, data from the ublox-neo-6 m GPS receiver are fused with INS data to obtain the position estimates.

As shown in Fig. 2.13b, the UAV flight trajectory is delineated by a red line, while the interruptions in the GPS signal, each lasting 35 s, are indicated by the yellow lines. To ensure the reliability of the GPS signal, the experiment is conducted in an open playground. Throughout the entire testing period, a minimum of 7 or more satellite signals are consistently available. Additionally, intentional GPS signal interruptions are induced by human-made GPS shielding.

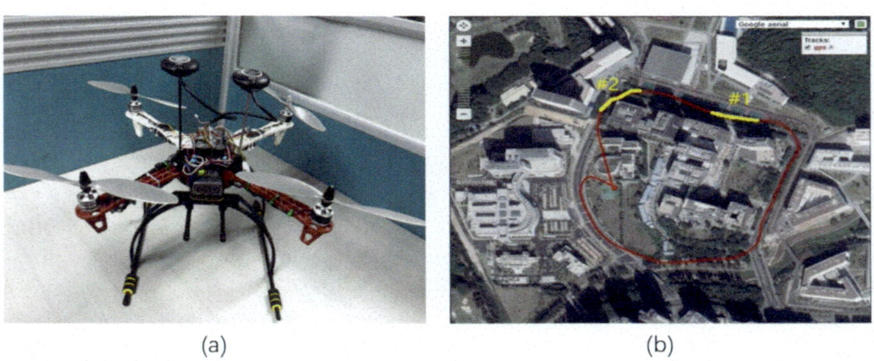

Fig. 2.13 The quadcopter equipped with PIXHAWK autopilot and flight trajectory. **a** The UAV. **b** The UAV flight trajectory

Figure 2.14 displays the northward and eastward position errors generated by different algorithms during the first GPS outage, while Table 2.2 summarizes the maximum errors in the eastward and northward positions during GPS outages. It is obvious that when a GPS outage occurs, the KF lacks the GPS update data, causing the INS to degrade into the inertial navigation system. Consequently, the navigation errors rapidly accumulate over time. However, when MLP, LSTM, or GI-NN is utilized for the estimation, navigation errors are somewhat suppressed. During the first GPS outage, the maximum eastward position errors for only INS, MLP, LSTM, and GI-NN are 107.8 m, 15.2 m, 11.8 m, and 2.4 m, respectively, while the maximum northward position errors are 36.1 m, 35.5 m, 12.9 m, and 10.6 m, respectively. Comparative analysis reveals that the maximum error of the GI-NN-assisted navigation method is smaller than that of MLP, LSTM, and only INS, demonstrating the effectiveness of the GI-NN-assisted integrated navigation method in improving the navigation accuracy during the GPS outages and its superior ability to suppress the inertial navigation errors compared to the traditional feedforward neural networks.

To further analyze the role of the GI-NN-assisted integrated navigation system during large-scale maneuvers of UAVs, the moment of the second GPS outage is chosen to coincide with the turning and direction change phase of the UAV. As depicted in Fig. 2.15, the eastward and northward position errors of INS, MLP, LSTM and GI-NN methods are compared and the results are already summarized

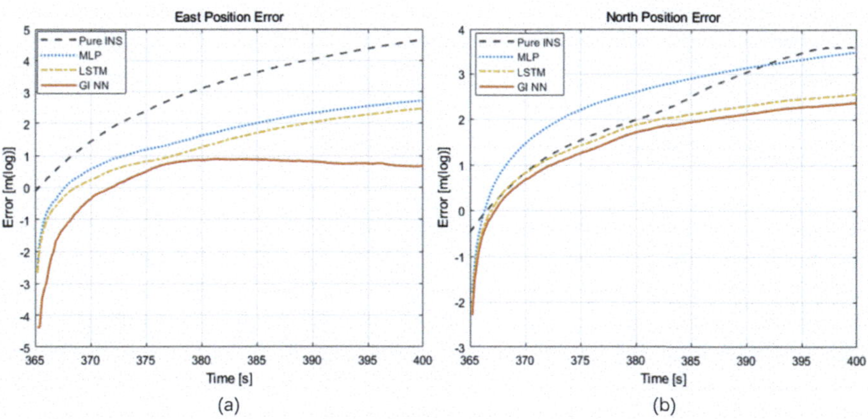

Fig. 2.14 The position errors among different algorithms during the period of 365–400 s in the real field test. **a** The east position error. **b** The north position error

Table 2.2 The max position error among different algorithms in the real field test

Max position error(m)	INS		MLP		LSTM		GI-NN	
	East	North	East	North	East	North	East	North
365–400 s	107.8	36.1	15.2	35.5	11.8	12.9	2.4	10.6
490–525 s	93.4	235.9	62.8	63.1	49.6	51.4	39.7	40.5

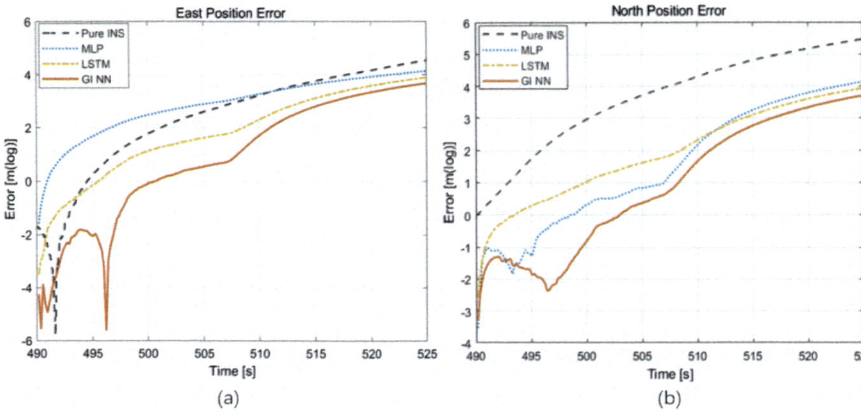

Fig. 2.15 The position errors among different algorithms during the period of 490–525 s in the real field test. **a** The east position error. **b** The north position error

in Table 2.2. During the second GPS outage period, compared to the single INS navigation mode without assistance, GI-NN reduces the positioning errors in the eastward and northward directions by 57.5% and 82.8%, respectively. Through the analysis of the actual UAV flight experiments, it is observed that the GI-NN-assisted integrated navigation method exhibits better position error suppression capability and higher navigation accuracy compared to the traditional neural networks.

2.5.3 Dataset Tests

The dataset experiment employs the publicly available INS/GPS integrated navigation dataset from the NaveGo toolbox [19], which was collected by Gonzalez and Dabove [20] through driving a vehicle equipped with Ekinox-D IMU and GNSS on the streets of Turin. Since the testing vehicle entered an underground parking lot, the dataset includes two 30-s GPS outage scenarios, used to evaluate the performance of the proposed GI-NN in assisting the INS. The trajectory of the NaveGo dataset is illustrated in Fig. 2.16, with the positions of the two GPS signal outages marked by red rectangles.

The results of the first GPS signal outage test are shown in Fig. 2.17 and Table 2.3: (1) The single INS mode exhibits the maximum position error. For instance, the maximum errors in the latitude and longitude positions reach 30.2 m and 19.1 m, respectively; (2) MLP outperforms the first method, with the maximum errors in the latitude and longitude reduced to 7.5 m and 9.1 m, respectively; (3) LSTM can achieve further reduction in the maximum errors of the latitude and longitude to 2.8 m and 3.7 m, respectively, compared to those of the MLP; (4) Among the four test results, the GI-NN method performs the best. Compared to the only inertial

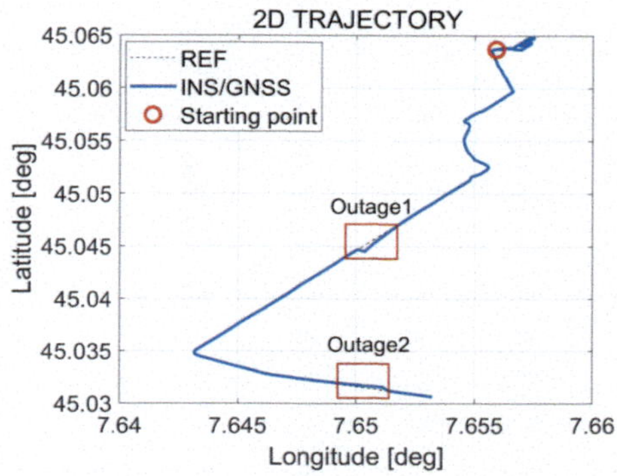

Fig. 2.16 The Navego dataset trajectory

mode, the maximum errors in the latitude and longitude are reduced by 92.7% and 90.1%, respectively.

To compare the performance of the Conv-LSTM neural network model and GI-NN, experiments are also conducted during the first GPS outage. The maximum errors in the latitude and longitude for the Conv-LSTM neural network model are 4.1 m and 4.5 m, respectively. From Fig. 2.18, it can be observed that GI-NN achieves higher accuracy than that of the Conv-LSTM.

From Fig. 2.19 and Table 2.3, it can be observed that although the four models exhibit different performance during the second GPS outage, they show similar

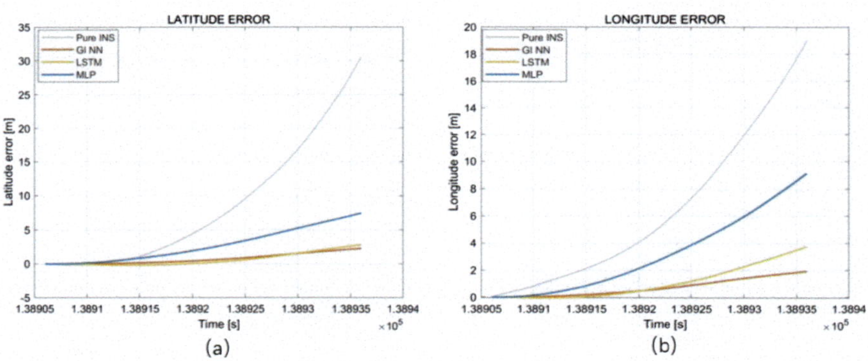

Fig. 2.17 The position errors among different algorithms during the first outage in the dataset test. **a** The latitude position error. **b** The longitude position error

Table 2.3 The max position error (m) among different algorithms in the dataset test

Max position error(m)	INS		MLP		LSTM		GI-NN	
	Lat	Lon	Lat	Lon	Lat	Lon	Lat	Lon
Outage 1	30.2	19.1	7.5	9.1	2.8	3.7	2.2	1.9
Outage 2	39.1	13.8	6.9	6.8	5.9	6.9	0.6	0.9

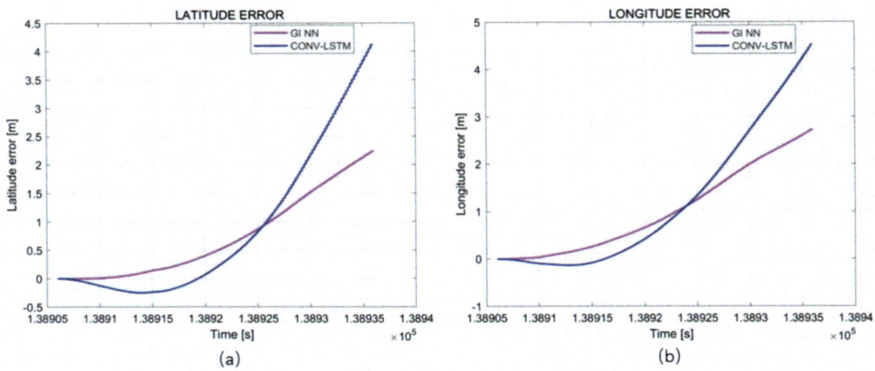

Fig. 2.18 The position errors between Conv-LSTM and GI-NN algorithms during the first outage in the dataset test. **a** The latitude position error. **b** The longitude position error

results to the first GPS outage: the single INS model has the largest errors, with a maximum latitude error of 39.1 m and a maximum longitude error of 13.8 m. Both LSTM and MLP outperform the only inertial model, with LSTM having maximum latitude and longitude errors of 5.9 m and 6.9 m, respectively, and MLP having latitude and longitude errors of 6.9 m and 6.8 m, respectively. Among them, GI-NN performs the best, with maximum errors in the latitude and longitude of 0.6 m and 0.9 m, respectively.

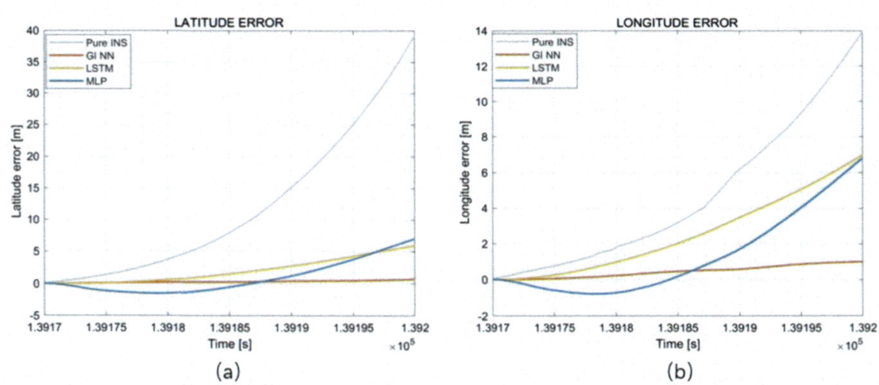

Fig. 2.19 The position errors among different algorithms during the second outage in the dataset test. **a** The latitude position error. **b** The longitude position error

2.6 Conclusion

This chapter proposes a novel INS/GPS integrated navigation system based on deep learning assistance to reduce the navigation error accumulation during GPS signal interruptions. The main advantages of this method are not only the extraction of feature representations of navigation sensors from measurement noise but also the automatic association of current inputs with historical model information. To validate the performance of the proposed method, numerical simulation, field experiments and public data experiments are conducted. The experimental results demonstrate that during a 35-s GPS outage, the navigation accuracy of the GI-NN algorithm is improved by 58% compared to the single INS algorithm, by 21% compared to the LSTM algorithm, and by 37% compared to the MLP algorithm. Therefore, the designed deep learning network, GI-NN, can estimate the intrinsic nonlinear relationship between INS outputs and GPS position increments, so as to provide accurate navigation information during GPS interruptions.

References

1. Abdel-Hafez MF, Saadeddin K, Jarrah M (2015) Constrained low-cost GPS/INS filter with encoder bias estimation for ground vehicles' applications. Mech Syst Signal Process 58:285–297
2. Noureldin A, Karamat TB, Eberts MD, El-Shafie A (2009) Performance enhancement of MEMS-based INS/GPS integration for low-cost navigation applications. IEEE Trans Veh Technol 58(3):1077–1096
3. Sebesta KD, Boizot N (2014) A real-time adaptive high-gain EKF, applied to a quadcopter inertial navigation system. IEEE Trans Industr Electron 61(1):495–503
4. Wang J et al (2008) Integration of GPS/INS/vision sensors to navigate unmanned aerial vehicles. Int Arch Photogram Remote Sens Spatial Inf Sci Conf 37:963–970
5. Kailath T (1968) An innovations approach to least-squares estimation–Part I: Linear filtering in additive white noise. IEEE Trans Autom Control 13(6):646–655
6. Angus JE (1992) Forecasting, structural time series and the Kalman filter. Technometrics 34(4):496–497
7. Kalman RE (1960) A new approach to linear filtering and prediction problems. J Basic Eng Trans ASME 82(1):35–45
8. Yan G, Wang J (2019) Strapdown inertial navigation algorithm and integrated navigation principle. Northwester Polytechnical University Press, Xi'an, China, in Chinese
9. Li J, Song N, Yang G, Li M, Cai Q (2017) Improving positioning accuracy of vehicular navigation system during GPS outages utilizing ensemble learning algorithm. Inf Fusion 35:1–10
10. Wagstaff B, Kelly J (2018) LSTM-based zero-velocity detection for robust inertial navigation. In: 2018 9th international conference on indoor positioning indoor navigation (IPIN). https://doi.org/10.1109/IPIN.2018.8533770
11. Nakashika T, Takiguchi T, Ariki Y (2014) Voice conversion using RNN Pre-Trained by recurrent temporal restricted Boltzmann machines. IEEE/ACM Trans Audio Speech Lang Proc 23(3):580–587
12. Li J et al (2016) Visualizing and understanding neural models in NLP. In: The North American chapter of the association for computational linguistics, 681–691

13. Chung J, Gulcehre C, Cho K, Bengio Y (2015) Gated feedback recurrent neural networks. Int Conf Mach Learn 37:2067–2075
14. Yao Y (2017) Xu X (2017) A RLS-SVM aided fusion methodology for INS during GPS outages. Sensors 17(3):432
15. Gonzalez R, Giribet JI, Patino HD (2015) NaveGo: a simulation framework for low-cost integrated navigation systems. Control Eng Appl Inf 17(2):110–120
16. Kingma DP, Ba J (2015) Adam: a method for stochastic optimization. In: 3rd international conference on learning representations (ICLR), pp 1–15
17. Loshchilov I, Hutter F (2017) SGDR: stochastic gradient descent with warm restarts. In: International conference on learning representations (ICLR 2017), pp 1–16
18. MRO Pixhawk Flight Controller (Pixhawk 1). Accessed: 2017. https://docs.px4.io/master/en/flightcontroller/mro pixhawk.html
19. NaveGo. Accessed: 2018. https://github.com/rodralez/NaveGo
20. Gonzalez R (2019) Dabove P (2019) Performance assessment of an ultra low cost inertial measurement unit for ground vehicle navigation. Sensors 19(18):3865

Chapter 3
Integrity Monitoring for GNSS Precision Positioning

Ling Yang, Jincheng Zhu, Yunri Fu, and Yangkang Yu

Abstract Integrity monitoring, originated from civil aviation and extending to various applications, is drawing increasing attentions in the navigation field, as the intelligent automation era approaches. With the improvement of GNSS infrastructures and the growing demands for precise positions, high-precision positioning technologies such as Precise Point Positioning (PPP), Real-Time Kinematic (RTK), and PPP-RTK have been well developed. Meanwhile, the integrity monitoring upon these high-precision GNSS positioning technologies is under developing. In this chapter, the demand for integrity monitoring in high-precision positioning is analyzed. The developments of integrity monitoring in civil aviation applications, and the key issues for high-precision GNSS positioning are briefly overviewed. Specifically, challenges on the Fault Detection and Exclusion (FDE) and Protection Level (PL) calculation procedures for high-precision positioning are comprehensively discussed. Furthermore, two preliminary studies that aim to extend the integrity monitoring algorithms to a generalized GNSS positioning procedure are presented. This chapter gives a whole picture of the GNSS integrity monitoring algorithms, and is expected to serve as a catalyst for further study on the integrity monitoring for GNSS high-precision positioning.

3.1 Introduction

With the improvement of GNSS infrastructure and the growing demand for precise applications, high-precision positioning technologies such as Precise Point Positioning (PPP) [1], Real-Time Kinematic (RTK) [2, 3], and PPP-RTK have deeply penetrated into people's daily lives [4, 5]. However, users often have complex requirements when using these technologies, not solely focusing on high precision but also pursuing high reliability. Autonomous driving is a typical application scenario that places additional emphases on reliability and safety [6–9].

L. Yang (✉) · J. Zhu · Y. Fu · Y. Yu
College of Surveying and Geo-informatics, Tongji University, Shanghai 200092, China
e-mail: lingyang@tongji.edu.cn

K. Yu (ed.), *Positioning and Navigation Using Machine Learning Methods*, Navigation: Science and Technology 14, https://doi.org/10.1007/978-981-97-6199-9_3

Integrity is a performance indicator that reflects the safety and reliability of positioning. Conducting integrity monitoring can effectively ensure the credibility of positioning, and therefore also become a necessitate to provide reliable services for high-precision positioning. Theoretical research on integrity monitoring algorithms for high-precision positioning is currently timely. The GNSS industry has already established a mature set of software and hardware processes for achieving high-precision positioning algorithms. The current challenge lies not only in providing high-precision positioning services but also in offering trustworthy high-precision positioning services. Integrity monitoring is currently a focal point of the industry, and it already has some technological foundations.

In this chapter, Sect. 3.1 analyzes the demand for integrity monitoring in high-precision positioning. Section 3.2 provides a brief overview about the developments of integrity monitoring for aviation, and presents the key issues of integrity monitoring for high-precision GNSS positioning. Section 3.3 comprehensively discussed the challenges on the FDE and PL calculation procedures for high-precision positioning. Section 3.4 presents two preliminary studies that aim to extend the integrity monitoring algorithms to a generalized GNSS positioning procedure. Section 3.5 concludes this chapter.

3.2 Developments of Integrity Monitoring

3.2.1 Overview of Integrity Monitoring for Aviation

Integrity reflects the trustworthiness of a navigation system, a concept initially introduced by the safety-centric civil aviation sector. Civil aviation conducts integrity monitoring to ensure flight operations remain unaffected by GNSS failures. Over several decades of development, currently, three augmentation systems, namely Ground Based Augmentation System (GBAS), Satellite Based Augmentation System (SBAS), and Airborne Based Augmentation System (ABAS), provide GNSS integrity monitoring services that comply with the standards set by the International Civil Aviation Organization (ICAO) [10, 11].

GBAS relies on local ground reference stations to monitor satellite signals and generate differential corrections as well as integrity information [12]. These messages are broadcasted to users via VHF Data Broadcasting (VDB) devices, thereby enhancing users' integrity. SBAS utilizes wide-area ground reference stations to monitor satellites and disseminates corrections and integrity information to users via Geostationary Earth Orbit (GEO) satellites, thereby improving positioning accuracy and integrity [13]. ABAS utilizes redundant information within user-end receivers to perform satellite fault detection and identification, exemplified by technologies such as Receiver autonomous integrity monitoring (RAIM) and Advanced RAIM (ARAIM) [14, 15].

In the operational environment of civil aviation aircraft, which typically operates in the open sky without any obstruction, these three types of augmentation systems can achieve sufficient positioning accuracy and integrity even when utilizing only pseudorange observations, without directly involving carrier-phase observations for positioning. Simultaneously, a vast amount of flight data has provided a fertile ground for the development and maturity of integrity monitoring models in this field. This has ultimately led to the establishment of standardized integrity monitoring systems as seen today. However, for high-precision applications that involve the simultaneous use of carrier-phase and pseudorange observations for positioning, the integrity monitoring methods in civil aviation are evidently not directly applicable to such scenarios. Challenges persist in achieving integrity monitoring for high-precision positioning.

3.2.2 Overview of Integrity Monitoring for Precise Positioning

RTK uses double-differencing (DD) carrier phase observations and combines ambiguity resolution to achieve instantaneous centimeter-level positioning accuracy [16, 17]. It is currently the most widely used high-precision positioning method. Double-differencing observations can eliminate the majority of errors, including satellite clock errors, atmospheric propagation errors, and receiver clock errors. However, due to the impact of multipath effects and carrier phase biases, the reliability of ambiguity resolution may be reduced, ultimately affecting positioning accuracy [18]. Therefore, the key factors for the integrity monitoring of RTK lie in the ambiguity resolution and biases [19]. For ambiguity resolution, the Carrier RAIM (CRAIM) is used to constrain the resolution of the L1 carrier phase ambiguity [20], and Solution Separation (SS)-based integrity monitoring method is employed to monitor the faulty ambiguity fixes [21]. An adaptive Kalman filtering algorithm with ambiguity success rate as a dynamic adjustment factor has also been proposed to enhance the efficiency of ambiguity resolution [22], and an integrity monitoring based ratio test is also used to replace the traditional ratio tests for ambiguity validation [23]. For biases, both cycle slips [24] and colored noise [25] in the observations need to be considered, and enhanced Fault Detection and Exclusion (FDE) algorithms should be used. As for the remaining atmospheric residuals and the multipath effect, the new weighting models can be employed to mitigate these influences [26].

PPP can achieve centimeter-level positioning accuracy using only a single receiver with the use of high-precision external augmentation products, most notably precise orbit and clock products, and it takes around half an hour to converge to centimeter-level accuracy. To achieve rapid convergence, PPP must be combined with Uncalibrated Phase Delays (UPD) [27] products and ambiguity resolution for ambiguity fixing, a technique known as PPP-AR [28]. PPP-AR, combined with high-precision atmospheric augmentation products calculated by Continuously Operating Reference

Station (CORS) network allows the realization of PPP-RTK [29]. Integrity monitoring for PPP/PPP-RTK should focus on three main aspects: server-end integrity monitoring, user-end integrity monitoring, and high-precision augmentation products integrity monitoring. Currently, the majority of research is focused on user-side integrity monitoring. This includes integrity monitoring algorithms based on Multi-Hypothesis Solution Separation (MHSS) in ARAIM [30–32], integrity monitoring algorithms that extend MHSS to Kalman Filters (KF) [33], as well as integrity monitoring algorithms based on MHSS that consider ambiguity fixing errors [34]. FDE algorithms for PPP/PPP-RTK have also been studied [35, 36]. To enhance the performance of Positioning, Navigation, and Timing (PNT) services in GNSS-denied environments, multi-sensor fusion technology has been proposed. GNSS/INS, as the most widely used integrated navigation system, can be divided into two types of integrity monitoring algorithms for its tightly coupled navigation system: innovation based or residual based methods [37] and MHSS based methods [38, 39]. There is also research that applies RAIM as integrity monitoring method for PPP/LiDAR loosely coupled SLAM [40].

3.3 Integrity Monitoring Challenges for High-Precision Positioning

Integrity monitoring in the civil aviation has developed to a highly mature stage. However, extending it to ground-based high-precision positioning still faces numerous significant challenges. Generally, the issues revolve around two main aspects: Fault Detection and Exclusion (FDE) and Protection Level (PL) calculation, which are also the two key tasks of integrity monitoring [21, 41].

3.3.1 FDE Algorithms

Integrity monitoring in the civil aviation is based on Single Point Positioning (SPP) using pseudorange measurements. Therefore, typical FDE algorithms such as RAIM and ARAIM generally focus on gross errors detection in pseudorange observations, often manifested as pseudorange outliers. Generally, only pseudorange observations from GPS or a combination of GPS/Galileo dual systems are used, and the positioning results are obtained using a snapshot weighted Least Squares (LS) algorithm [14, 42].

On the contrary, high-precision positioning algorithms such as RTK/PPP/PPP-RTK not only use pseudorange observations but also use carrier phase observations. Different types of data introduce different outliers, particularly manifested as cycle slips and faulty-fixed ambiguities on carrier phase observations [43, 44]. The introduction of new observation types renders LS algorithms inadequate, and currently, high-precision positioning relies on KF algorithms. Additionally, to ensure the stable performance of high-precision positioning algorithms in complex environments, observations from multi-systems and multi-frequencies are used [45].

Therefore, traditional FDE algorithms are not directly applicable to high-precision positioning. FDE algorithms for high-precision positioning need to consider at least the following three aspects: (1) Applicability to multi-frequency and multi-system observations, (2) Meeting the requirements for cycle slip detection and ambiguity resolution validation, and (3) Suitability for KF algorithm.

While multi-frequency and multi-system observations can effectively increase measurement redundancy, they may also lead to a higher number of potential outliers. Existing FDE algorithms, such as RAIM, initially only used observations from the GPS L1 frequency band [8]. Although ARAIM can handle dual-frequency observations, it is limited to the GPS and Galileo dual-system [14]. Moreover, the core technique for outlier detection in FDE, hypothesis testing (e.g., w-test), can only detect one outlier at a time [46, 47]. To improve the efficiency of outlier detection, traditional binary hypothesis testing has been extended to multi-hypothesis testing [48]. However, the significant increase in computational complexity associated with multi-hypothesis testing remains a challenge to be addressed [49].

Cycle slips and ambiguities are unique to carrier phase observations, and FDE algorithms based on SPP cannot effectively handle these outliers. Cycle slip detection and repair are essential components of GNSS data preprocessing [44]. Currently, there are algorithms based on DD observations [50] and algorithms based on undifferenced observations for handling cycle slips. The Turbo Edit algorithm [51], as the first undifferenced observations based algorithm proposed, is adopted by many international precision positioning software [52]. The LAMBDA algorithm is the most widely used RTK ambiguity resolution algorithm [53], and PPP/PPP-RTK ambiguity resolution involves Narrow-Lane (NL) and Wide-Lane (WL) ambiguities as well as UPD products [43]. Incorporating the impact of residual cycle slips and faulty ambiguity fixing into FDE algorithms remains a challenging issue.

High-precision positioning employs KF for position estimation, hence, FDE algorithms should be applicable to filtering modes rather than snapshot modes. Both the function model and stochastic model of the filtering process change with time, and there is a need to modify FDE algorithms to integrate them effectively with filtering. Detection, Identification and Adaptation (DIA) is widely applied for the FDE purpose [54, 55], and there have been initial explorations of its application within the framework of KF [17]. However, challenges persist in terms of how to fix the detected cycle slips to whole-cycle numbers for ambiguity resolution.

3.3.2 User-End PL Calculation

The concept of PL in civil aviation integrity monitoring is a measure of the robustness and reliability of the positioning results, ensuring the system security under specified operation conditions. Mathematically, PL can be simplified as the percentile of positioning error at an extreme large cumulative distribution probability. Due to the influence of various risk sources, the observation errors are actually non-Gaussian. As a result, the positioning errors do not rigorously obey the Gaussian distribution. Theoretically, determining the percentiles of non-Gaussian random variables is the core challenges for PL calculation. If the probability density function of the positioning errors can be determined, the percentile of an arbitrary probability can be obtained by integral. The corresponding PL can then be obtained by inverse calculation using numerical methods such as the binary searching. Although the exact value of PL can be obtained via the Monte Carlo simulation, this is not applicable for real-time situation. In ICAO regulations and algorithms, the PL is particularly defined with accordance to the Probability of Hazardous Misleading Information (PHMI) and Time-to-Alert (TTA), which requires for integral calculation in multiple intervals within a certain time window, and thus further increasing the computational complexity.

Calculating PL is the main task of integrity monitoring. Due to the difficulty of integral calculation upon a non-Gaussian probability density function, the overbounding technologies is generally used in current integrity monitoring algorithms for aviation applications. With the FDE procedure, significant outlying observations are firstly excluded. Then, by overbounding each error term of the reserved observations, distribution of the final positioning error can also be overbounded by a Gaussian distribution. As a valid parameter, the PL value should balance between being overly conservative and too permissive. The primary requirement for PL is that it must successfully envelop the positioning error. This ensures that the system integrity is not compromised by excessive errors in positioning. However, an overly conservative PL value, on the other hand, would lead to a decrease in system availability.

For those high-precise positioning modes, PL calculation should be designed according to the specific positioning and FDE algorithms. Two common issues should always be considered: (1) the stochastic model that describes the observation error characteristics should be capable to overbound the actual observation errors; (2) impacts of the undetected faults, mainly including pseudorange outliers, wrongly fixed ambiguities, carrier phase cycle slips, should be considered.

Firstly, compared with the functional model, the impacts of stochastic model on positioning are usually considered to be less significant, since an inaccurate stochastic model would only lead to an more dispersive estimation, but still unbiased. However, for integrity monitoring, the impacts of stochastic model play the same role as the function model. An inaccurate stochastic model would not only destroy the validity of hypothesis testing in the FDE procedure, but also destroy the validity of PL for the final system availability determination. The stochastic model for high-precise positioning algorithms should consider their discrepancies on dealing with each error term, i.e., the impacts of residual tropospheric and ionospheric delays should be modelled as the code and phase observation error terms, and compensated for in the stochastic model, if these residuals have not been parameterized in the RTK or PPP/PPP-AR/PPP-RTK functional model. Otherwise, if these residual effects have been parametrized in the function model, prior values with accurate variance information would be introduced as constraints to mitigate overparametrization effects. In both cases, conservative but tight variances of each factor are the core for producing valid PLs in the positioning domain.

Secondly, even after some cautious FDE procedures, undetected outlying observations may still be reserved. Because the FDE procedure always relies on hypothesis testing, by which there are always some probabilities of committing missed detection and wrong identification under a certain fault mode. For high-precision GNSS positioning, carrier phase observations play a dominant role, and meanwhile it also introduces some specific undetected faults. Both the wrongly fixed ambiguities and undetected cycle slips would behave as integer biases on the carrier phase observations. Although this kind of effects can be mostly eliminated via conducting an overly strict FDE procedure which abandons all those doubtful observations, the system availability would adversely reduce. Generally, the corresponding impacts on the positioning biases should be modelled and considered during the PL calculation. This leads to an intrinsic combination between testing and estimation, based on a mixed-integer model.

3.4 Preliminary Studies on Generalization of Integrity Monitoring Algorithms

In this section, two preliminary studies that aim to extend the integrity monitoring algorithms to a generalized GNSS positioning procedure have been summarized. First, a DIA-MAP method that can be used for FDE purpose is presented. Second, an overbounding method that stochastically characterizes the residual tropospheric delays is presented.

3.4.1 The DIA-MAP Method for FDE Purpose

For FDE purpose, while various test statistics have been developed to address scenarios involving multiple outliers [56–58], directly comparing individual test statistics becomes impractical when the suspected outlier count varies outlier number [17]. Consequently, the Detection and Identification Adaptation (DIA) for multiple outliers, based solely on residuals, is hindered by masking and swamping effects. In cases where the maximum suspected outlier count is high, statistical tests tend to falsely declare more outliers than actually present [59]. Moreover, the presence of multiple outliers can easily mask each other, complicating the detection and identification process [60].

To solve this problem, the DIA based on maximum a posteriori estimate (DIA-MAP) is proposed [61]. By leveraging the prior distribution of gross errors, DIA-MAP selects the hypothesis with the maximum posterior probability for outlier detection and identification. With the priors of gross error, DIA-MAP provides a unified

DIA process for both single and multiple outliers. Also, the prior can be flexibly adjusted rather than fixed to be uniform, so that the DIA method can be adapted to different application cases.

In the detection step, the validity of the null hypothesis is checked by a global test, without parameterizing particular alternative hypotheses. Generally, the overall test statistic is formed as

$$T_0 = \boldsymbol{y}^{\mathrm{T}}\boldsymbol{\Sigma}^{-1}\boldsymbol{\Sigma}_{\hat{\boldsymbol{e}}\hat{\boldsymbol{e}}}\boldsymbol{\Sigma}^{-1}\boldsymbol{y} \tag{3.1}$$

where $\boldsymbol{\Sigma}$ is the variance–covariance matrix of the observations, and $\boldsymbol{\Sigma}_{\hat{\boldsymbol{e}}\hat{\boldsymbol{e}}}$ is the variance–covariance matrix of the residual vector. Since $T_0 \sim \chi^2(m-n, 0)$, the acceptance of H_0 can be defined as

$$\text{accept } H_0 \text{ if } T_0 \leq k_\alpha \tag{3.2}$$

with

$$k_\alpha = \chi^2_\alpha(m-n) \tag{3.3}$$

Here $\chi^2_\alpha(m-n)$ denotes the corresponding critical value of the Chi-square distribution that is dependent on the confidence level α and the degree of freedom $m-n$.

In the identification step, following the MAP principle, if the observation vector $\boldsymbol{y}$ is contaminated by outliers, then the optimal estimates of the hypothesis H_i, the gross error $\widehat{\nabla}_i$, and the unknown parameter $\hat{\boldsymbol{x}}_i$ are the ones that maximize the posterior probability distribution:

$$H_i, \widehat{\nabla}_i, \hat{\boldsymbol{x}}_i = \underset{H_j, \nabla_j, \boldsymbol{x}}{\operatorname{argmax}}\, p\big(H_j, \nabla_j, \boldsymbol{x}|\boldsymbol{y}\big), j \in \{1, \ldots, N\} \tag{3.4}$$

Then, one can accept the model as:

$$\text{accept } H_i \text{ if } \mathrm{Post}_i = \max_{j\in\{1,\ldots,N\}}\big\{\mathrm{Post}_j\big\} \tag{3.5}$$

Here Post_i represents the magnitude relationship of the posterior probability of the hypotheses, which is given by:

$$\begin{aligned}\mathrm{Post}_i &= q_i \ln\varepsilon + (m-q_i)\ln(1-\varepsilon) - \frac{1}{2}\boldsymbol{y}^T\boldsymbol{\Sigma}^{-1}\boldsymbol{\Sigma}_{\hat{\boldsymbol{e}}\hat{\boldsymbol{e}}}\boldsymbol{\Sigma}^{-1}\boldsymbol{y} \\ &+ \frac{1}{2}\boldsymbol{y}^T\boldsymbol{\Sigma}^{-1}\boldsymbol{\Sigma}_{\hat{\boldsymbol{e}}\hat{\boldsymbol{e}}}\boldsymbol{\Sigma}^{-1}\boldsymbol{C}_i\Big(\boldsymbol{C}_i^T\boldsymbol{\Sigma}^{-1}\boldsymbol{\Sigma}_{\hat{\boldsymbol{e}}\hat{\boldsymbol{e}}}\boldsymbol{\Sigma}^{-1}\boldsymbol{C}_i\Big)^{-1}\boldsymbol{C}_i^T\boldsymbol{\Sigma}^{-1}\boldsymbol{\Sigma}_{\hat{\boldsymbol{e}}\hat{\boldsymbol{e}}}\boldsymbol{\Sigma}^{-1}\boldsymbol{y}\end{aligned} \tag{3.6}$$

where q_i is the outlier number in H_i and ε is the preset outlier rate. The estimate of gross errors and unknown parameters are respectively described as:

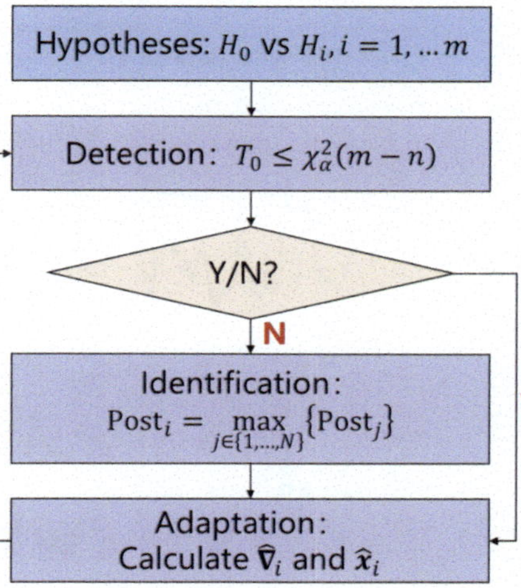

Fig. 3.1 The flow chart of DIA-MAP

$$\widehat{\nabla}_i = \left(\boldsymbol{C}_i^T\boldsymbol{\Sigma}^{-1}\boldsymbol{\Sigma}_{\hat{e}\hat{e}}\boldsymbol{\Sigma}^{-1}\boldsymbol{C}_i\right)^{-1}\boldsymbol{C}_i^T\boldsymbol{\Sigma}^{-1}\boldsymbol{\Sigma}_{\hat{e}\hat{e}}\boldsymbol{\Sigma}^{-1}\boldsymbol{y} \tag{3.7}$$

$$\widehat{\nabla}_i = \left(\boldsymbol{C}_i^T\boldsymbol{\Sigma}^{-1}\boldsymbol{\Sigma}_{\hat{e}\hat{e}}\boldsymbol{\Sigma}^{-1}\boldsymbol{C}_i\right)^{-1}\boldsymbol{C}_i^T\boldsymbol{\Sigma}^{-1}\boldsymbol{\Sigma}_{\hat{e}\hat{e}}\boldsymbol{\Sigma}^{-1}\boldsymbol{y} \tag{3.7}$$

and

$$\hat{\boldsymbol{x}}_i = \left(\boldsymbol{A}^T\boldsymbol{\Sigma}^{-1}\boldsymbol{A}\right)^{-1}\boldsymbol{A}^T\boldsymbol{\Sigma}^{-1}\left(\boldsymbol{y} - \boldsymbol{C}_i\widehat{\nabla}_i\right) \tag{3.8}$$

where $\boldsymbol{A}$ is the design matrix of unknown parameters, $\boldsymbol{C}_i$ is the design matrix of gross error size under H_i. The flow chart of DIA-MAP is illustrated in Fig. 3.1.

Figure 3.2 shows the positioning RMSEs of the five methods (LS, DIA-datasnooping, iterative DIA-datasnooping, DIA-p-value, and DIA-MAP) in one-day data of BeiDou Navigation Satellite System (BDS) SPP examples. Table 3.1 summarizes the average RMSEs of all epochs for each method. One can find that, due to lack of robustness, the RMSE of LS always shows an increasing tendency when the gross error enlarges, regardless of the value of ε. As for the DIA-datasnooping, iterative DIA-datasnooping, and DIA-p-value, the RMSEs remain in a stable trend with the growth of gross error size when $\varepsilon = 10^{-3}$. It means that when the gross error rate is relatively small, the conventionally used methods all show satisfactory performance on robustness. However, when ε is relatively large these methods cannot always bound the estimation errors. But for DIA-MAP, no matter how many outliers occur, the RMSE of DIA-MAP can always decline to a lower bound as the gross error enlarges.

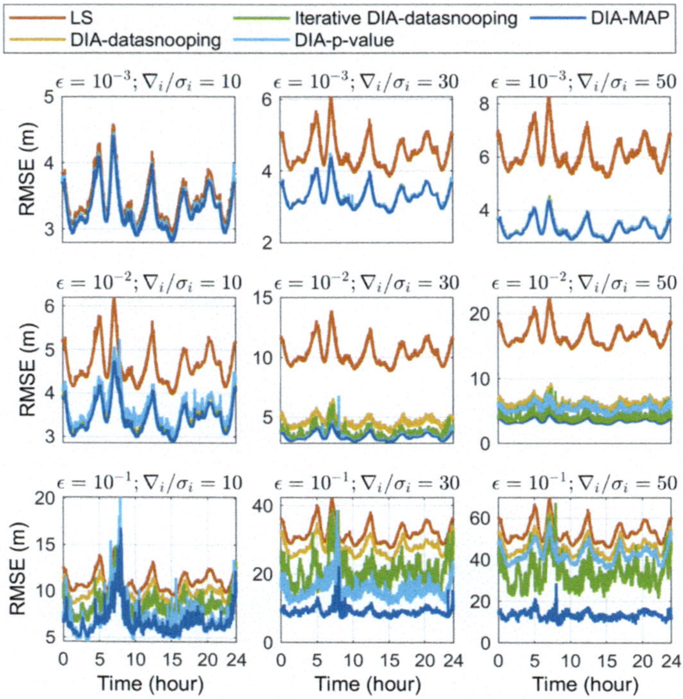

Fig. 3.2 Positioning RMSEs of five methods under different geometry

3.4.2 Stochastic Modelling for the Residual Tropospheric Delays

Tropospheric delay due to the neutral atmosphere is a major error source for GNSS positioning. Although Various Zenith Total Delay (ZTD) models have been proposed and implemented, residual tropospheric delay can still lead to meter-level biases in positioning [62]. The Minimum Operational Performance Standards by Radio Technical Committee for Aeronautics (RTCA MOPS) uses a constant standard deviation (STD) of 0.12 m to characterize the residual ZTDs [14], equivalent to an upper bound of 0.64 m for residual ZTDs. However, this value is too conservative, and does not consider the spatiotemporal-varying characteristic of residual ZTDs [63]. Therefore, a global and spatiotemporal-varying overbounding method was proposed to evaluate the performance of a ZTD model and apply it for GNSS positioning and integrity monitoring [64]. The proposed method is applied to three conventionally used ZTD models, providing a tighter but still conservative overbounding model of residual ZTDs.

The flowchart of stochastic modelling for the residual tropospheric delays is shown in Fig. 3.3. First, to analyze the residual ZTDs, the global grid-wise VMF3 products

Table 3.1 Average positioning RMSEs of five methods under different rates and sizes of gross errors (Unit: m)

Methods	$\varepsilon = 10^{-3}$			$\varepsilon = 10^{-2}$			$\varepsilon = 10^{-1}$		
	$\nabla_i/\sigma_i = 10$	$\nabla_i/\sigma_i = 30$	$\nabla_i/\sigma_i = 50$	$\nabla_i/\sigma_i = 10$	$\nabla_i/\sigma_i = 30$	$\nabla_i/\sigma_i = 50$	$\nabla_i/\sigma_i = 10$	$\nabla_i/\sigma_i = 30$	$\nabla_i/\sigma_i = 50$
LS	3.44	4.52	6.14	4.65	10.40	16.77	11.09	31.96	53.09
DIA-datasnooping	3.30	3.30	3.32	3.55	4.60	6.26	9.98	28.12	46.66
Iterative DIA-datasnooping	3.30	3.29	3.29	3.46	3.64	4.03	8.74	20.71	33.36
DIA-p-value	3.34	3.33	3.33	3.65	3.41	5.47	7.09	15.59	42.79
DIA-MAP	3.30	3.29	3.29	3.43	3.34	3.33	6.68	9.17	13.39

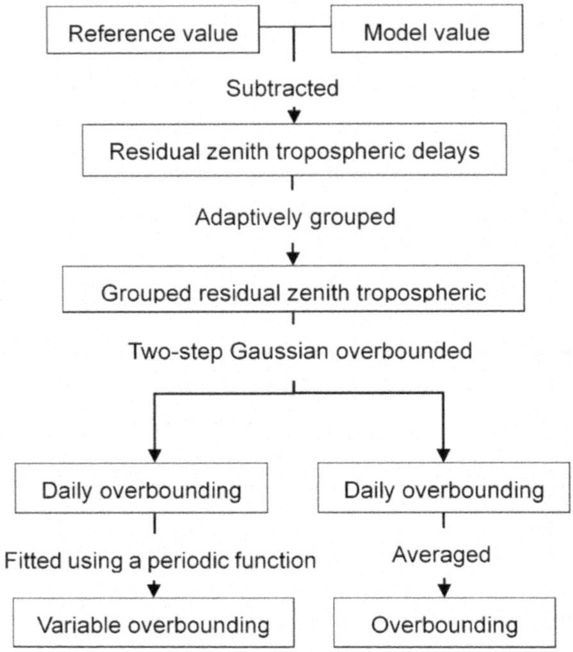

Fig. 3.3 Flowchart of stochastic modelling

were selected as the reference, and the residual ZTDs were obtained by subtracting the ZTDs calculated by a model from the corresponding reference value at each grid point. Then the residual ZTDs are adaptively banded using the hierarchical clustering method, which uses two criteria: samples with similar seasonal variations should be clustered together, and samples at adjacent latitudes should be clustered together.

After the adaptive banding, the residuals at the same latitude band are stacked along the date axis, and the two-step gaussian overbounding method [65] was used to calculate the overbounding STD and bias at each DOY. Then the overbounding biases were averaged, and the overbounding STDs were fitted using a periodic function. The overbounding models for residual ZHDs and ZWDs are established separately, and the overbounding model for the total residual ZTDs is established by combining the two. The upper bound $\Delta_{\max}(DOY)$ of the residual ZTDs can be calculated by:

$$\Delta_{\max}(DOY) = \sigma(DOY) \times K_{IR} + b \tag{3.9}$$

where $\sigma(DOY)$ is the fitted function of the overbounding STD, b is the averaged overbounding bias, and K_{IR} represents the right tail quantile of the PDF of a standard normal distribution at the extreme probability level corresponding to the preset integrity risk (IR).

As an illustration, Fig. 3.4 shows the overbounding results for the UNB3 model, which show global and annual characteristics, with overbounding biases and daily STD values varying at different latitude bands. Seasonal variations are observed in

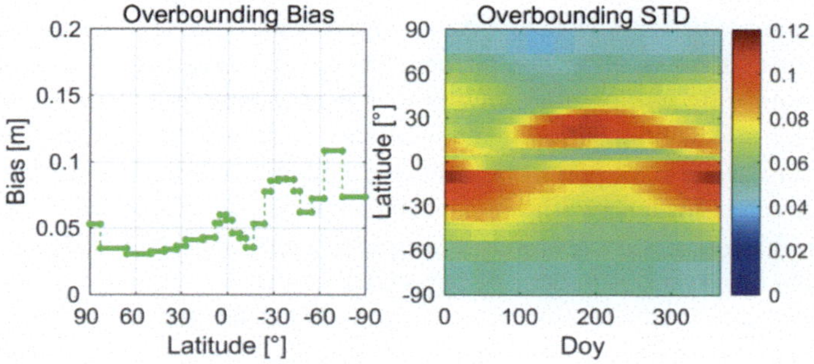

Fig. 3.4 Overbounding model results for UNB3 model

Fig. 3.5 Residual ZTDs overbounding results for UNB3 model

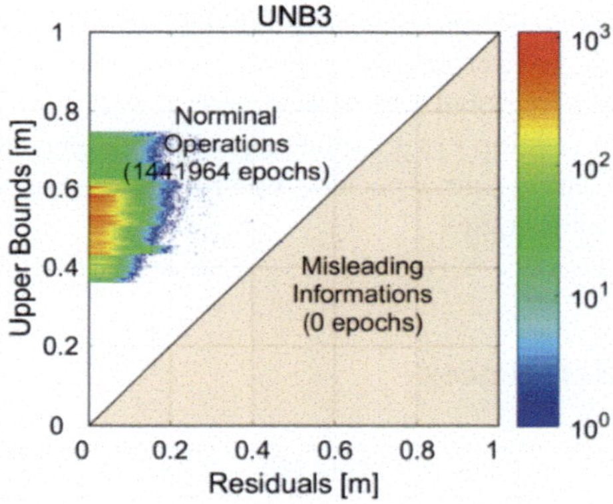

low and middle latitudes and are asymmetrical in the northern and southern hemispheres. The overbounding model varies across seasons and latitudes, making it neither overly conservative nor overly optimistic.

Figure 3.5 shows the residual ZTDs overbounding results for UNB3 model using the external IGS ZTD products, which shows that the calculated upper bound could successfully overbound the residual ZTDs, and is tighter than 0.64 m recommended by RTCA in most cases. Therefore, adopting spatiotemporal-varying STDs can improve positioning and integrity monitoring performance.

3.5 Concluding Remarks

The demands of integrity monitoring for GNSS positioning have been steadily growing, as the developments of intelligent transport systems (ITS), and other safety-of-life (SoL) applications. Initially in civil aviation, three augmentation systems, namely GBAS, SBAS, and ABAS, mainly utilizing pseudorange or phase-smoothing-pseudorange observations, have been well developed, to provide GNSS integrity monitoring services that comply with the standards set by the ICAO. While for those high-precision positioning technologies such as PPP, RTK and PPP-RTK, significant challenges still persist in achieving integrity monitoring service.

Issues revolve around two main aspects: FDE and PL calculation. In general, the FDE for high-precision positioning should be applicable to multi-frequency and multi-system observations under a KF procedure, as well as should fulfill cycle slip detection and ambiguity resolution validation. Also, the PL should be calculated based on a stochastical model that can overbound the actual code and phase errors, as well as by considering impacts of the undetected pseudorange outliers, carrier phase cycle slips, and wrongly fixed ambiguities.

To generalize the FDE procedure, the authors have proposed a DIA-MAP method which can provides a unified DIA process for multiple outliers by utilizing the priors of gross errors. To generalize one term of the stochastical model for PL calculation, a global and spatiotemporal-varying overbounding method has been proposed to envelop the residual tropospheric delays for different ZTD models. These investigations are expected to be further extended upon those high-precision positioning technologies.

References

1. Zumberge JF, Heflin MB, Jefferson DC, Watkins MM, Webb FH (1997) Precise point positioning for the efficient and robust analysis of GPS data from large networks. J Geophys Res Solid Earth 102:5005–5017. https://doi.org/10.1029/96JB03860
2. Feng Y (2008) GNSS three carrier ambiguity resolution using ionosphere-reduced virtual signals. J Geod 82:847–862. https://doi.org/10.1007/s00190-008-0209-x
3. Li B, Shen Y, Feng Y, Gao W, Yang L (2014) GNSS ambiguity resolution with controllable failure rate for long baseline network RTK. J Geod 88:99–112. https://doi.org/10.1007/s00190-013-0670-z
4. Teunissen PJG, Khodabandeh A (2015) Review and principles of PPP-RTK methods. J Geod 89:217–240. https://doi.org/10.1007/s00190-014-0771-3
5. Zhang B, Hou P, Odolinski R (2022) PPP-RTK: from common-view to all-in-view GNSS networks. J Geod 96:102. https://doi.org/10.1007/s00190-022-01693-y
6. Du Y, Wang J, Rizos C, El-Mowafy A (2021) Vulnerabilities and integrity of precise point positioning for intelligent transport systems: overview and analysis. Satell Navig 2:3. https://doi.org/10.1186/s43020-020-00034-8
7. Hassan T, El-Mowafy A, Wang K (2021) A review of system integration and current integrity monitoring methods for positioning in intelligent transport systems. IET Intell Transp Syst 15:43–60. https://doi.org/10.1049/itr2.12003
8. Yang L, Sun N, Rizos C, Jiang Y (2022) ARAIM stochastic model refinements for GNSS positioning applications in support of critical vehicle applications. Sensors 22:9797. https://doi.org/10.3390/s22249797
9. Zhu N, Marais J, Bétaille D, Berbineau M (2018) GNSS position integrity in urban environments: a review of literature. IEEE Trans Intell Transp Syst 19:2762–2778. https://doi.org/10.1109/TITS.2017.2766768
10. ICAO (2018) ICAO Standards and recommended practices (SARPs), Annex 10 vol I, Radio navigation aids

11. RTCA (2017) DO-253D, minimum operational performance standards for GPS local area augmentation system airborne equipment
12. McGraw GA, Murphy T, Brenner M, Pullen S, Dierendonck AJV (2000) Development of the LAAS accuracy models, pp 1212–1223
13. Roturier B, Chatre E, Ventura-Traveset J (2001) The SBAS integrity concept standardised by ICAO. Application to EGNOS
14. Blanch J, Walker T, Enge P, Lee Y, Pervan B, Rippl M, Spletter A, Kropp V (2015) Baseline advanced RAIM user algorithm and possible improvements. IEEE Trans Aerosp Electron Syst 51:713–732. https://doi.org/10.1109/TAES.2014.130739
15. Hewitson S, Wang J (2006) GNSS receiver autonomous integrity monitoring (RAIM) performance analysis. GPS Solut 10:155–170. https://doi.org/10.1007/s10291-005-0016-2
16. Parkins A (2011) Increasing GNSS RTK availability with a new single-epoch batch partial ambiguity resolution algorithm. GPS Solut 15:391–402. https://doi.org/10.1007/s10291-010-0198-0
17. Teunissen PJ, Montenbruck O (2017) Springer handbook of global navigation satellite systems. Springer Cham
18. Li L, Shi H, Jia C, Cheng J, Li H, Zhao L (2018) Position-domain integrity risk-based ambiguity validation for the integer bootstrap estimator. GPS Solut 22:39. https://doi.org/10.1007/s10291-018-0703-4
19. Wang K, El-Mowafy A (2021) Effect of biases in integrity monitoring for RTK positioning. Adv Space Res 67:4025–4042. https://doi.org/10.1016/j.asr.2021.02.032
20. Feng S, Ochieng W, Moore T, Hill C, Hide C (2009) Carrier phase-based integrity monitoring for high-accuracy positioning. GPS Solut 13:13–22. https://doi.org/10.1007/s10291-008-0093-0
21. Gao Y, Jiang Y, Gao Y, Huang G, Yue Z (2023) Solution separation-based integrity monitoring for RTK positioning with faulty ambiguity detection and protection level. GPS Solut 27:140. https://doi.org/10.1007/s10291-023-01472-y
22. Wang Z, Hou X, Dan Z, Fang K (2022) Adaptive Kalman filter based on integer ambiguity validation in moving base RTK. GPS Solut 27:34. https://doi.org/10.1007/s10291-022-01367-4
23. Li L, Li Z, Yuan H, Wang L, Hou Y (2016) Integrity monitoring-based ratio test for GNSS integer ambiguity validation. GPS Solut 20:573–585. https://doi.org/10.1007/s10291-015-0468-y
24. Kim D, Song J, Yu S, Kee C, Heo M (2018) A new algorithm for high-integrity detection and compensation of dual-frequency cycle slip under severe ionospheric storm conditions. Sensors 18:3654. https://doi.org/10.3390/s18113654
25. Gao Y, Gao Y, Liu B, Jiang Y (2021) Enhanced fault detection and exclusion based on Kalman filter with colored measurement noise and application to RTK. GPS Solut 25:82. https://doi.org/10.1007/s10291-021-01119-w
26. Wang K, El-Mowafy A, Rizos C, Wang J (2020) Integrity monitoring for horizontal RTK positioning: new weighting model and overbounding CDF in open-sky and suburban scenarios. Remote Sens 12:1173. https://doi.org/10.3390/rs12071173
27. Geng J, Chen X, Pan Y, Mao S, Li C, Zhou J, Zhang K (2019) PRIDE PPP-AR: an open-source software for GPS PPP ambiguity resolution. GPS Solut 23:91. https://doi.org/10.1007/s10291-019-0888-1
28. Li X, Li X, Yuan Y, Zhang K, Zhang X, Wickert J (2018) Multi-GNSS phase delay estimation and PPP ambiguity resolution: GPS, BDS, GLONASS, Galileo. J Geod 92:579–608. https://doi.org/10.1007/s00190-017-1081-3
29. Lyu Z, Gao Y (2022) PPP-RTK with augmentation from a single reference station. J Geod 96:40. https://doi.org/10.1007/s00190-022-01627-8
30. Gunning K, Blanch J, Walter T, Groot LD, Norman L (2018) Design and evaluation of integrity algorithms for PPP in kinematic applications. Florida, Miami, pp 1910–1939
31. Zhang J, Zhao L, Yang F, Li L, Liu X, Zhang R (2022) Integrity monitoring for undifferenced and uncombined PPP under local environmental conditions. Meas Sci Technol 33:065010. https://doi.org/10.1088/1361-6501/ac4b12

32. Zhang W, Wang J, El-Mowafy A, Rizos C (2023) Integrity monitoring scheme for undifferenced and uncombined multi-frequency multi-constellation PPP-RTK. GPS Solut 27:68. https://doi.org/10.1007/s10291-022-01391-4
33. Wang S, Zhan X, Xiao Y, Zhai Y (2022) Integrity Monitoring of PPP-RTK based on multiple hypothesis solution separation. In: Yang C, Xie J (eds) China satellite navigation conference (CSNC 2022) proceedings. Springer Nature, Singapore, pp 321–331
34. Zhang W, Wang J (2023) GNSS PPP-RTK: integrity monitoring method considering wrong ambiguity fixing. GPS Solut 28:30. https://doi.org/10.1007/s10291-023-01572-9
35. Elsayed H, El-Mowafy A, Wang K (2023) A new method for fault identification in real-time integrity monitoring of autonomous vehicles positioning using PPP-RTK. GPS Solut 28:32. https://doi.org/10.1007/s10291-023-01569-4
36. Innac A, Gaglione S, Troisi S, Angrisano A (2018) A proposed fault detection and exclusion method applied to multi-GNSS single-frequency PPP. In: 2018 European navigation conference (ENC), pp 129–139
37. Tanil C, Khanafseh S, Joerger M, Kujur B, Kruger B, Groot LD, Pervan B (2019) Optimal INS/GNSS coupling for autonomous car positioning integrity, pp 3123–3140
38. Gunning K, Blanch J, Walter T, Groot LD, Norman L (2019) Integrity for tightly coupled PPP and IMU. Florida, Miami, pp 3066–3078
39. Wang S, Zhai Y, Zhan X (2023) Implementation of solution separation-based Kalman filter integrity monitoring against all-source faults for multi-sensor integrated navigation. GPS Solut 27:103. https://doi.org/10.1007/s10291-023-01423-7
40. Li T, Pei L, Xiang Y, Wu Q, Xia S, Tao L, Guan X, Yu W (2021) P3-LOAM: PPP/LiDAR loosely coupled SLAM with accurate covariance estimation and robust RAIM in Urban Canyon environment. IEEE Sens J 21:6660–6671. https://doi.org/10.1109/JSEN.2020.3042968
41. El-Mowafy A (2019) On detection of observation faults in the observation and position domains for positioning of intelligent transport systems. J Geod 93:2109–2122. https://doi.org/10.1007/s00190-019-01306-1
42. Blanch J, Walter T, Enge P, Lee Y, Pervan B, Rippl M, Spletter A (2012) Advanced RAIM user algorithm description: integrity support message processing, fault detection, exclusion, and protection level calculation, pp 2828–2849
43. Ge M, Gendt G, Rothacher M, Shi C, Liu J (2008) Resolution of GPS carrier-phase ambiguities in precise point positioning (PPP) with daily observations. J Geod 82:389–399. https://doi.org/10.1007/s00190-007-0187-4
44. Zhao Q, Sun B, Dai Z, Hu Z, Shi C, Liu J (2015) Real-time detection and repair of cycle slips in triple-frequency GNSS measurements. GPS Solut 19:381–391. https://doi.org/10.1007/s10291-014-0396-2
45. Wu Z, Wang Q, Hu C, Yu Z, Wu W (2022) Modeling and assessment of five-frequency BDS precise point positioning. Satell Navig 3:8. https://doi.org/10.1186/s43020-022-00069-z
46. Baarda W (1967) Statistical concepts in geodesy. Nederlandse Commissie Voor Geodesie
47. Baarda W (1968) A testing procedure for use in geodetic networks. Netherland Geodetic Commission
48. Yang L, Wang J, Knight NL, Shen Y (2013) Outlier separability analysis with a multiple alternative hypotheses test. J Geod 87:591–604. https://doi.org/10.1007/s00190-013-0629-0
49. El-Mowafy A, Wang K (2022) Integrity monitoring for kinematic precise point positioning in open-sky environments with improved computational performance. Meas Sci Technol 33:085004. https://doi.org/10.1088/1361-6501/ac5d75
50. Gao Y, Li Z (1999) Cycle slip detection and ambiguity resolution algorithms for dual-frequency GPS data processing. Mar Geod 22:169–181. https://doi.org/10.1080/014904199273443
51. Blewitt G (1990) An automatic editing algorithm for GPS data. Geophys Res Lett 17:199–202. https://doi.org/10.1029/GL017i003p00199
52. Liu J, Ge M (2003) PANDA software and its preliminary result of positioning and orbit determination. Wuhan Univ J Nat Sci 8:603–609. https://doi.org/10.1007/BF02899825
53. Teunissen PJG (1995) The least-squares ambiguity decorrelation adjustment: a method for fast GPS integer ambiguity estimation. J Geod 70:65–82. https://doi.org/10.1007/BF00863419

54. Teunissen PJG (2018) Distributional theory for the DIA method. J Geod 92:59–80. https://doi.org/10.1007/s00190-017-1045-7
55. Yang L, Shen Y, Li B, Rizos C (2021) Simplified algebraic estimation for the quality control of DIA estimator. J Geod 95:14. https://doi.org/10.1007/s00190-020-01454-9
56. Ding X, Coleman R (1996) Multiple outlier detection by evaluating redundancy contributions of observations. J Geod 70:489–498. https://doi.org/10.1007/BF00863621
57. Kok JJ (1984) On data snooping and multiple outlier testing. National Geodetic Survey
58. Teunissen PJG (2006) Testing theory: an introduction. VSSD Press
59. Beckman RJ, Cook RD (1983) Outlier ... ... s. Technometrics 25:119–149. https://doi.org/10.1080/00401706.1983.10487840
60. Rousseeuw PJ, Leroy AM (1987) Robust regression and outlier detection
61. Yu Y, Yang L, Shen Y, Sun N (2023) A DIA method based on maximum a posteriori estimate for multiple outliers. GPS Solut 27:199. https://doi.org/10.1007/s10291-023-01534-1
62. Yang L, Wang J, Li H, Balz T (2021) Global assessment of the GNSS single point positioning biases produced by the residual tropospheric delay. Remote Sens 13:1202. https://doi.org/10.3390/rs13061202
63. McGraw GA (2012) Tropospheric error modeling for high integrity airborne GNSS navigation. In: Proceedings of the 2012 IEEE/ION position, location and navigation symposium, pp 158–166
64. Yang L, Fu Y, Zhu J, Shen Y, Rizos C (2023) Overbounding residual zenith tropospheric delays to enhance GNSS integrity monitoring. GPS Solut 27:76. https://doi.org/10.1007/s10291-023-01408-6
65. Blanch J, Walter T, Enge P (2019) Gaussian bounds of sample distributions for integrity analysis. IEEE Trans Aerosp Electron Syst 55:1806–1815. https://doi.org/10.1109/TAES.2018.2876583

Chapter 4
Machine Learning-Aided Tropospheric Delay Modeling over China

Hongxing Zhang and Luohong Li

Abstract Real-time precise tropospheric corrections are critical for global navigation satellite system (GNSS) data processing. This chapter aims to develop a new tropospheric delay model over China with advanced machine learning method. Compared with previous models, the new model has features such as high accuracy, a small number of coefficients and good continuity of service, showing a good performance in severe weather conditions. The new model utilizes the complementary advantages of numerical weather prediction (NWP) forecasts and real-time GNSS observations with the aid of machine learning, which alleviates the high-dependency on the dense GNSS network and allows for the ease of generating tropospheric corrections. The results can provide a new insight into augmenting tropospheric delays for BeiDou Satellite-Based PPP service across China.

4.1 Introduction

Tropospheric effect is a major error source in microwave-based space geodetic techniques, which makes the signals through the troposphere are delayed and bent [1, 2]. Accurate external prior zenith tropospheric delay (ZTD) can effectively constraint the algorithm model, improve positioning precision and reduce convergence time in Precise Point Positioning (PPP) because ZTD is correlated to station coordinates and receiver clock error. Besides, the real-time tropospheric delay also can be used in water vapour monitoring [3–6]. The troposphere delay can be calculated using Numerical Weather Prediction (NWP) models and GNSS network [7]. Meanwhile, the tropospheric corrections also can be considered in satellite-based PPP services,

H. Zhang (✉) · L. Li
State Key Laboratory of Geodesy and Earth's Dynamics, Innovation Academy for Precision Measurement Science and Technology, Chinese Academy of Sciences, Wuhan, China
e-mail: caszhx@apm.ac.cn

L. Li
College of Earth and Planetary Sciences, University of Chinese Academy of Sciences, Beijing, China

K. Yu (ed.), *Positioning and Navigation Using Machine Learning Methods*, Navigation: Science and Technology 14, https://doi.org/10.1007/978-981-97-6199-9_4

broadcasting in real-time as one of augmentation information, which providing an atmosphere constraint to improve real-time PPP performance [8–11].

There are two common methods to mitigate tropospheric delays in GNSS data processing. We can obtain the initial value from the empirical tropospheric models or the external products [12]. Besides, we also consider the tropospheric delay as unknown parameters and estimated in algorithm model [13]. However, estimating zenith wet delay (ZWD) parameters in PPP not only increase the convergence time and but also enhance the correlation among other parameters [14–16]. The traditional empirical models are also imperfect in temporal and spatial variations in tropospheric delay, which hardly provide accurate corrections for real-time PPP.

However, the accurate tropospheric delay corrections can effectively enhance the GNSS positioning model, especially in real-time case [2, 12, 17]. The corrections also can be provided as a spatial grid for wide area PPP users, but the users suffer from a significant burden to receive and store them [4, 18–20]. The huge number of tropospheric corrections are compressed using a polynomial approximation, accompanied by a certain loss of accuracy. In this way, we can only use few coefficients to express the tropospheric delay over China instead of spatial grid with amounts of data.

There are two types of datasets used in modeling the tropospheric delay, which are ground-based observations and NWP model. Ground-based meteorological or GNSS stations offer precise and timely site-wise ZTD. However, their effectiveness is hindered by uneven distribution, particularly in challenging terrain like unman mountainous regions. Conversely, NWP can forecast ZTD with extensive spatial coverage and continuous service. But NWP models must be carefully calibrated by ground-based data to address inherent biases. We can make use of the combination of NWP and ground-based data as a promising alternative, pursing a balance between accuracy and spatiotemporal resolution in tropospheric correction modeling.

For example, Dousa et al. [16] proposed a novel correction model by combining Numerical Weather Model (NWM) forecasts and precise GNSS ZTDs data. Yu et al. [21] developed a tropospheric correction model for interferometric synthetic aperture radar (InSAR) data by weighting the near real-time NWM-ZTDs from European Centre for Medium-Range Weather Forecasts (ECMWF) and GNSS-ZTDs. Wilgan and Geiger [22] combined the COSMO-1 model and GNSS data to generate post-processing tropospheric corrections for alpine regions in Switzerland. Most previous works have been conducted by artificially weighting GNSS and NWP tropospheric data in accordance with their specific characteristics (accuracy/spatiotemporal resolution, etc.). However, the quality of NWP data varies by region and is subject to changes in weather conditions and underlying NWM [16, 23–25], while the density and geometry of GNSS networks also differ significantly across regions. These variables can impact modeling performance, thus potentially posing challenges for manual weighting strategies due to their limited flexibility and hardly to implement. Moreover, to date, there have been few reports on the joint use of numerical weather forecasts and real-time GNSS tropospheric data across China, indicating that the optimal fusion strategy remains unclear.

After tropospheric corrections are generated, broadcasting these corrections effectively is of great importance. While gridded corrections offer high accuracy, their size of transmission data increases significantly with larger regions, especially in a vast country like China. Conversely, polynomial approximation can represent the gridded correction with fewer coefficients, albeit with a slight loss in accuracy. Therefore, this approach shows potential to obtain a balance between accuracy and broadcasting burden, which can be considered in satellite-based correction services.

For example, Shi et al. [26] proposed a method of determining the optimal fitting coefficients (OFCs) for polynomial approximation and tested it on a small GNSS network over Wuhan. Oliveira et al. [27] examined the OFC-based polynomial approximation method for wet delay over France and evaluated its performance in PPP-RTK. Zheng et al. [5] applied OFC-based polynomial approximation in tropospheric modeling using more than 240 GPS stations across China in a simulated real-time mode but failed to achieve satisfactory results. It still is a challenging problem to implement polynomial approximation to model tropospheric corrections across China. The potential of the polynomial approximation approach in tropospheric modelling over a wide area with highly variable topographic relief has not yet been fully investigated.

The BDS-3 now provide PPP-B2b service to Asia by three BDS-3 GEO (Geostationary Earth Orbit) satellites. The GPS/BDS-3 orbit/clock corrections and differential code bias products are included. The accurate tropospheric correction over China is potential information to be considered implemented in future. In this study, we seek to address the above challenges concerning both the tropospheric datasets and modeling methods, aiming to make new light on the generation of wide-area precise tropospheric corrections (WAPTCs) across China.

Regarding the tropospheric dataset, in order to solve the problem of sparse spatial coverage of GNSS networks (which often affects modeling performance), we first propose a fusion approach of real-time GNSS-ZTD and NWP-ZTD forecasts data to obtain better accuracy and spatial and temporal resolution of tropospheric delay corrections. To combine the advantages of the good spatial coverage of NWP-ZTD and the high accuracy of GNSS-ZTD, a machine learning technique called general regression neural network (GRNN) is applied to the joint use of GNSS-ZTD and NWP-ZTD [28]. Many studies have shown that GRNN is a powerful and convenient tool for fusion of multi-source atmospheric datasets [29, 30]. Compared with the manual weighting strategies used in previous studies, machine learning can better exploit the huge potential of relationships between variables for their joint use.

As for modeling method, using a combination of the improved tropospheric dataset and our empirical IGGtrop model [31], we apply polynomial approximation to express tropospheric corrections across China by optimizing polynomial coefficients. This approach reduces high-volume gridded corrections to only a small number of coefficients for easy transmission via satellite commutation. Furthermore, unlike many previous studies that used post-processing or simulated real-time mode, our study was conducted in a "real" real-time mode and therefore provides valuable experience and insights into the enhancement performance of atmospheric corrections on real-time PPP applications.

This chapter is organized as follows: Sect. 4.2 describes the datasets used in this study. Section 4.3 introduces the methodology for generating the WAPTCs. Section 4.4 presents evaluations on the WAPTCs. Section 4.5 summarizes the conclusions.

4.2 Data Description

Real-time data of 264 GNSS stations across China from two national monitoring networks are used, including the China Crustal Movement Observation Network (CMONOC) and the Beidou Ground-Based Augmentation System (GBAS) network. The altitudes of GNSS stations range from sea level to as high as 4570 m and the distribution of GNSS stations is shown in Fig. 4.1. The GPS/GLONASS dual-frequency observation are tracked by all stations with the sampling interval of 1 s. The period of GNSS data range from August 1 to December 31, 2020, covering summer and winter.

The Global Forecast System (GFS) is a numerical weather prediction model, which is developed and operated by the National Centers for Environmental Prediction (NCEP) in American. The GFS model divides the atmosphere parameters into 26

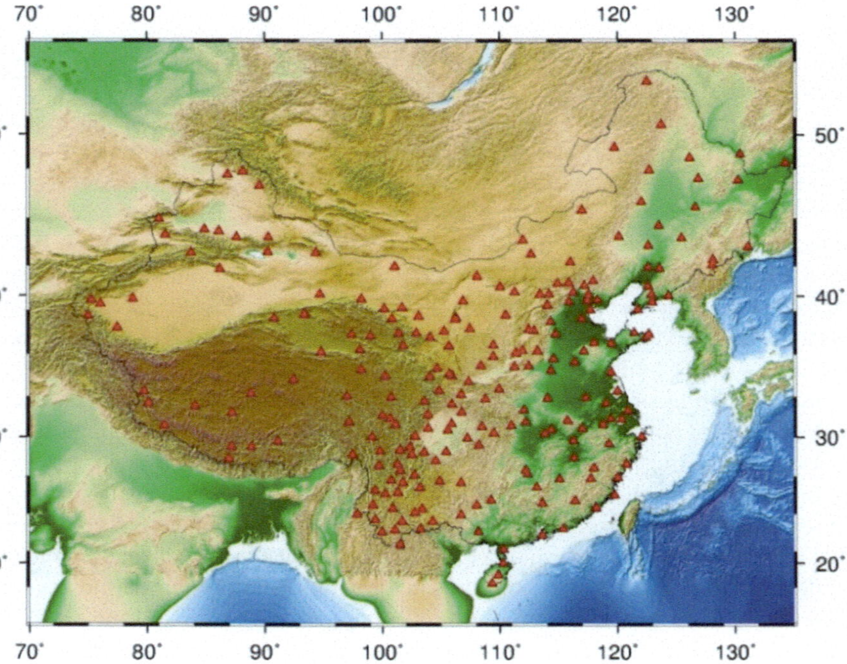

Fig. 4.1 Geographic distribution of the 264 GNSS stations used in this study

Table 4.1 Characteristics of the data used in this study

Data	Source	Station number/Spatial resolution	Temporal resolution
GNSS stream	CMONOC/GBAS	264	1 s
SSR stream	CNES	–	5 s
NWP model	NCEP-GFS	$0.25^\circ \times 0.25^\circ$	1 h

isobaric levels with a spatial resolution of $0.25^\circ \times 0.25^\circ$ every hour, such as temperature, pressure, specific humidity, and other variables at each grid point. The ZTDs can be estimated from the hourly GFS model with integral method, namely GFS ZTD. We extracted the GFS-ZTDs with spatial resolution of $1.0^\circ \times 1.0^\circ$ over China, because the resampled result is sufficient for our study with reduced calculation burden.

The GFS operates and forecasts every 6 h at 00:00, 06:00, 12:00, and 18:00 UTC for the first 120 h. The forecasts are successively uploaded with a latency of approximately 3–5 h. Thus, the forecasts with short-range (5–10 h) and GNSS ZTDs of 1s interval are combined to generate WAPTCs model. The WAPTCs model has a temporal resolution of 5 min. The characteristics of the data used are summarized in Table 4.1.

4.3 Methodology

The flowchart of generating and implementation of the WAPTCs is presented in Fig. 4.2. The main features and procedures of this method are briefly summarized as follows:

(1) GFS-forecasted and real-time GNSS-derived ZTD corrections to the IGGtrop-based background are calculated.
(2) A GRNN model is trained based on the site-wise GFS-forecasted and real-time GNSS-derived corrections.
(3) An improved ZTD correction grid is predicted using the trained GRNN model.
(4) Polynomial approximation is conducted to generate WAPTCs based on the improved ZTD correction grid.
(5) The WAPTCs are encoded and broadcasted to wide-area users.

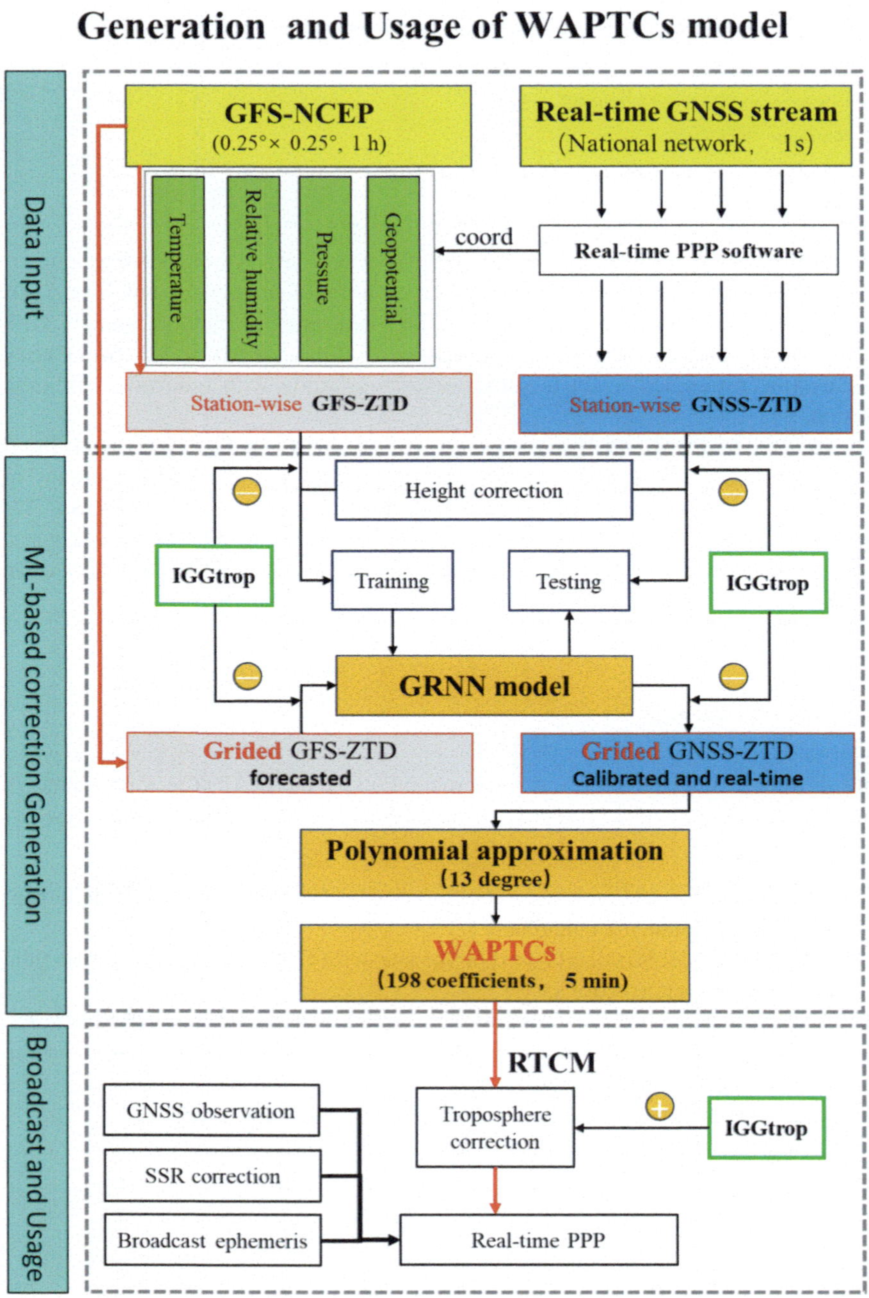

Fig. 4.2 Flowchart for the generation of WAPTCs

4.3.1 ZTD Calculation Using Real-Time GNSS and GFS Forecasts

Real-time GNSS-derived ZTDs

The real-time GNSS-derived ZTDs are estimated by self-developed real-time PPP (RTPPP) software while post-processing results are treated as reference with high-accuracy precise orbit/clock products and robust processing strategies. As for the accuracy of post-processing approach, ZTD with the root mean square (RMS) of around 5 mm is confirmed in [32]. The real-time and postprocessing strategies used are summarized in Table 4.2.

GFS-forecasted ZTDs

The numerical integration method is used to calculate the ZTD from GFS-forecasted atmospheric profiles data as follows:

$$\begin{cases} ZTD = 10^{-6} \int\limits_{h_o} N dh \\ N = \frac{k_1(P-e)}{T} + \frac{k_2 e}{T} + \frac{k_3 e}{T^2} \end{cases} \tag{4.1}$$

Table 4.2 Strategies used in PPP software

Item	Strategies	
	Postprocessing mode	Real-time mode
Observations	GPS/GLONASS raw pseudorange and phase observables	
Frequency selection	L1 & L2	
Orbit and clock	IGS final products	CNES (SSRA00CNE0)
Sampling rate	30 s	1 s
Ambiguities	Float	
Estimator	Kalman filter (smooth)	Kalman filter (forwards)
Weighing strategy	$\sigma_0 = 0.60$ m and $\sigma_0 = 3.0$ m for GPS and GLONASS pseudoranges, respectively; $\sigma_0 = 0.60$ m for GPS/GLONASS phase $\sigma = \sigma_0\sqrt{1 + 4\cos^8 e}$ [32]	
Elevation cut-off angle	7°	
Phase centre variation	igs14.atx	
Ionospheric delay	Estimated as white noise	
Station coordinates	Estimated as constant	
Zenith wet delay	Estimated as random-walk noise (10^{-8} m^2/s)	
Tropospheric gradients	Estimated every 24 h	
Mapping function		GMF

where N is the atmospheric refractivity index, h_0 is the station height, P is the pressure, $k_1 = 77.689 \pm 0.0094$; $k_2 = 71.295 \pm 1.3$; $k_3 = (3.75463 \pm 0.0076)10^5$. T is the temperature and e is the water vapour pressure calculated as follows:

$$e = 0.01RH \cdot e_s \cdot f_w \tag{4.2}$$

where RH is the relative humidity, e_s is the saturation water vapour pressure, and f_w is an enhancement factor. It should be noted that hydrostatic delay above top level of GFS atmosphere profiles is calculated with Saastamoinen model while the wet delay is too small so it can be ignored. In addition, to stay consistency with the heights system in GNSS, the geopotential heights used in GFS are transformed into ellipsoidal heights using Earth Gravitational Model 2008.

Topography and climate significantly affect tropospheric delay over China, due to diverse topography and dynamic weather patterns, which lead to considerable spatial and temporal variation in tropospheric delays. To simplify modeling complexity, the advanced empirical model IGGtrop is introduced as the background reference to mitigate the main spatiotemporal variation of tropospheric delay. The users are also easier to retrieve ZTD based on the IGGtrop mode and several coefficients for tropospheric corrections. This approach would significantly reduce redundancy and complexity in tropospheric correction modeling and approximation.

4.3.2 Calibrating GFS Forecasts Using Real-Time GNSS Dataset

Machine learning is a powerful tool to makes complex models easier to build, which has been developed for tropospheric modeling. The GRNN model is introduced to build a flexible model, mapping GNSS-ZTD to GFS-predicted ZTD, and calibrate it with real-time GNSS ZTD.

The GRNN model has a strong nonlinear mapping ability. The training set includes values of inputs x and values of an output y corresponding to each component of x. The model consists of four layers: input layer, pattern layer, summation layer, and output layer. The neurons in summation layer can be divided into two types. One calculates the algebraic sum of the neurons in the pattern layer, which is called the denominator unit. Another calculates the weighted sum of the neurons in the pattern layer, which is called the molecular unit. The output layer merely divides the denominator unit by the molecular unit to yield the desired estimate of y.

In this study, it can be determined by GRNN model that nonlinear relationship between GFS-ZTD and GNSS-ZTD corrections using sparse data samples. The structure of the GRNN model is shown in Fig. 4.3. The input variables consist of four parameters, namely site GFS-ZTD correction, site latitude, longitude, and height. The output variables are the site-wise ZTD corrections estimated by real-time PPP.

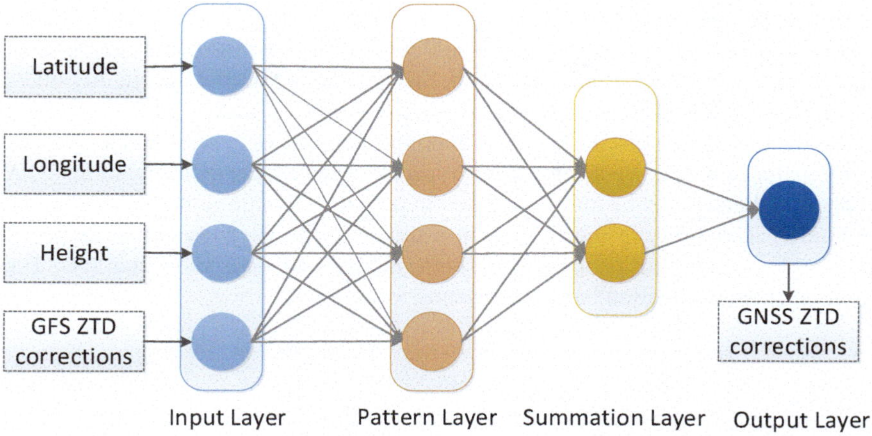

Fig. 4.3 Architecture of the GRNN used in this study

It should be noted that GNSS-ZTD estimates after convergence are used to avoid the adverse impacts of inaccurate samples on training.

The training process only involves one parameter (i.e., the spread parameter σ) that needs to be manually set in advance. In this study, we adapt a sample-based tenfold cross-validation technique to determine the optimal spread parameter for each epoch. This technique is essentially a resampling process, in which all samples are used as test data once, which performs particularly well in a limited number of data samples. Firstly, we divide all data samples (i.e., site-wise ZTD correction) into ten groups randomly and uniformly. Next, each group will be used as the test sample, while the remaining nine groups will be used as learning samples to train the model. Then, we repeated the training of the model using learning samples and evaluated it using test samples until each group of 10 samples was used as a test sample once. Finally, the optimal σ can be determined based on the results of ten rounds of evaluation. Once the GRNN model is properly trained, it can be used as a tool for obtaining improved ZTD datasets, using gridded GFS-ZTD correction as input.

4.3.3 Polynomial Approximation

The polynomial approximation is adapted to express tropospheric corrections with a cost-effective way. However, there are no studies achieved satisfactory accuracies by applying this technique across China, especially in satellites corrections. The reasons for this lie in the following facts: (1) The nonuniform and sparse modelling datasets (GNSS-ZTDs only) used in previous studies always result in poor spatial representation, causing the modelling performances to degrade. (2) The improper coordinate parameters used in previous studies can lead to poor numerical stability in the estimation of the polynomial coefficients. (3) The insufficient polynomial degrees used in previous studies also limit their modelling accuracies. In this study, we address the above issues by conducting the following tasks. (1) An improved tropospheric dataset represented on a finer surface grid is generated and used as the modelling dataset, which significantly improves the spatial representation. (2) A proper coordinate scale factor is employed in polynomial approximation to improve the numerical stability of the coefficient estimation. (3) The polynomial degree is optimized according to feedback from both internal and external evaluations. The polynomial approximation has the following form:

$$T_{cor} = \sum_{i=0}^{n}\sum_{j=0}^{n} a_{ij}(kdL)^i(kdB)^j + \sum_{t=1}^{m} b_t h^t \tag{4.5}$$

where T_{cor} is the ZTD correction; n is the maximum degree of the polynomial function accounting for ZTD corrections in the horizontal direction; k is the coordinate scale factor; dL and dB (in rad) are the latitude and longitude differences, respectively, between the sample point and centre point of the region; and h is the height (in km) of the sample point. The second term of the equation accounts for the vertical variations of the ZTD corrections. Since the vertical ZTD variations are already well characterized by the IGGtrop model a priori, $m = 2$ is sufficient for modelling. a_{ij} and b_t are the polynomial coefficients, which are estimated using the least-squares technique. The optimal determination of n and k is described in the following sections.

The GRNN is a one-pass learning algorithm that features a fast-learning process, making it suitable for real-time applications. Moreover, it also provides flexibility to include a time-varying number of training samples, which can overcome the issue of the time-varying number of valid GNSS stations during training due to accident outages and breaks in the real-time stream. We briefly review the operation of GRNN as follows.

The pattern layer of GRNN generally adopts a Gaussian function, which can approximate a continuous value with arbitrary precision. Each neural unit in the pattern layer has a basis function, and these basis functions are linearly combined through the weights. Infinite approximation makes the output of the neural network approach a certain value or causes it to no longer change, so that the GRNN model is stabilized. The GRNN model draws the function estimate directly from training data and does not need an iterative training procedure; thus, GRNN does not have a local minimum issue.

4.4 Results and Discussion

4.4.1 Evaluation of the GFS-Forecasted ZTDs

Overall accuracies of GFS-ZTD and RT GNSS-ZTD

The quality of the GFS-ZTDs and real-time GNSS-ZTDs is of great importance in the GRNN model and the combination results. Thus, we evaluate the site-wise forecasting GFS-ZTDs and estimted GNSS-ZTD in real-time using the post-processing ZTD as reference. The biases, standard deviation (STD) and RMS values of GFS-ZTD (left panels) and real-time GNSS-ZTD (right panels) at 264 stations are shown in Fig. 4.4, where red and blue colours indicate the results for summer (August) and winter (December) months, respectively.

Three characteristic can be drawn from Fig. 4.4 as follows: (1) The accuracy of GFS-ZTD is uneven in space and time domin, exhibiting significant latitude and seasonal dependencies, where lower values appear in low latitudes and warm seasons. The tropospheric delays at lower latitudes are usual larger and exhibit greater variations, which making GFS-ZTD prediction more difficult. (2) Overall positive biases

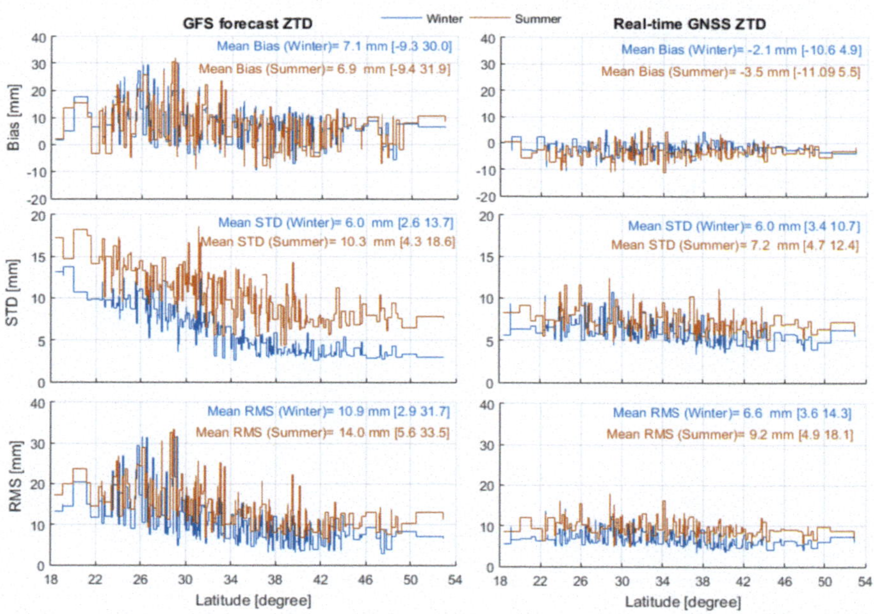

Fig. 4.4 Bias, STD and RMS error values of the GFS-ZTDs (left panels) and real-time GNSS-ZTDs (right panels) in summer (in red) and winter (in blue) months at the 264 GNSS stations with respect to the GNSS-ZTDs from the postprocessing run. The values in brackets represent the minimum and maximum values of the bias, STD and RMS for all stations

with around 7.0 mm appears in GFS-ZTD, indicating that GFS-ZTDs are overestimated in China, especially in low latitude regions. (3) Real-time GNSS-ZTD outperforms GFS-ZTD in terms of biases, STD, and RMS. It is thus feasible to use the real-time GNSS-ZTD to calibrate GFS-ZTDs.

To further illustate the accuracy of the GFS-ZTD over China, we evaluate the GFS-ZTD with respect to the ZTD derived from ERA5 reanalysis products, namely ERA5-ZTD. Zue et al. (2015) distributed a global tropospheric grid products in a horizontal resolution of $1^\circ \times 1^\circ$ based on ERA5 at GFZ. We use the same orography model ($1^\circ \times 1^\circ$) and refractivity coefficients adpated in GFZ products when gernerating the GFS-ZTD. Therefore, we compare the GFS-ZTD grid to the ERA5-ZTD grid (from GFZ) directly without horizontal or vertical adjustment for ZTD. The bias and RMS of the ZTD differences between the GFS-ZTD and ERA5-ZTD are presented in Fig. 4.5. A significant spatial variation in biases is observed on GFS-ZTD over China, with most regions showing positive values, indicating that GFS-ZTD is often overestimated. In addition, RMS values are related to latitude and are larger in low latitude regions of China. The same results are confirmed by another dataset as reference.

As for the spatial characteristics, a significant negative bias appears in GFS-ZTD over Sichuan Basin, showing a notable dark blue at this area in Fig. 4.5a (bias). This relates to the unique basin topography. Moreover, large RMS values are seen in Fig. 4.5b (RMS), predominantly along the coast. This is mainly attributed to the influences of the Asian and Pacific monsoons, which bring the amount of water vapour over the southern coast of China from ocean. Thus, the variation of water vapour would be larger than that over other regions, increasing uncertainties and leading to worse consistency with the ERA5-ZTD. Therefore, the quality of the GFS-ZTDs is related not only to latitude but also to topography and climate zones.

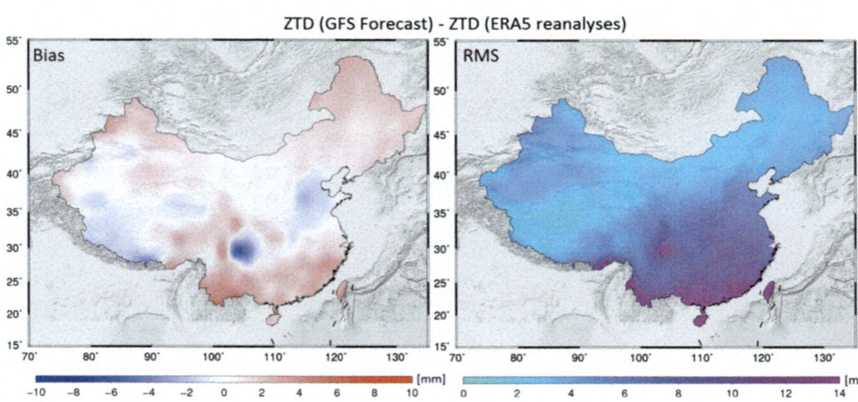

Fig. 4.5 Bias (left panel) and RMS (right panel) of the GFS-ZTDs with respected to the ERA5-derived ZTDs over China from 1 August to 31 December in 2020

GFS-ZTD accuracy degradation with an increasing forecast horizon

To assess how the accuracy of GFS-ZTD forecasts diminishes over time, we computed the GFS-ZTD errors at 264 GNSS stations for various forecast horizons: 5 h (5, 11, 17, 23 UTC), 6 h (6, 12, 18, 00 UTC), 7 h (7, 13, 19, 01 UTC), 8 h (8, 14, 20, 02 UTC), 9 h (9, 15, 21, 03 UTC), and 10 h (10, 16, 22, 04 UTC). Post-processing GNSS-ZTDs served as the reference.

Figure 4.6 illustrates the average RMS errors for forecast horizons of 5, 6, 7, 8, 9, and 10 h during both summer and winter. Notably, the accuracy of GFS-ZTD exhibits a slight decline as the forecast horizon increases, albeit at a marginal rate of approximately 0.1 mm/h within the 5–10 h forecast window, as evidenced by the slope of the linear line. Consequently, we assert that GFS-ZTDs demonstrate no significant deterioration in accuracy within the short-term forecast horizons (5–10 h) examined in this study.

Accuracy of the WAPTCs

We utilized the enhanced ZTD correction grid to produce WAPTCs, employing the polynomial approximation settings defined earlier. We first derive the ZTD correction grid from the original GFS-ZTDs in WAPTC model, which is labeled as WAPTC (GFS-only). The GNSS calibrated GFS-ZTD named WAPTC (GNSS/GFS) can be obtained as follows: we calculate the site-wise ZTD corrections from the original GFS forecasts and then we calibrate the site-wise GFS-ZTD with real-time GNSS data.

To evaluate how effectively sparse real-time GNSS-ZTDs mitigate the spatially varying biases of GFS-ZTDs across China, we compared the ZTDs obtained from WAPTC (GFS-only) and WAPTC (GNSS/GFS) at 1 h intervals against their respective post-processing results to calculate ZTD errors. Figure 4.7 depicts histograms

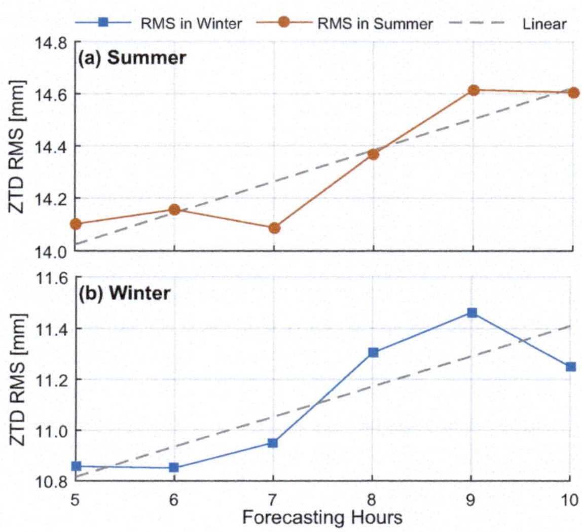

Fig. 4.6 Mean RMS errors of the GFS-ZTDs at the 264 GNSS stations with forecast horizons of 5, 6, 7, 8, 9 and 10 h in summer (**a**, in red) and winter (**b**, in blue) months. The GNSS-ZTDs from the postprocessing run are used as references

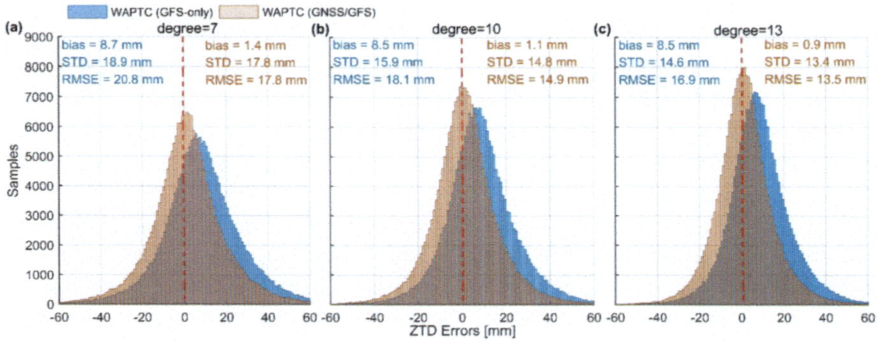

Fig. 4.7 Histogram of the ZTD errors obtained from the WAPTC (GFS-only) and WAPTC (GNSS/GFS) at all GNSS stations across China from 1 August to 31 December 2020. The GNSS-ZTDs from the postprocessing run are used as references. The statistical results for the errors in terms of bias, STD and RMS values are also depicted

of the ZTD errors across all GNSS stations from August 1st to December 31st, 2020. Three cases with varying polynomial degrees (7, 10, and 13) are illustrated in Fig. 4.7a, b, c, respectively, aiming to showcase the influence of polynomial degree on modeling accuracy.

Two key observations are summaried from Fig. 4.7 as follows: Firstly, both WAPTC (GFS-only) and WAPTC (GNSS/GFS) demonstrate improved accuracy with higher polynomial degrees. Secondly, WAPTC (GNSS/GFS) outperforms WAPTC (GFS-only) with smaller biases. This highlights the superior performance of WAPTC (GNSS/GFS) in mitigating biases compared to solutions relying solely on GFS data. The GRNN model trained on sparse GNSS stations contributes to the production of tropospheric datasets with enhanced accuracy compared to GFS-only solutions.

An external validation was conducted to verify the accuracy of WAPTC (GNSS/GFS) using post-processing GNSS-ZTDs. The validation methodology is briefly described as follows: WAPTC (GNSS/GFS) was generated epoch by epoch. For each epoch, 90% of available GNSS stations were randomly selected to train the GRNN for calibrating the GFS-ZTD correction grid, while the remaining 10% were designated as external test stations. ZTDs obtained from WAPTC (GNSS/GFS) at the test stations were then compared against post-processing GNSS-ZTDs to calculate errors. Simultaneously, WAPTC (GFS-only) was evaluated using the same selected test stations. Additionally, we generated and evaluated WAPTCs by using only the sitewise real-time GNSS-ZTDs simultaneously for comparison, which is denoted as WAPTC (GNSS-only). The evaluation spanned a five-month period from August 1st to December 31st, 2020, with 1 h intervals. Since test stations were randomly selected in each epoch, the evaluation results were deemed sufficiently representative.

Figure 4.8 presents mean bias and RMS error values for WAPTC (GNSS-only), WAPTC (GFS-only), and WAPTC (GNSS/GFS) across all randomly selected test stations during summer and winter seasons, with polynomial degree (n) ranging from 3 to 16. WAPTC (GNSS-only) consistently reached its minimum RMS error

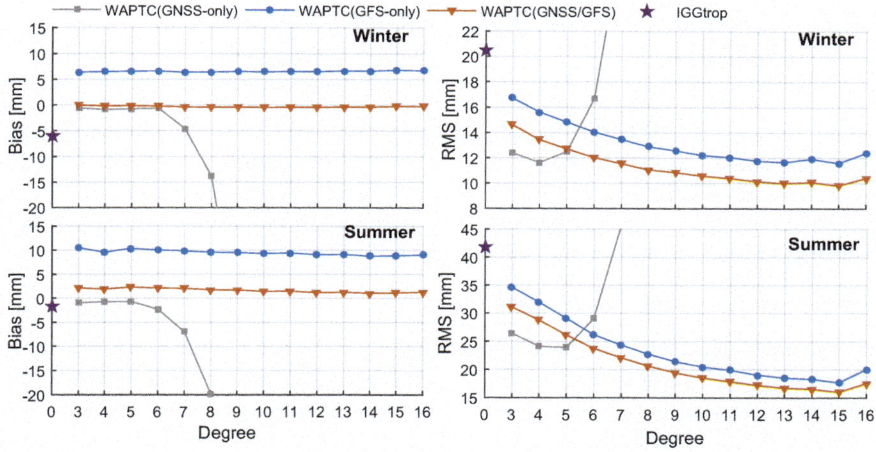

Fig. 4.8 Mean biases and RMS errors of the ZTDs obtained from the WAPTC (GNSS-only), WAPTC (GFS-only) and WAPTC (GNSS/GFS) with the polynomial degree varying from 3 to 16 at the 10% randomly selected test stations during winter (upper panels) and summer (lower panels). The accuracy of the IGGtrop model is also shown for comparison. The GNSS-ZTDs from the post-processing are used as references

(11.9 mm in winter and 23.8 mm in summer) at relatively low degrees due to the limited spatial coverage of GNSS stations. This scarcity makes estimating excessive coefficients challenging and burdensome in GNSS-only conditions. WAPTC (GNSS-only) exhibited no conspicuous bias at lower degrees before RMS error began to rise. In contrast, increasing the degree led to decreased RMS error for WAPTC (GFS-only), primarily attributable to the extensive spatial coverage of the GFS-ZTD grid. However, WAPTC (GFS-only) displayed noticeable bias. Ultimately, WAPTC (GNSS/GFS) leveraged the complementary advantages of both WAPTC(GNSS-only) and WAPTC (GFS-only) to achieve optimal overall performance.

Furthermore, the mean RMS error of WAPTC (GNSS/GFS) decreases to 10.0 mm (in winter) and 16.0 mm (in summer) when $n = 13$ and shows no significant further decrease at higher degrees. This finding aligns with the outcomes depicted in Figs. 4.7 and 4.8, where fitting errors serve as feedback. Hence, a balance between accuracy and efficiency can be attained by setting n as 13. With this optimal value of n, WAPTC (GNSS/GFS) enhances ZTD accuracy by approximately 16% (in winter) and 23% (in summer) compared to the WAPTC (GNSS-only) scenario, by about 16% (in winter) and 17% (in summer) compared to the WAPTC (GFS-only) scenario, and by roughly 50% (in winter) and 62% (in summer) compared to the empirical IGGtrop model for China.

4.4.2 Performance of the WAPTCs Under Challenging Conditions

To gain insight into the performance of the proposed WAPTCs, we conducted evaluations focusing on challenging conditions, particularly complex terrain and severe weather events. These scenarios are known to pose significant challenges to tropospheric models. By scrutinizing WAPTCs under such conditions, we can better understand their capabilities and advantages.

Complex terrain: Hengduan Mountains

We evaluated WAPTCs in the challenging terrain of the Hengduan Mountains, selecting 30 GNSS stations across the region. These stations, situated between 150 and 3500 m elevation, represent areas with complex topography. The spatial distribution of these 30 stations is shown in Fig. 4.9. This assessment aims to gauge the performance of WAPTCs in demanding conditions, particularly in regions with intricate terrain and low latitudes over China.

The ZTDs obtained from WAPTC (GNSS/GFS) at these test stations were compared against the corresponding post-processing GNSS-ZTDs to calculate errors. The biases and RMS errors at all 30 stations are presented in Fig. 4.10. Importantly, only errors generated when the stations served as test stations (not used in GRNN training) are included in these results. The experimental period spans from August 1st to December 31st, 2020, with 1 h intervals. Each station functions as a test station more than 670 times (equivalent to 4 weeks) during this period, ensuring the external evaluation's representativeness. This process was repeated using WAPTC (GFS-only), and corresponding evaluation results are depicted in the left panels of Fig. 4.10 for comparison.

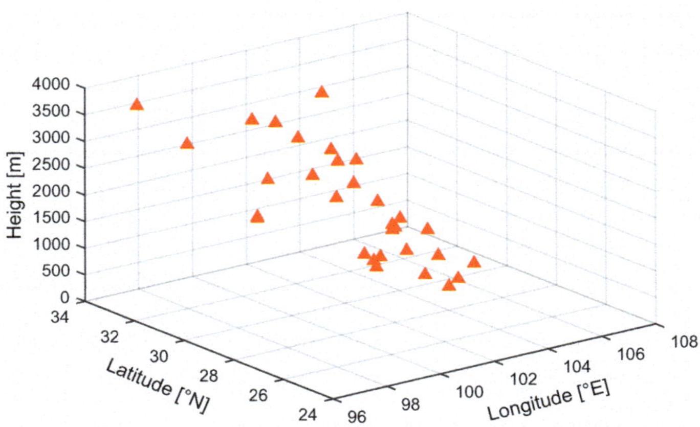

Fig. 4.9 Spatial distribution of the 30 test stations in the Hengduan Mountains

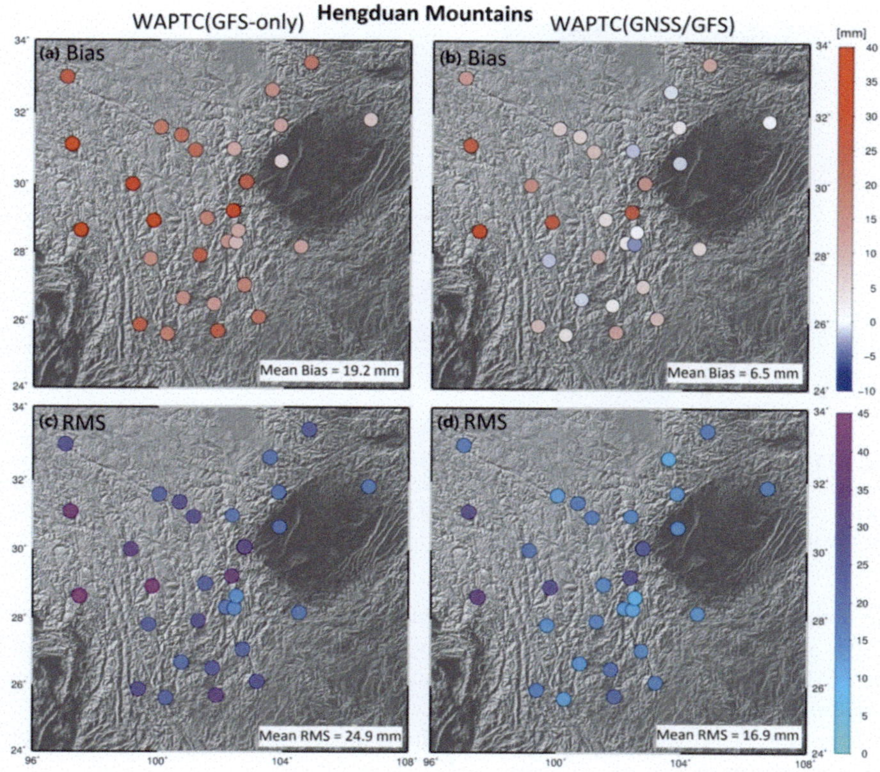

Fig. 4.10 Biases and RMS errors of the ZTDs obtained from the WAPTC (GFS-only) and WAPTC (GNSS/GFS) over a mountainous region from 1 August to 31 December 2020. The GNSS-ZTDs from the postprocessing run are used as references

In Fig. 4.10a, the WAPTC (GFS-only) results have notable positive biases across most stations with a mean value of 19.2 mm. This highlights a tendency for ZTDs derived from WAPTC (GFS-only) to be consistently overestimated. However, the implementation of WAPTC (GNSS/GFS) can effectively mitigate these biases, reducing the mean bias value to 6.5 mm. This outcome underscores the successful calibration of GFS-ZTD biases through the trained GRNN model. Furthermore, the mean RMS error value of WAPTC (GNSS/GFS) at these 30 test stations stands at 16.9 mm, marking a notable 21% enhancement in accuracy compared to WAPTC (GFS-only).

Severe weather: Typhoon Maysak

To illustrate the potentials of the WAPTC under severe weather conditions, Fig. 4.11 shows a case of Typhoon Maysak. The path information of Maysak is shown in Fig. 4.11a. The error time series of the ZTDs obtained from WAPTC (GNSS/GFS) at the two test stations (i.e., JLCB and HRBN, not used in training GRNN), which

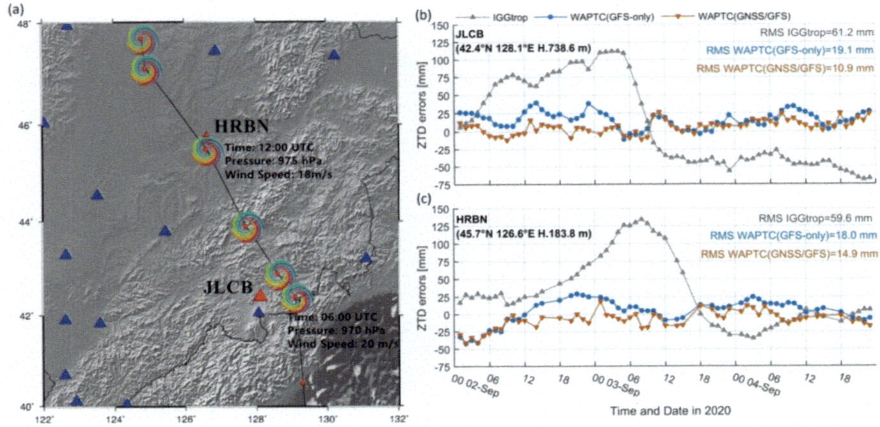

Fig. 4.11 **a** The path track of Typhoon Maysak on 3 September 2020. **b–c** Time series of ZTD errors obtained from IGGtrop, WAPTC (GFS-only) and WAPTC (GNSS/GFS) at stations JLCB (**b**) and HRBN (**c**) before, during and after the typhoon event. The absolute ZTDs from postprocessing run are also presented for references

are both near the path of Maysak, over a 3 day period (from 2 to 4 in September of 2020), are shown in Fig. 4.11b, c, respectively. The errors of the ZTDs obtained from the WAPTC (GFS-only) and the IGGtrop model are also presented for comparison purposes.

From Fig. 4.11, we can see that the IGGtrop model exhibit error peaks at both two stations (even exceeds 125 mm at HRBN) when the Maysak is approaching. This indicates that Typhoon Maysak causes rapid change in ZTD, which cannot be characterized by the empirical IGGtrop model and thus resulting in large errors. The ZTD errors obtained from WAPTC (GFS-only) are generally stable before, during and after the typhoon event with a mean RMS error of 19.1 mm at JLCB and 18.0 mm at HRBN. The WAPTC (GNSS/GFS) jointly using the GNSS and GFS further decreases the mean RMS error to 10.9 mm at JLCB station and 14.9 mm at HRBN station, which suggests the superiority of the WAPTC (GNSS/GFS) over the WAPTC (GFS-only) under severe weather conditions.

Performance of WAPTC in augmenting RTPPP

To showcase the potential of WAPTC under severe weather conditions, Fig. 4.11 presents a case study of Typhoon Maysak. The path information of Maysak is depicted in Fig. 4.11a. Figure 4.11b, c display the error time series of ZTDs obtained from WAPTC (GNSS/GFS) at two test stations (JLCB and HRBN), both located near the path of Maysak, over a 3 day period (from September 2nd to 4th, 2020). These stations were not used in the GRNN training set.

From Fig. 4.11, it’s obvious that the IGGtrop model exhibits error peaks at both stations (even exceeding 125 mm at HRBN) as Typhoon Maysak approaches. This suggests that the rapidly changing ZTD induced by the typhoon cannot be adequately characterized by the empirical IGGtrop model, resulting in significant errors. Conversely, ZTD errors obtained from WAPTC (GFS-only) remain relatively stable before, during, and after the typhoon event, with a mean RMS error of 19.1 mm at JLCB and 18.0 mm at HRBN. Furthermore, WAPTC (GNSS/GFS), which integrates both GNSS and GFS data, reduces the mean RMS error to 10.9 mm at the JLCB station and 14.9 mm at the HRBN station. This indicates the superiority of WAPTC (GNSS/GFS) over WAPTC (GFS-only) under severe weather conditions, demonstrating its ability to provide more accurate tropospheric corrections during extreme events like Typhoon Maysak.

To demonstrate the advantages of utilizing WAPTCs in BDS-3 positioning, Fig. 4.12 presents a case study of BDS-3 RTPPP testing using BDS-3 PPP-B2b corrections. The data were collected on September 12, 2023, during a summer day with active atmospheric conditions.

For comparison, standard PPP and WAPTCs-augmented PPP were simultaneously conducted at the two stations. Table 4.3 summarizes the strategies employed in these two PPP scenarios, differing only in how tropospheric delays were handled. Other settings, such as sampling rate, estimator, elevation cut-off angle, and mapping functions, remained consistent with those outlined in Table 4.2 (for GPS real-time mode). Additionally, real-time WAPTC messages were encoded in a self-defined RTCM3 format and broadcasted to users via Networked Transport of RTCM via Internet Protocol (NTRIP). This broadcasting approach was validated in the process, ensuring our experiments were conducted under a "true" stream-based real-time mode.

Figure 4.12 presents the time series of ZTD and positioning errors in up component from the two PPP scenarios throughout a whole-day period at the two test stations. In Fig. 4.12, we observed that WAPTCs-augmented PPP solutions tend to exhibit greater stability than standard PPP solutions in both delay and positioning

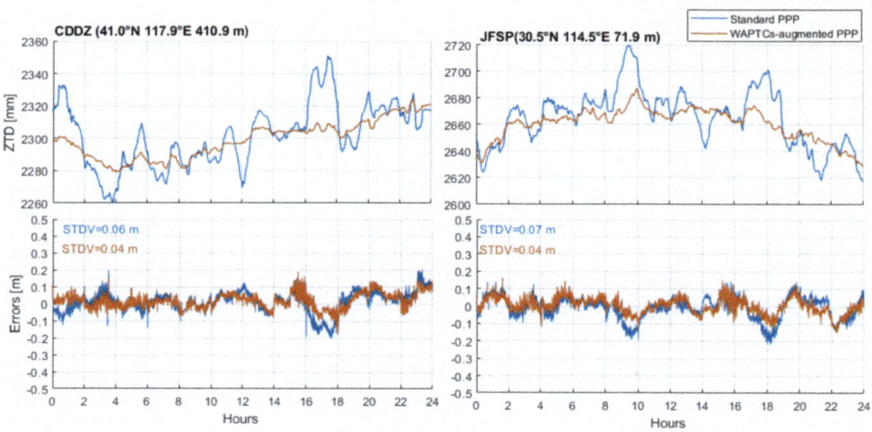

Fig. 4.12 Real-time kinematic PPP ZTD (top panels) and positioning errors (bottom panels) from Standard PPP and WAPTCs-augmented PPP at the test stations CDDZ and JFSP over a 24 h period

Table 4.3 Strategies used in the four BDS-3 RTPPP scenarios

Item	Strategies	
	Standard PPP	WAPTCs-augmented PPP
Observations	BDS-3 ionosphere-free combinations (B1I/B3I)	
SSR corrections	BDS-3 PPP-B2b	
ZTD	Estimated	Constrained to WAPTCs
Station coordinates	Estimated as white noise	

domains, suggesting that WAPTCs can enhance the performance of real-time PPP. The average STD error of the WAPTCs-augmented PPP solution in the up component is approximately 4.0 cm at the two test stations. In contrast, with standard PPP, this value increases to 6.0–7.0 cm. Consequently, the positioning accuracy in the up component improves by 23.8% when employing WAPTCs.

4.5 Conclusion

In this study, we developed a novel method for generating real-time WAPTCs in China, leveraging NCEP-GFS forecasts and sparse real-time GNSS stations. Our approach capitalizes on the spatial resolution of GFS forecasts and the accuracy of GNSS-derived ZTDs, employing a machine learning technique called GRNN. Additionally, we introduced a new polynomial approximation method to express tropospheric corrections across China, significantly reducing the volume of gridded corrections while maintaining satisfactory accuracy. These WAPTCs alleviate dependence on dense GNSS networks and facilitate troposphere augmentation for BDS/GNSS satellite-based PPP services.

Firstly, we addressed spatially varying biases in NCEP-GFS forecasts and proposed a ML-based strategy to calibrate these biases in GRNN model using real-time GNSS-ZTD from sparse stations. This approach improved the quality of tropospheric datasets for generating WAPTCs, reducing mean bias from 8.5 to 0.9 mm.

Subsequently, based on the improved tropospheric dataset, we employed a new polynomial approximation method to express tropospheric corrections across China. This method significantly reduced the number of model coefficients, allowing for easier transmission via satellite. Incorporating our modern IGGtrop model as background further enhanced the performance of polynomial approximation.

External evaluations of the proposed WAPTCs demonstrated overall accuracies of 10.0 mm (winter) and 16.0 mm (summer) across China. Case studies confirms the ability of our real-time WAPTCs to capture turbulent variability in tropospheric delays over complex terrain and under severe weather conditions, underscoring the robustness of our method.

When applied in BDS-3 PPP-B2b service, WAPTCs improved user-end positioning accuracy in the up component by 23.8%. Our approach can be extended to other regions worldwide by combining available NWP models and GNSS stations. Future work will focus on utilizing higher-quality regional NWP models and advanced machine learning techniques to achieve more robust performance.

References

1. Boehm J, Werl B, Schuh H (2006) Troposphere mapping functions for GPS and very long baseline interferometry from European centre for medium-range weather forecasts operational analysis data. J Geophys Res-Solid Earth 111(B2)
2. Hobiger T, Ichikawa R, Koyama Y, Kondo T (2008) Fast and accurate ray-tracing algorithms for real-time space geodetic applications using numerical weather models. J Geophys Res-Atmos 113(D20)
3. Li XX, Zhang XH, Ge MR (2011) Regional reference network augmented precise point positioning for instantaneous ambiguity resolution. J Geodesy 85(3):151–158
4. Wilgan K, Hadas T, Hordyniec P, Bosy J (2017) Real-time precise point positioning augmented with high-resolution numerical weather prediction model. GPS Solutions 21(3):1341–1353
5. Zheng F, Lou YD, Gu SF, Gong XP, Shi C (2018) Modeling tropospheric wet delays with national GNSS reference network in China for BeiDou precise point positioning. J Geodesy 92(5):545–560
6. Bisnath S and IEEE (2020) PPP: Perhaps the natural processing mode for precise GNSS PNT, presented at the 2020 IEEE/ION Position, Location and Navigation Symposium (Plans)
7. Lu C, Zhong Y, Wu Z, Zheng Y, Wang Q (2023) A tropospheric delay model to integrate ERA5 and GNSS reference network for mountainous areas: application to precise point positioning. GPS Solutions 27(2)
8. European Union (2022) Galileo High Accuracy Service Signal-In-Space Interface Control Document (HAS SIS ICD) Issue 1.0
9. European Union (2023) Galileo High Accuracy Service Service Definition Document (HAS SDD) Issue 1.0
10. Cabinet Office (2020) Quasi-Zenith Satellite System Interface Specification Centimeter Level Augmentation Service (IS-QZSS-L6-003)
11. CSNO (2020) BeiDou Navigation Satellite System Signal in space interface control document Precise Point Positioning Service Signal PPP-B2b (Version 1.0)
12. Hadas T, Kaplon J, Bosy J, Sierny J, Wilgan K (2013). Near-real-time regional troposphere models for the GNSS precise point positioning technique. Meas Sci Technol 24(5)
13. Bock O, Tarniewicz J, Thom C, Pelon J, Kasser M (2001) Study of external path delay correction techniques for high accuracy height determination with GPS. Phys Chem Earth Part A-Solid Earth Geodesy 26(3):165–171
14. Yao YB, Xu XY, Xu CQ, Peng WJ, Wan YY (2019) Establishment of a real-time local tropospheric fusion model. Remote Sens 11(11)
15. Hadas T, Teferle FN, Kazmierski K, Hordyniec P, Bosy J (2017) Optimum stochastic modeling for GNSS tropospheric delay estimation in real-time. GPS Solutions 21(3):1069–1081
16. Dousa J, Elias M, Václavovic P, Eben K, Krc P (2018) A two-stage tropospheric correction model combining data from GNSS and numerical weather model. GPS Solutions 22(3)
17. Dousa J, Vaclavovic P (2014) Real-time zenith tropospheric delays in support of numerical weather prediction applications. Adv Space Res 53(9):1347–1358
18. Li XX, Ge MR, Dousa J, Wickert J (2014) Real-time precise point positioning regional augmentation for large GPS reference networks. GPS Solutions 18(1):61–71

19. Lu CX et al (2017) Improving BeiDou real-time precise point positioning with numerical weather models. J Geodesy 91(9):1019–1029
20. Vaclavovic P, Dousa J, Elias M, Kostelecky J (2017) Using external tropospheric corrections to improve GNSS positioning of hot-air balloon. GPS Solutions 21(4):1479–1489
21. Yu C, Li ZH, Penna NT, Crippa P (2018) Generic Atmospheric correction model for interferometric synthetic aperture radar observations. J Geophys Res-Solid Earth 123(10):9202–9222
22. Wilgan K, Geiger A (2019) High-resolution models of tropospheric delays and refractivity based on GNSS and numerical weather prediction data for alpine regions in Switzerland. J Geodesy 93(6):819–835
23. Andrei CO, Chen RZ (2009) Assessment of time-series of troposphere zenith delays derived from the global data assimilation system numerical weather model. GPS Solutions 13(2):109–117
24. Lu CX et al (2016) Tropospheric delay parameters from numerical weather models for multi-GNSS precise positioning. Atmos Meas Tech 9(12):5965–5973
25. Zhang WX et al. (2020) Rapid troposphere tomography using adaptive simultaneous iterative reconstruction technique. J Geodesy 94(8)
26. Shi JB, Xu CQ, Guo JM, Gao Y (2014) Local troposphere augmentation for real-time precise point positioning. Earth Planets Space 66
27. Oliveira PS et al (2017) Modeling tropospheric wet delays with dense and sparse network configurations for PPP-RTK. GPS Solutions 21(1):237–250
28. Specht DF (1991) A general regression neural network. IEEE Trans Neural Netw 2(6):568–576
29. Yuan QQ, Xu HZ, Li TW, Shen HF, Zhang LP (2020) Estimating surface soil moisture from satellite observations using a generalized regression neural network trained on sparse ground-based measurements in the continental U.S. J Hydrol 580
30. Zhang B, Yao YB (2021) Precipitable water vapor fusion based on a generalized regression neural network. J Geodesy 95(3)
31. Li W et al (2015) New versions of the BDS/GNSS zenith tropospheric delay model IGGtrop. J Geodesy 89(1):73–80
32. Hadas T, Hobiger T, Hordyniec P (2020) Considering different recent advancements in GNSS on real-time zenith troposphere estimates. GPS Solutions 24(4)

Chapter 5
Deep Learning Based GNSS Time Series Prediction in Presence of Color Noise

Hongkang Chen, Xiaoxing He, and Tieding Lu

Abstract Global Navigation Satellite System (GNSS) time series prediction plays a significant role in monitoring crustal plate motion, landslide detection, and maintenance of the global coordinate framework. Long Short-Term Memory (LSTM), a deep learning model has been widely applied in the field of high-precision time series prediction especially when combined with Variational Mode Decomposition (VMD) to form the VMD-LSTM hybrid model. To further improve the prediction accuracy of the VMD-LSTM model, this paper proposes a dual variational modal decomposition long short-term memory (DVMD-LSTM) model to effectively handle the noise in GNSS time series prediction. This model extracts fluctuation features from the residual terms obtained after VMD decomposition to reduce the prediction errors associated with residual terms in the VMD-LSTM model. Daily E, N, and U coordinate data recorded at multiple GNSS stations between 2000 and 2022 are used to validate the performance of the proposed DVMD-LSTM model. The experimental results demonstrate that compared to the VMD-LSTM model, the DVMD-LSTM model achieves significant improvements in prediction performance across all measurement stations. The average root mean squared error (RMSE) is reduced by 9.86%, and the average mean absolute error (MAE) is reduced by 9.44%, and the average R2 increased by 17.97%. Furthermore, the average accuracy of the optimal noise model for the predicted results is improved by 36.50%, and the average velocity accuracy of the predicted results is enhanced by 33.02%. These findings collectively attest to the superior predictive capabilities of the DVMD-LSTM model, thereby enhancing the reliability of the predicted results.

H. Chen · T. Lu
School of Geodesy and Geomatics, East China University of Technology, Nanchang 341000, China
e-mail: chk@ecut.edu.cn

T. Lu
e-mail: tdlu@ecut.edu.cn

X. He (✉)
School of Civil and Surveying and Mapping Engineering, Jiangxi University of Science and Technology, Ganzhou 341000, China
e-mail: xxh@jxust.edu.cn; hjh@mail.jxust.edu.cn

K. Yu (ed.), *Positioning and Navigation Using Machine Learning Methods*, Navigation: Science and Technology 14, https://doi.org/10.1007/978-981-97-6199-9_5

5.1 Introduction

Over the past three decades, owing to the rapid advancements in satellite navigation technology, a global network of Global Navigation Satellite System (GNSS) continuously operational reference stations has been established. These stations play a pivotal role as primary sources of information for various purposes, such as monitoring crustal plate movements [1], detecting landslides [2], monitoring deformations in structures like bridges or dams [3], and maintaining regional or global coordinate frameworks [4]. By analyzing extensive time series data collected from these GNSS stations, it becomes possible to predict changes in coordinates at regular intervals, forming a fundamental basis for identifying patterns in motion. This carries significant practical and theoretical implications in the fields of geodesy and geodynamics research [5].

Time series prediction techniques can be broadly categorized into two main groups: physical simulation and numerical simulation [6]. Traditional methods in both physical and numerical simulation rely on geophysical principles, linear components, periodic elements, and gap filling to construct models. However, these models often struggle to capture intricate nonlinear data, requiring manual selection of feature information and model parameters, which can lead to systematic biases and limitations [7]. In contrast, deep learning, an emerging technology, has the capability to automatically extract relevant information by constructing deep network architectures. Deep learning demonstrates robust learning abilities and excels in handling extensive and high-dimensional data [8, 9].

Long Short-Term Memory (LSTM), a significantly improved variant of Recurrent Neural Networks (RNN), effectively tackles the challenges of gradient vanishing, gradient exploding, and limited long-term memory commonly encountered in conventional RNNs [10]. Due to its remarkable abilities in long-range time series forecasting, LSTM has found extensive application across various domains of time series prediction, including satellite navigation. For instance, Kim et al. enhanced the precision and stability of absolute positioning solutions in autonomous vehicle navigation by employing a multi-layer LSTM model [11]. Tao et al. adopted a CNN-LSTM approach to extract deep multipath features from GNSS coordinate sequences, thereby mitigating the impact of multipath effects on positioning accuracy [12]. Additionally, Xie et al. [13] achieved accurate predictions of landslide periodic components using the LSTM model, establishing a landslide hazard warning system [13].

Variational Mode Decomposition (VMD) is a signal processing methodology rooted in variational inference principles. It decomposes signals into distinct mode components called Intrinsic Mode Functions (IMF), each with varying frequencies achieved through an optimization process. This process effectively extracts time–frequency local characteristics from signals, enabling efficient signal decomposition and analysis [14–18].

The integration of LSTM with VWD, i.e., the VMD-LSTM model has gained widespread adoption in various fields for time series prediction. However, most studies typically follow a common approach: VMD is used to decompose the original data, predict each Intrinsic Mode Function (IMF) and the residual term separately, and then combine these predictions to obtain the final result. While this method yields good results for each IMF, it encounters challenges in effectively capturing the fluctuation characteristics of the residual term, leading to notable prediction errors in the model. Additionally, existing methods primarily focus on the accuracy of prediction results, but often overlook the inherent noise characteristics within the data. In light of these limitations, this chapter introduces a novel hybrid model known as the Dual VMD-LSTM (DVMD-LSTM) model, which takes the characteristics of noise in the data into consideration. By applying VMD decomposition to the residual components derived from the initial VMD decomposition, the proposed model adeptly extracts the fluctuation features within the residuals, thereby enabling high-precision prediction of GNSS time series data.

In this chapter, by fully utilizing the multi-site GNSS coordinate data, the proposed hybrid deep learning model is first evaluated and compared with multiple deep learning models using traditional accuracy evaluation metrics which are Root Mean Square Error (RMSE), Mean Absolute Error (MAE), and coefficient of determination (R^2). Subsequently, a noise model is introduced and the prediction results of each model are analyzed in comparison with the optimal noise model and speed calculated by the noise model. These multi-level comparison methods fully demonstrate the excellent performance of the hybrid deep learning model algorithm proposed in this chapter, which provides an innovative hybrid deep learning model in the field of GNSS time series prediction, and provides an important support for research and application in related fields.

5.2 Principle and Method

5.2.1 Variational Modal Decomposition (VMD)

Variational Mode Decomposition (VMD) is an adaptive and entirely non-recursive technique used to address issues related to modal variation and signal processing. GNSS time series data inherently exhibit non-stationarity. Through decomposition, VMD effectively separates the non-stationarity data into stationary signals. This procedure enables the extraction of the inherent fluctuation characteristics within GNSS time series data, forming a robust foundation for predictive modeling.

The VMD model is essentially a set of multiple adaptive Wiener filters and involves the following four important concepts in processing the signal [19]: Intrinsic

Mode Functions, Total Practical IMF Bandwidth, Hilbert Transforms and Analytic Signal.

(1) Intrinsic Mode Function

In the VMD algorithm, the IMF is defined as an amplitude-modulated-frequency-modulated (AM-FM) signal with the following expression:

$$u_k(t) = A_k(t)\cos(\varphi_k(t)) \tag{5.1}$$

where $A_k(t)$ is the instantaneous amplitude of $u_k(t)$ and the instantaneous frequency of $u_k(t)$ is given by:

$$w_k(t) = \varphi_k'(t) = \frac{d\varphi(t)}{dt} \geq 0 \tag{5.2}$$

$A_k(t)$ and $w_k(t)$ exhibit slow variations with respect to the phase, denoted by $\phi_k(t)$, occurring within the interval of $[t-\delta, t+\delta]$ (where $\delta \approx 2\pi/\phi_k'(t)$), $u_k(t)$ can be regarded as a harmonic signal with amplitude $A_k(t)$ and frequency $w_k(t)$.

(2) Total Practical IMF Bandwidth

The signal $u_k(t)$ obtained by decomposition typically comprises two main frequency components: the instantaneous frequency and the carrier frequency. If w_k is the average frequency of a mode, its actual bandwidth BW_{FM} increases with the maximum deviation of the instantaneous frequency from the center frequency and the rate of offset. Based on Carson's rule [20], we have:

$$BW_{FM} = 2(\Delta f + f_{FM}) \tag{5.3}$$

where Δf denotes the instantaneous frequency of the maximum deviation to the center, and f_{FM} represents the rate of frequency shift.

During VMD decomposition, the bandwidth of the instantaneous amplitude is also related to its own maximum frequency, thus the total real bandwidth of the IMF is defined as:

$$BW_{AM-FM} = 2(\Delta f + f_{FM} + f_{AM}) \tag{5.4}$$

where f_{AM} is the maximum frequency of the envelope $A_k(t)$.

(3) Hilbert Transform

The Hilbert transform is an all-pass filter characterized by a transfer function. Assuming the existence of a real-valued signal $f(t)$, the Hilbert transform can be expressed as:

$$H[f(t)] = f(t) * \frac{1}{\pi t} = \frac{1}{\pi}\int_{-\infty}^{+\infty} \frac{f(\xi)}{t-\xi} d\xi \tag{5.5}$$

where $H[\cdot]$ stands for Hilbert transform; * stands for convolution. From the above equation, it can be seen that the process of Hilbert transform can be regarded as that the signal f (t) passes through a filter and the impulse response is $h(t) = \frac{1}{\pi t}$.

Since the convolution of $h(t) = \frac{1}{\pi t}$ is non-accumulatable, the Cauchy principal value is introduced for the solution:

$$p.v. \int f(t)dt = \lim_{\varepsilon \to 0^+} \left(\left(\int_{-\infty}^{-\varepsilon} f(t)dt + \int_{\varepsilon}^{+\infty} f(t)dt \right) \right) \tag{5.6}$$

where the Hilbert transform of a signal is obtained as the Cauchy principal value (denoted *p.v.*) of the convolution integral.

So the Hilbert transform of the signal f (t) can be expressed as:

$$H[f(t)] = \frac{1}{\pi} p.v. \int_R \frac{f(u)}{t-u} du \tag{5.7}$$

The amplitude of the signal f (t) does not change after the Hilbert transform, and the most prominent use of Hilbert is to construct a purely real signal into a complex-valued analytic signal.

The analytical equation obtained for the real signal f (t) after the Hilbert transform is defined as:

$$f_A(t) = f(t) + jH[f(t)] = A(t)e^{j\varphi(t)} \tag{5.8}$$

where $j^2 = -1$; the complex exponential term $e^{j\varphi(t)}$ denotes the amount of variation of the complex signal rotated in time; $\varphi(t)$ is the phase, and $A(t)$ denotes the time-domain amplitude. For signals of the form (5.1), the analyzed signal can be expressed as the same amplitude function:

$$u_{k,A}(t) = A_k(t)(\cos(\varphi(t)) + j\sin(\varphi(t))) = A_k(t)e^{j\varphi(t)} \tag{5.9}$$

(4) Analytic Signal

The combination of nonlinear signals with different frequencies generates a crossed frequency term, and then the two real signals with frequencies w_1 and w_2 are multiplied together to obtain a mixed frequency signal as follows:

$$2\cos(w_1 t)\cos(w_2 t) = \cos((w_1 + w_2)t) + \cos(w_1 - w_2)t) \tag{5.10}$$

From the above equation, the multiplication of two analytic signals automatically become signals composed of two frequencies (i.e. addition and subtraction of the original two frequencies) when they are mixed. However, in the Fourier transform, there are pairs of transforms as shown in the following equation:

$$f_A(t)e^{-jw_0t} \Leftrightarrow \tilde{f}_A(w) * \delta(w + w_0) = \tilde{f}_A(w + w_0) \tag{5.11}$$

where δ is a Dirac function, so multiplication of an analytic signal with a purely exponential function causes the signal to shift in frequency.

VMD is mainly to compute the optimal solution of variational equations through iteration, and finally decompose the signal data into IMFs with different bandwidths and center frequencies. The VMD algorithm consists of two steps: variational construction and variational solution. The steps of the specific variational construction are as follows:

(1) The Hilbert transform is applied to each modal component $u_K(t)$ to derive the corresponding analytic signal, facilitating the acquisition of its unilateral spectrum:

$$\left[\delta(t) + \frac{j}{\pi t}\right] * u_K(t) \tag{5.12}$$

(2) By multiplying the exponential term with each converted mode component, the center frequency $e^{-j\omega_K t}$ of each mode is estimated and the spectral components of each mode are modulated to align with their respective fundamental bands:

$$\left[\left(\delta(t) + \frac{j}{\pi t}\right) * u_K(t)\right] e^{-j\omega_K t} \tag{5.13}$$

(3) The bandwidth ω_K calculation, i.e., the calculation of the squared gradient paradigm for each mode, is based on Gaussian smoothing demodulation.

$$\begin{cases} \min\limits_{\{\mu_K\},\{\omega_K\}} \left\{ \sum\limits_K \left\| d_t\left[\left(\delta(t) + \frac{j}{\pi t}\right) * u_K(t)\right] e^{-j\omega_K t} \right\|_2^2 \right\} \\ s.t. \sum\limits_K u_K = x(t) \end{cases} \tag{5.14}$$

where $x(t)$ represents the original signal. The procedure for solving the variational problem is described as follows.

We introduce quadratic penalty factor denoted as α and the Lagrange multiplier operator denoted as λ_t to convert the constrained variational problem into an unconstrained variational problem. The extended Lagrange expression can be formulated as follows:

$$L(\{u_K\}, \{\omega_K\}, \lambda) = \alpha \sum_K \left\| \partial_t\left[\left(\delta(t) + \frac{j}{\pi t}\right) * \mu_K(t)\right] e^{-j\omega_K t} \right\|_2^2 + \left\| f(t) - \sum_K u_K(t) \right\|_2^2 + \langle \lambda(t), f(t) - \sum_K u_K(t) \rangle \tag{5.15}$$

where $\left\|f(t)-\sum_k u_k(t)\right\|_2^2$ is a quadratic penalty term that speeds up convergence.

To address this unconstrained variational problem, the Alternating Direction Method of Multipliers (ADMM) is harnessed. The focus is on locating the saddle point through iterative updates of u_K^{n+1}, ω_K^{n+1} and λ^{n+1}, thereby seeking the optimal solution for the constrained variational model as articulated in Eq. (5.15).

That is, the modal component is determined by the minimization of the extended Lagrange formula:

$$u_k^{n+1} = \arg\min\left\{\alpha\left\|\partial_t\left[\left(\delta(t)+\frac{j}{\pi t}\right)*u_k(t)\right]\exp\left(-jw_k^{n+1}\right)\right\|_2^2 + \left\|f(t)-\sum_{i\neq k}u_i(t)^{n+1}+\frac{\lambda(t)}{2}\right\|_2^2\right\} \tag{5.16}$$

Applying the Fourier transform to the above equation yields the frequency domain expressions for the modal components and the center frequency, respectively:

$$\hat{u}_k^{n+1}(w) = \frac{\hat{f}(w)-\sum\limits_{i\neq k}\hat{u}_i(w)+\frac{\hat{\lambda}(w)}{2}}{1+2\alpha(w-w_k)^2} \tag{5.17}$$

$$w_k^{n+1} = \frac{\int_0^\infty w|\hat{u}_k(w)|^2 dw}{\int_0^\infty |\hat{u}_k(w)|^2 dw} \tag{5.18}$$

where $\hat{u}_k^{n+1}(w)$ is the Wiener filter of $\hat{f}(w)-\sum\limits_{i\neq k}\hat{u}_i(w)$,$w_k^{n+1}$ is the modal center frequency, n denotes the number of iterations, $\hat{u}_k(w)$ denotes the Fourier inverse transform, and the real part of the final result is $u_k(t)$.

The steps for the complete implementation of VMD are:

Step 1. Initialize $\{\hat{u}_k^1\}$, $\{w_k^1\}$, $\hat{\lambda}^1$ and n;

Step 2. Update the values of $\hat{u}_k$ and w_k according to Eqs. (5.17) and (5.18);

Step 3. Update the Lagrange multiplier operator $\hat{\lambda}^{n+1}(w)$ according to

$$\hat{\lambda}^{n+1}(w) = \hat{\lambda}^n(w)+\tau\left[\hat{f}(w)-\sum_k \hat{u}_k^{n+1}(W)\right] \tag{5.19}$$

Step 4: Given the discriminant accuracy $\varepsilon > 0$, end the iteration and output the result if the iteration termination condition $\frac{\sum_k \|\hat{u}_k^{n+1}-\hat{u}_k^n\|_2^2}{\sum_k \|\hat{u}_k^n\|_2^2} < \varepsilon$ is satisfied; otherwise, return to step 2 and continue the iteration again.

5.2.2 Long Short Term Memory (LSTM)

LSTM is an advanced iteration of recurrent neural networks (RNN) designed to overcome the challenge of handling long-term dependencies. LSTM achieves this by incorporating memory cells, effectively addressing issues related to vanishing and exploding gradients. In contrast to traditional neural networks, LSTM demonstrates robust capabilities in tackling long-term sequence prediction tasks and has found widespread application in domains such as time series forecasting and fault detection [21]. The LSTM architecture consists of input layers, hidden layers, and output layers. Each hidden layer incorporates input gates, forget gates, and output gates for data storage and access, as visually represented in Fig. 5.1.

As shown in Fig. 5.1, LSTM is accomplished through the operation of three gates and the procedure can be summarized as follows:

(1) Within the framework of LSTM, the forget gate, marked as f_{t-1}, is responsible for deciding whether to discard or preserve information related to X_t and h_{t-1}. This decision is determined by the activation factor σ, which is associated with the forget gate.

$$f_{t-1} = \sigma(W_f \cdot [h_{t-1}, X_t] + b_f) \tag{5.20}$$

where W represents the weight matrices, b represents the biases, and f_{t-1} is a vector with elements ranging from 0 to 1. Each element in the vector indicates the degree of

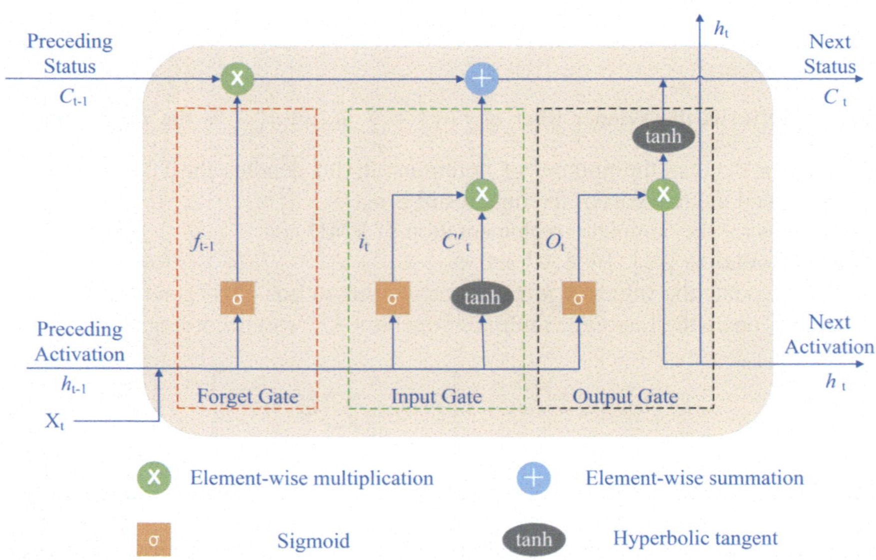

Fig. 5.1 Basic structure of an LSTM

information preservation in the cell state C_{t-1}. A value of 0 indicates no preservation, while a value of 1 indicates complete preservation.

(2) The update process of the cell state is facilitated by the input gate. This process involves passing two components, X_t and h_{t-1}, through an activation function σ to determine the information update. Additionally, X_t and h_{t-1} are processed through a hyperbolic tangent (tanh) function to generate a new candidate vector, which is a vector with element values ranging from -1 to 1. Finally, the output from the tanh operation is scaled by the multiplication factor σ

$$i_t = \sigma(W_i \cdot [h_{t-1}, X_t] + b_i) \tag{5.21}$$

$$C'_t = \tanh(W_c \cdot [\mathrm{h}_{t-1}, X_t] + b_c) \tag{5.22}$$

(3) The cell state from the previous layer undergoes an element-wise multiplication with the forget vector, which is then added to the output of the input gate. This process results in the updated cell state:

$$C_t = f_{t-1} * C_{t-1} + i_t * C'_t \tag{5.23}$$

where $f_{t-1} * C_{t-1}$ determines how much information from the previous memory cell state C_{t-1} is forgotten, while $i_t * C'_t$ determines how much information from C'_t is added to the new memory cell state C_t.

(4) The value of the subsequent hidden state O_t is determined through the output gate h_t, which incorporates information from previous inputs:

$$O_t = \sigma(W_O \cdot [h_{t-1}, X_t] + b_O) \tag{5.24}$$

$$h_t = O_t * \tanh(C_t) \tag{5.25}$$

5.2.3 *Dual Variational Mode Decomposition Long-Short Term Memory Network Model (DVMD-LSTM)*

The VMD-LSTM model, a well-established hybrid deep learning framework, has found extensive applications in various time series prediction tasks, particularly in areas such as load forecasting and wind speed prediction, consistently demonstrating impressive predictive accuracy [22]. This model utilizes Variational Mode Decomposition (VMD) to partition the original dataset into a collection of Intrinsic Mode Functions (IMFs) and a residual component denoted as "r". Subsequently, each IMF,

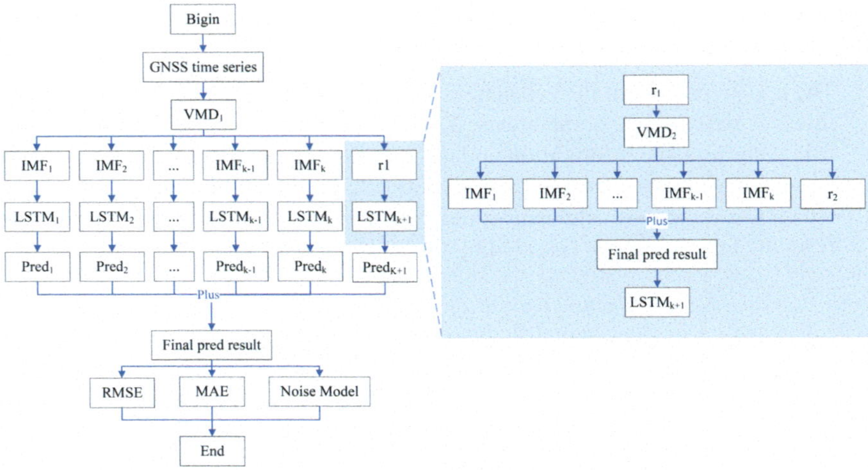

Fig. 5.2 DVMD-LSTM hybrid model prediction process

along with the residual component, undergoes individual prediction, and their predictions are cumulatively combined to derive the final output of the model. It is essential to emphasize that IMFs, characterized by their stationary nature, achieve superior predictive accuracy when addressed independently, thus significantly enhancing the overall predictive capability of the VMD-LSTM model. Notably, the specific prediction process, as depicted on the left-hand side of Fig. 5.2, does not involve any decomposition of the residual value.

However, it is crucial to acknowledge that the residual component resulting from the VMD decomposition of real-world data retains certain fluctuation characteristics and non-white noise elements, such as high-frequency noise. In response to this, the model proceeds to conduct further decomposition of the residual terms using VMD and predicts the decomposed modal components to mitigate the influence of incomplete VMD decomposition. The DVMD-LSTM model enhances overall prediction accuracy by replacing the predicted outcomes of the original residual terms with the combined modal components. This strategic adjustment effectively reduces the impact of residual terms on prediction accuracy. The detailed workflow is elucidated in Fig. 5.2.

The precise prediction procedure of the DVMD-LSTM model can be delineated as follows:

Step 1: Commence with the preprocessing of GNSS time series data, encompassing tasks such as the removal of outliers, interpolation, and other data preprocessing techniques. Subsequently, feed the preprocessed data into the VMD for decomposition.

Step 2: Further break down the residue component denoted as "r_1", derived from the initial VMD operation, into individual modal components. Concurrently, conduct another round of VMD to obtain a new residue component, designated as "r_2".

Step 3: Aggregate the modal components derived from the VMD decomposition of "r_1" to formulate the Fused Intrinsic Mode Function (Fuse-IMF). This Fuse-IMF is utilized as a predictive feature within the LSTM model.

Step 4: Employ the individual modal components, extracted from the VMD decomposition of the original GNSS time series, as distinct features. These features are input separately into the LSTM model for prediction, yielding K prediction outcomes, where K signifies the count of modal components generated during VMD decomposition.

Step 5: Sum the K prediction outcomes generated in Step 4 with the prediction outcome of the Fuse-IMF to acquire the ultimate prediction result of the DVMD-LSTM model.

Step 6: Compute the performance metrics to evaluate the model's performance across various noise models.

5.2.4 *Precision Evaluation Index*

To evaluate the prediction accuracy and analyze the noise characteristics of the hybrid model, this research employs several evaluation metrics, including Root Mean Square Error (RMSE), Mean Absolute Error (MAE), and the coefficient of determination (R^2). Additionally, the Bayesian Information Criterion (BIC_tp) is utilized to determine the most appropriate noise model for both the original GNSS time series and the predicted time series under each model. This aids in assessing whether the prediction results adequately account for colored noise patterns [23]. The specific definitions of these three evaluation metrics are given as follows:

(1) RMSE:

$$RMSE = \sqrt{\frac{1}{n}\sum_{i=1}^{n}(y_i - \hat{y}_i)^2} \tag{5.26}$$

(2) MAE:

$$MAE = \frac{1}{n}\sum_{i=1}^{n}|(y_i - \hat{y}_i)| \tag{5.27}$$

(3) R^2:

$$R^2 = 1 - \frac{\sum_{i=1}^{n}(y_i - \hat{y}_i)^2}{\sum_{i=1}^{n}(y_i - \overline{y})^2} \tag{5.28}$$

where y_i is the actual GNSS data values, $\overline{y}$ is the mean of the actual GNSS data values, $\hat{y}_i$ is the predicted results generated by each model, and n denotes the total number of GNSS data points.

The values of RMSE and MAE serve as crucial evaluation metrics for assessing model prediction accuracy. Smaller values of RMSE and MAE indicate a higher level of prediction accuracy in the model, while larger values signify reduced prediction accuracy.

The coefficient of determination R^2 falls within the range of 0 to 1. When R^2 approximates 1, it signifies that the prediction model effectively explains the variability observed in the dependent variable. Conversely, when R^2 approaches 0, it suggests that the predictive model exhibits weak explanatory power.

(4) BIC_tp

$$BIC\_tp = -2\log(L) + \log(\frac{n}{2\pi})v \tag{5.29}$$

where L is likelihood function.

To visually demonstrate the improvement achieved by the hybrid model for each evaluation metric, this study introduces the concept of the Improvement Ratio (I). The Improvement Ratio quantifies the degree of enhancement in each accuracy evaluation metric. By calculating the I value, one can precisely assess the extent to which the hybrid model has improved accuracy. The formula for calculating the Improvement Ratio is expressed as follows:

$$I_{y\hat{y}} = \frac{y - \hat{y}}{y} \tag{5.30}$$

where y is the accuracy for the initial model's predictions, whereas $\hat{y}$ stands for the accuracy for the predictions generated by the hybrid model.

The magnitude of $I_{y\hat{y}}$ indicates the degree of improvement observed in the evaluation metric achieved by the hybrid model. In other words, a larger value of $I_{y\hat{y}}$ implies a larger enhancement in the evaluation metric, while a smaller value suggests a smaller degree of improvement.

5.3 Data and Experiments

5.3.1 Data Sources

In this chapter, we have utilized daily time series data consisting of three directions (i.e. position coordinates), obtained from 8 GNSS stations affiliated with the Enhanced Solid Earth Science (ESDR) System. The daily solutions were computed using GAMIT and GIPSY with loose constraints, and subsequently, we employed the open-source Quasi-Observation Combination Analysis (QOCA) software, developed by the Jet Propulsion Laboratory (JPL), to generate consolidated daily GNSS coordinate time series data for the selected sites. Detailed station-specific information is presented in Table 5.1, and the geographical distribution of these stations is visually illustrated in Fig. 5.3.

To mitigate the influence of missing data on the estimation and prediction outcomes of the noise model, the station selection process adhered to the following guidelines:

Table 5.1 Information of eight GNSS stations

Site	Longitude (°)	Latitude (°)	Time span (year)	Date missing rate (%)
ALBH	−123.49	48.39	2000–2022	0.61
BURN	−117.84	42.78	2000–2022	1.27
CEDA	−112.86	40.68	2000–2022	2.74
FOOT	−113.81	39.37	2000–2022	3.40
GOBS	−120.81	45.84	2000–2022	3.65
RHCL	−118.03	34.02	2000–2022	1.79
SEDR	−122.22	48.52	2000–2022	0.49
SMEL	−112.84	39.43	2000–2022	0.79

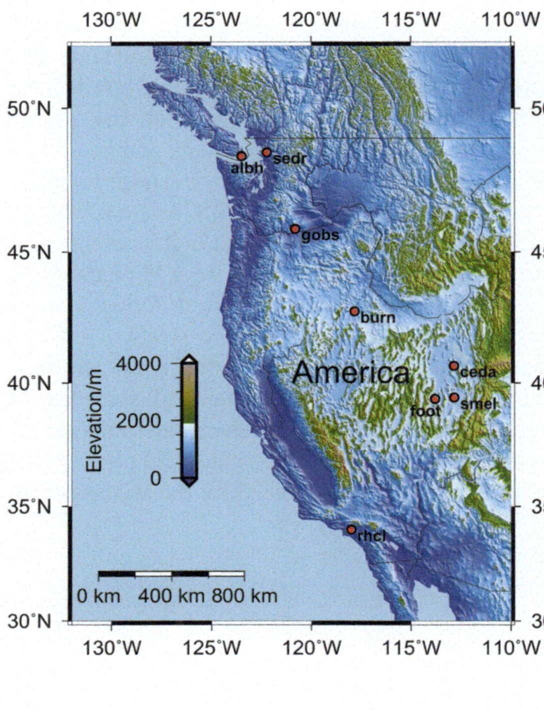

Fig. 5.3 Distribution of eight GNSS stations

(1) Temporal Consistency: The chosen station coordinate time series were required to encompass data spanning from the year 2000 to 2022. The selection of the same long-term time series data was essential to maintain experiment consistency and facilitate the optimal noise model to calculate reliable speed parameter estimates.
(2) Limited Missing Data: Within the time frame spanning from 2000 to 2022, the selected station data were expected to exhibit an average missing data rate not exceeding 5%. This stipulation was put in place to ensure the reliability of the predictive results by minimizing data gaps.
(3) Spatial Evenness: In order to mitigate the influence of inter-regional correlations on the velocity parameters and noise modeling, the selected sites were deliberately distributed in an evenly spread manner.

5.3.2 Data Preprocessing

For data preprocessing, the Hector software was used to identify and eliminate outliers through detecting any step discontinuities present in the raw data. When step discontinuities were identified, a correction process was applied using the least squares fitting method. Subsequently, the rectified data underwent interpolation, which was achieved using the Regularized Expectation Maximization (RegEM) algorithm [24].

The RegEM algorithm combines the Expectation Maximization (EM) algorithm with regularization techniques. This combination allows for the simultaneous optimization of the likelihood function while considering the model's smoothness and facilitating noise reduction. As a result, the algorithm effectively addresses the challenge of interpolating missing data points.

It is important to note that, due to space limitations, only a comparison of the interpolation results for the "GOBS" station, which exhibited the highest missing rate in the E (East), N (North), and U (Up) components, is presented in Fig. 5.4.

As illustrated in the figure, the RegEM method generated favorable interpolation results, and the obtained interpolation followed a rough trend, especially for data points with missing regions. Impressively, it effectively preserves the underlying sequence trend even when faced with a significant amount of continuous missing data. This accomplishment highlights the RegEM method's ability to overcome the limitations associated with traditional linear interpolation, particularly in areas with continuous data gaps. Additionally, the RegEM method provides high-quality continuous time series data, which is crucial for the success of subsequent experiments.

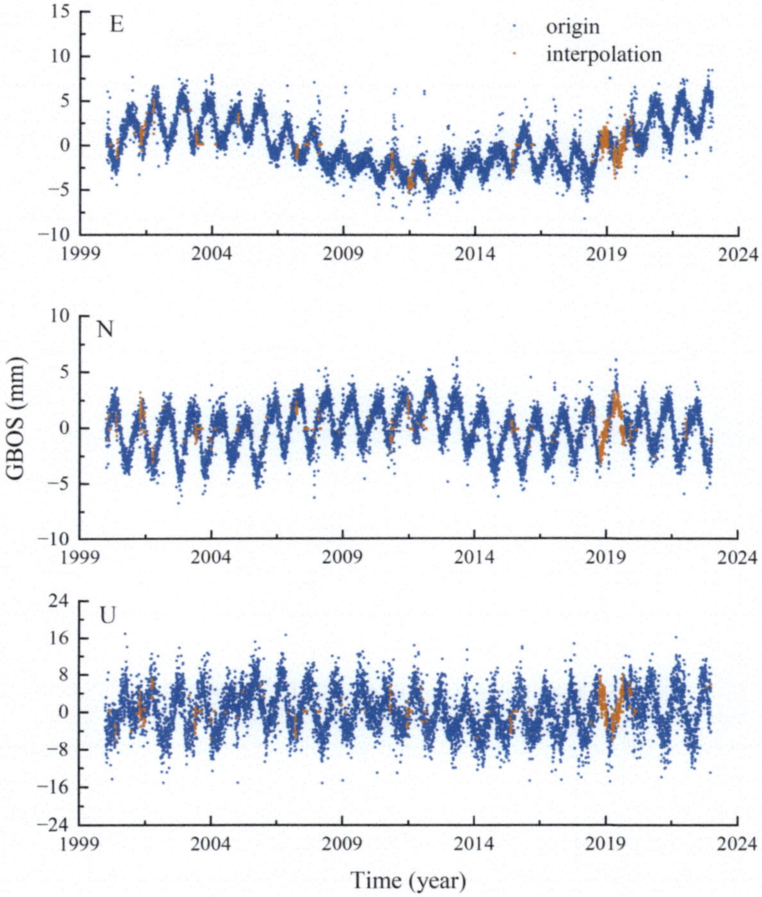

Fig. 5.4 Three-direction interpolation comparison chart of GBOS station

5.3.3 VMD Parameter Discussion

When conducting data decomposition using VMD, the selection of an appropriate number of mode components, denoted as K, is crucial in achieving high-quality decomposition results. An excessively large K can lead to over-decomposition, while a small K may result in insufficient decomposition of the data. To determine the optimal K value for the E (East), N (North), and U (Up) time series across different stations, this study utilizes the signal-to-noise ratio (SNR) as a metric to assess the quality of the decomposition results. A higher SNR indicates a more distinct signal decomposition and a superior denoising effect. Through a series of comprehensive experiments and guided by empirical guidelines, this research confines the K value to a range spanning from 2 to 10. Within this specified range, the K value that results

Table 5.2 Results of K value selection in three directions at each site

Site	Direction		
	N	E	U
ALBH	3	6	3
BURN	4	4	3
CEDA	4	4	3
FOOT	3	8	5
GOBS	3	6	5
RHCL	7	3	3
SEDR	3	5	7
SMEL	7	3	5

in the highest SNR for each time series is selected as the optimal K value [25]. The SNR is calculated as follows:

$$\mathrm{SNR} = 10\lg \frac{\sum_{i=1}^{N} f^2(i)}{\sum_{i=1}^{N} [f(i) - g(i)]^2} \tag{5.31}$$

where $f(i)$ is the original signal, $g(i)$ is the reconstructed signal, and N is the length of the time series.

The choice of the penalty factor, denoted as α, also exerts a certain influence on the outcomes of the VMD data decomposition process. Considering the empirical guideline that suggests selecting a penalty factor approximately 1.5 times the magnitude of the decomposed data is optimal, this study maintains experimental consistency by setting a penalty factor of 10,000 for all the decomposition procedures.

The outcomes of the K value selection for the three directions at each site are tabulated in Table 5.2, providing valuable insights into the decomposition process.

5.4 Experimental Results and Analysis

5.4.1 DVMD-LSTM Prediction Results Analysis

To ensure fairness and consistency in the experiments, all deep learning models employed in this study adhered to a uniform dataset division scheme. The dataset was segregated into three distinct subsets: the training set (Jan 2000–Sept 2011), the validation set (Jan 2012–Sept 2014), and the test set (Jan 2015–Sept 2022). Each subset had a specific role in the modeling process:

Training Set: This set was dedicated to training the model parameters and enabling the model to learn the underlying data features.

Validation Set: It served as fine-tuning the model's hyperparameters and conducting an intermediate evaluation of the model performance.

Test Set: This set played a crucial role in the final assessment of the model's performance, serving as the basis for evaluating its effectiveness in practical applications. Additionally, by obtaining substantial prediction results on the test set, it becomes possible to evaluate the optimal noise model for prediction accuracy.

The primary aim of this dataset partitioning scheme was to ensure that the model had access to an adequate amount of training data, enabling it to effectively capture and comprehend the data's distinctive features. To visually illustrate the differences in prediction outcomes between the DVMD-LSTM model and the VMD-LSTM model, this study conducts a comparative analysis of the prediction results for the decomposed IMFs and residual terms generated by the two hybrid models. Due to space constraints, this chapter only presents the prediction results of the IMFs and residual terms in the U (Up) direction for the SEDR station, as shown in Fig. 5.5.

It can be seen from Fig. 5.5 that the LSTM models excel in delivering commendable prediction results for each IMF component. However, a noteworthy distinction emerges when addressing the residual terms. The VMD-LSTM model faces challenges in effectively capturing the fluctuation characteristics within the residual terms, which lack apparent regularity. Consequently, this difficulty in modeling the residual terms leads to lower prediction accuracy, ultimately impacting the overall performance of the VMD-LSTM model.

To address this issue, the proposed DVMD-LSTM model conducts a secondary VMD decomposition on the residual terms obtained after the initial VMD decomposition. This additional decomposition extracts further fluctuation information within the residual terms, resulting in a substantial improvement in prediction accuracy. Compared to the VMD-LSTM model, the overall prediction results of the DVMD-LSTM model show a 17.30% increase in RMSE and a 17.65% increase in MAE.

To assess the potential benefits of conducting multiple VMD decompositions, an analysis is carried out on the residual terms following the second decomposition. However, it is observed that these terms lack conspicuous fluctuation characteristics. Consequently, incorporating these results into the model for prediction fails to yield significant improvements, and, in some cases, prediction accuracy even decreases. This suggests that increasing the number of decompositions on the residual terms may not necessarily enhance the model's prediction accuracy.

As a result, in this study, the data after the secondary VMD decomposition is used as the feature input for the subsequent deep learning experiments, as it has proven to be an effective representation for achieving high prediction accuracy.

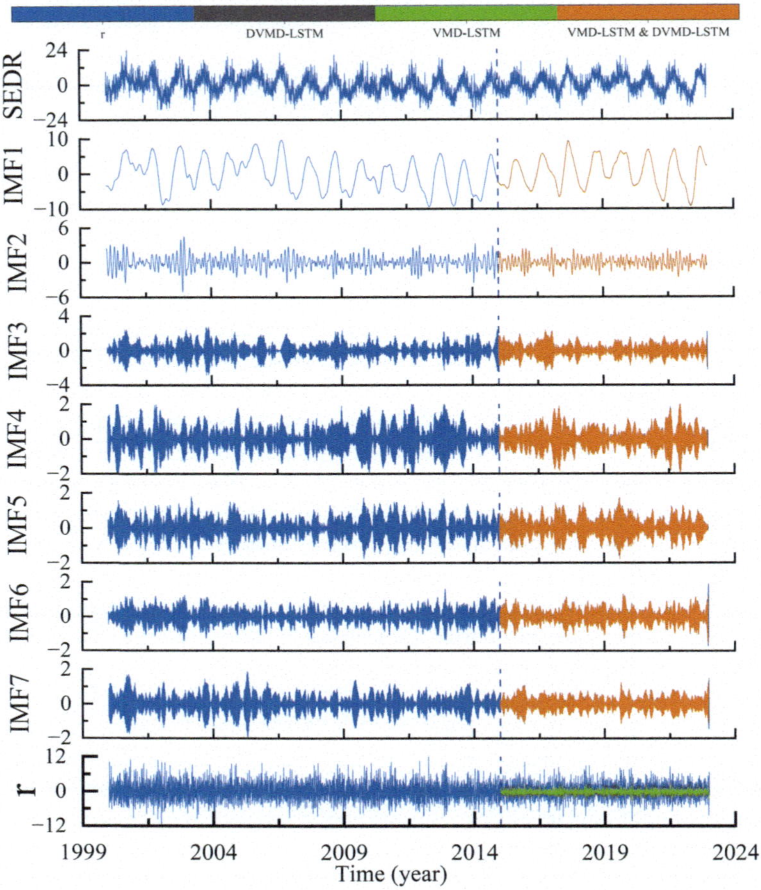

Fig. 5.5 Prediction results of each IMF and residual under different models after VMD decomposition in U direction of SEDR station (The blue curve represents the original data as well as the IMF components and residual terms obtained from VMD decomposition. The orange curve represents the same prediction results of IMF components by DFVMD-LSTM and VMD-LSTM models, the green curve represents the prediction results of residual terms by VMD-LSTM model, and the black curve represents the prediction results of residual terms by DVMD-LSTM model)

5.4.2 DVMD-LSTM Model Prediction Results and Precision Analysis

To evaluate and compare the improvement in predictive accuracy achieved by the DVMD-LSTM model and the VMD-LSTM model in contrast to the LSTM model, this study conducted a series of experiments using datasets from diverse stations. Due to space limitations, only the predicted coordinates and prediction errors at the SEDR station are presented, as shown in Fig. 5.6.

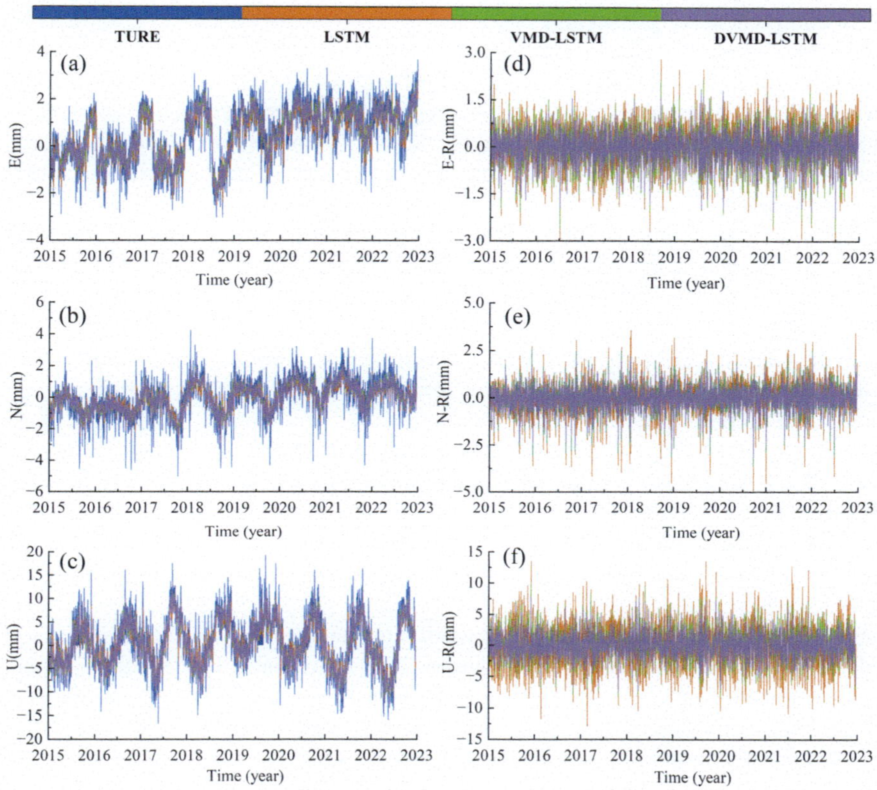

Fig. 5.6 Comparison of predictions and prediction errors of position coordinates at SEDR station under three different models (sub-figures **a**, **b**, **c** show the coordinate prediction results, and sub-figures **d**, **e**, **f** show prediction errors)

As the fluctuation amplitude of the original data increases, the prediction errors of the various models also exhibit varying degrees of escalation. The U (Up) direction consistently presents the largest errors among the three directions. Compared to the baseline LSTM model, the VMD-LSTM model excels in capturing the fluctuation trends and amplitudes of the data. It also demonstrates smaller variations and extremes in the prediction error. This suggests that after undergoing VMD decomposition, the VMD-LSTM model efficiently captures the inherent fluctuation characteristics of the original data, leading to more accurate predictions.

Both the VMD-LSTM and DVMD-LSTM models display similar patterns in prediction fluctuations and trends. However, the DVMD-LSTM model exhibits notably smaller prediction errors, indicating that it not only retains the advantages of the VMD-LSTM model in forecasting fluctuation trends and amplitudes but also achieves a higher level of prediction accuracy.

To comprehensively assess the applicability and robustness of the DVMD-LSTM model, this study conducted predictions in the E (East), N (North), and U (Up) directions for each GNSS station using the LSTM, VMD-LSTM, and DVMD-LSTM models. The prediction accuracy and the degree of improvement achieved by each model are summarized in Table 5.3. In this table, "I" denotes the extent of accuracy improvement of the hybrid model compared to the single LSTM model across various accuracy metrics.

The results presented in Table 5.3 reveal significant improvements in prediction accuracy achieved by the VMD-LSTM model in comparison to the standalone LSTM model. The improvements are particularly notable in terms of RMSE, MAE, and R^2 across different directions:

In the E (East) direction, the VMD-LSTM model demonstrates an average reduction of 19.77% in RMSE, an average reduction of 20.31% in MAE, and an average increase of 43.66% in R^2.

In the N (North) direction, the VMD-LSTM model showcases an average reduction of 26.83% in RMSE, an average reduction of 27.12% in MAE, and an average increase of 43.47% in R^2.

In the U (Up) direction, the VMD-LSTM model manifests an average reduction of 19.31% in RMSE, an average reduction of 19.48% in MAE, and an average increase of 44.54% in R^2.

These outcomes underscore the substantial enhancement in prediction accuracy delivered by the VMD-LSTM model compared to the baseline LSTM model. Furthermore, while the degree of improvement in R^2 may vary across different stations, it is notably more pronounced in stations where the LSTM model exhibited lower initial R^2 values. This observation implies that the VMD-LSTM model yields predictions that closely align with observed values and offers improved fitting results.

In comparison to the VMD-LSTM model, the DVMD-LSTM model delivers the following improvements in prediction accuracy across different directions:

In the E (East) direction, the DVMD-LSTM model showcases an average reduction of 9.71% in RMSE, an average reduction of 9.17% in MAE, and an average increase of 20.68% in R^2.

In the N (North) direction, the DVMD-LSTM model demonstrates an average reduction of 8.84% in RMSE, an average reduction of 8.55% in MAE, and an average increase of 12.18% in R^2.

In the U (Up) direction, the DVMD-LSTM model manifests an average reduction of 11.02% in RMSE, an average reduction of 10.61% in MAE, and an average increase of 21.03% in R^2. The overall average R^2 value reaches 0.78, indicating a strong correlation between the DVMD-LSTM model's prediction results and the original data, accompanied by enhanced fitting performance.

These findings underscore that the DVMD-LSTM model achieves a notable enhancement in accuracy when compared to the VMD-LSTM model, with particularly substantial improvements observed in R^2. Notably, the U direction benefits the most, suggesting that the DVMD-LSTM model excels in handling time series with significant fluctuations. This advantage stems from the larger residual terms obtained

Table 5.3 Comparison of the prediction results of each GNSS station in the three directions of E, N, and U under different models (The units of RMSE and MAE in the table are mm)

(a) Comparison of the prediction results of each GNSS station in the direction of E under different models

Model		ALBH	BURN	CEDA	FOOT	GOBS	RHCL	SEDR	SMEL
LSTM	RMSE	0.89	1.40	1.73	0.58	1.00	1.62	0.68	0.57
	MAE	0.65	1.10	1.35	0.44	0.70	1.28	0.53	0.44
	R^2	0.65	0.51	0.70	0.13	0.86	0.61	0.66	0.40
VMD-LSTM	RMSE	0.76	1.16	1.37	0.51	0.86	1.07	0.58	0.40
	I/%	13.91	17.00	20.75	12.91	13.74	34.08	15.00	30.80
	MAE	0.55	0.92	1.06	0.38	0.58	0.83	0.45	0.30
	I/%	14.03	16.70	21.18	13.51	16.08	34.78	15.13	31.08
	R^2	0.74	0.66	0.81	0.34	0.90	0.83	0.76	0.71
	I/%	13.75	30.37	16.00	157.60	4.10	35.51	14.23	77.69
DVMD-LSTM	RMSE	0.67	1.02	1.21	0.45	0.77	0.94	0.50	0.34
	I/%	24.56	27.00	29.82	22.12	23.53	41.63	27.07	40.11
	MAE	0.49	0.82	0.94	0.34	0.52	0.74	0.39	0.26
	I/%	24.31	25.78	30.32	22.27	24.50	41.91	26.76	39.98
	R^2	0.80	0.74	0.85	0.47	0.92	0.87	0.82	0.79
	I/%	22.89	45.61	21.83	256.70	6.66	41.40	24.00	95.60

(b) Comparison of the prediction results of each GNSS station in the direction of N under different models

Model		ALBH	BURN	CEDA	FOOT	GOBS	RHCL	SEDR	SMEL
LSTM	RMSE	0.73	1.39	1.38	0.59	0.86	3.14	0.85	0.55
	MAE	0.57	1.11	1.1	0.43	0.63	2.54	0.63	0.42
	R^2	0.62	0.55	0.46	0.48	0.78	0.46	0.44	0.45
VMD-LSTM	RMSE	0.55	1.07	1.05	0.39	0.63	1.71	0.66	0.47
	I/%	24.53	22.74	23.54	33.45	26.95	45.59	22.23	15.62
	MAE	0.43	0.85	0.83	0.29	0.46	1.31	0.5	0.35
	I/%	24.23	23.37	24.05	31.81	26.6	48.53	21.79	16.54
	R^2	0.78	0.73	0.68	0.77	0.88	0.84	0.66	0.61
	I/%	26.18	32.59	48.72	59.65	13.33	81.39	50.49	35.42
DVMD-LSTM	RMSE	0.49	0.95	0.9	0.34	0.56	1.58	0.56	0.41
	I/%	32.77	31.65	34.5	41.35	34.86	49.55	34.15	26.53
	MAE	0.38	0.76	0.72	0.26	0.41	1.21	0.42	0.3
	I/%	32.53	32.13	34.33	39.95	34.1	52.28	33.1	26.91
	R^2	0.83	0.79	0.77	0.82	0.91	0.86	0.76	0.7
	I/%	33.33	43.08	66.97	70.25	16.46	86.19	72.34	56.6

(continued)

Table 5.3 (continued)

(c) Comparison of the prediction results of each GNSS station in the direction of U under different models

Model		ALBH	BURN	CEDA	FOOT	GOBS	RHCL	SEDR	SMEL
LSTM	RMSE	3.38	2.3	2.65	2.39	2.92	2.45	3.33	2.36
	MAE	2.6	1.78	2.03	1.83	2.22	1.9	2.62	1.87
	R^2	0.58	0.53	0.51	0.31	0.62	0.31	0.65	0.32
VMD-LSTM	RMSE	2.89	1.94	2.27	1.87	2.28	2.1	2.37	1.84
	I/%	14.57	15.78	14.48	21.89	22.17	14.5	28.68	22.38
	MAE	2.25	1.49	1.73	1.43	1.72	1.63	1.87	1.43
	I/%	13.77	16.29	15.08	22.23	22.48	14.04	28.79	23.12
	R^2	0.69	0.66	0.64	0.58	0.77	0.49	0.82	0.59
	I/%	19.4	26.08	25.63	88.11	24.56	60.46	26.63	85.49
DVMD-LSTM	RMSE	2.51	1.66	1.96	1.6	1.99	1.87	1.96	1.58
	I/%	25.74	27.82	26.09	32.94	32.04	23.68	41.19	33.17
	MAE	1.96	1.29	1.49	1.23	1.53	1.46	1.54	1.24
	I/%	0.83	0.79	0.77	0.82	0.91	0.86	0.76	0.7
	R^2	0.77	0.75	0.73	0.69	0.82	0.6	0.88	0.7
	I/%	32.21	42.98	43.28	124.3	33.52	93.85	35.44	118.9

after VMD decomposition in time series with pronounced fluctuations, which contain more distinctive fluctuation characteristics.

In summary, the DVMD-LSTM model preserves the advantages of the VMD-LSTM model in predicting fluctuation trends and frequencies while attaining higher prediction accuracy. The results across different directional components and stations substantiate the model's applicability and robustness, affirming its potential for extensive utilization in the domain of high-precision time series forecasting.

5.4.3 *Optimal Noise Model Research*

5.4.3.1 Comparison of Optimal Noise Models Under Each Prediction Model

To further assess whether the DVMD-LSTM model effectively accounts for the noise characteristics in various datasets during the prediction process, it is essential to consider the prevailing beliefs among domestic and international scholars regarding optimal noise models for GPS coordinate time series. Currently, two primary models are widely considered for describing the noise characteristics of GPS coordinate time series: White noise + Flicker Noise (FN + WN) and a minor amount of random walk noise + flicker noise (RW + FN). Furthermore, some scholars have proposed that the noise in GPS coordinate time series can be considered as power law noise (PL) and the Gaussian Markov model (GGM).

In this study, we focused on GNSS reference stations in North America with the same time span. Four combined noise models were considered for analysis, namely random walk noise + flicker noise + white noise (RW + FN + WN), flicker noise + white noise (FN + WN), power law noise + white noise (PL + WN), and Gaussian Markov + white noise (GGM + WN). The training and test data of each station were examined, and ultimately, eight stations sharing the same optimal noise model were selected for experimentation. The optimal noise model for each prediction model, concerning the prediction results for each station, was then determined. The specific outcomes are presented in Table 5.4.

Table 5.4 The optimal noise model of each station under different models in the three directions of E, N, and U

Site	ENU	Optimal noise model			
		TURE	LSTM	VMD-LSTM	DVMD-LSTM
ALBH	E	RW + FN + WN	PL + WN	RW + FN + WN	RW + FN + WN
BURN		RW + FN + WN	PL + WN	PL + WN	RW + FN + WN
CEDA		RW + FN + WN	PL + WN	PL + WN	RW + FN + WN
FOOT		PL + WN	GGM + WN	FN + WN	PL + WN
GOBS		RW + FN + WN	PL + WN	RW + FN + WN	RW + FN + WN
RHCL		RW + FN + WN	GGM + WN	PL + WN	RW + FN + WN
SEDR		RW + FN + WN	PL + WN	PL + WN	RW + FN + WN
SMEL		FN + WN	PL + WN	FN + WN	FN + WN
ALBH	N	RW + FN + WN	PL + WN	RW + FN + WN	RW + FN + WN
BURN		FN + WN	PL + WN	PL + WN	PL + WN
CEDA		RW + FN + WN	PL + WN	PL + WN	RW + FN + WN
FOOT		FN + WN	GGM + WN	FN + WN	FN + WN
GOBS		RW + FN + WN	PL + WN	RW + FN + WN	RW + FN + WN
RHCL		RW + FN + WN	RW + FN + WN	PL + WN	PL + WN
SEDR		FN + WN	GGM + WN	RW + FN + WN	FN + WN
SMEL		FN + WN	PL + WN	FN + WN	FN + WN
ALBH	U	PL + WN	PL + WN	RW + FN + WN	FN + WN
BURN		PL + WN	GGM + WN	PL + WN	PL + WN
CEDA		PL + WN	PL + WN	RW + FN + WN	PL + WN
FOOT		PL + WN	PL + WN	FN + WN	FN + WN
GOBS		PL + WN	GGM + WN	PL + WN	FN + WN
RHCL		FN + WN	PL + WN	RW + FN + WN	FN + WN
SEDR		PL + WN	PL + WN	PL + WN	PL + WN
SMEL		PL + WN	PL + WN	FN + WN	PL + WN

From Table 5.4, it becomes evident that different stations exhibit different optimal noise models, indicating the presence of inconsistent noise characteristics in the data. The LSTM model shows significant disparities between its prediction results and the optimal noise models associated with the original data, with an average accuracy of only 25% across all three directions. Additionally, it is noteworthy that the predominant optimal noise models tend to be PLWN and GGMWN. This suggests that the LSTM model does not adequately consider the inherent noise characteristics of GNSS time series during the prediction process.

On the contrary, the VMD-LSTM model demonstrates improved accuracy in capturing the optimal noise models, achieving an average accuracy of 42.67%. This indicates that VMD decomposition effectively captures the noise characteristics within the IMF components, although the noise characteristics in the residual component are not fully addressed, resulting in relatively lower overall accuracy.

The proposed DVMD-LSTM model further enhances the consideration of noise characteristics within the residual component by applying VMD decomposition once again. As a result, the DVMD-LSTM model achieves an impressive average accuracy of 79.17% in capturing the optimal noise models. In summary, the DVMD-LSTM model effectively takes into account the noise characteristics of the data during the prediction process by processing both the original data and the decomposed residual component.

5.4.3.2 Velocity Estimation Impact Analysis

To evaluate the predictive quality of each deep learning model, this study initially employs these models to predict the original data. Subsequently, the optimal noise model and corresponding velocities are computed for the prediction results of each model. These velocities are then compared with the velocities calculated using the optimal noise model of the original data with the assistance of the Hector software. By calculating the absolute error between the prediction results of each model and the original velocities at various measurement stations, we can obtain the average absolute error between the velocities computed from each deep learning model's prediction results and the velocities derived from the original data. This process allows us to assess the accuracy of the model's prediction. The velocities computed from the prediction results of each deep learning model under the optimal noise model at different measurement stations are presented in Table 5.5.

It can be seen from Table 5.5 that average absolute error of the LSTM model in velocity prediction varies across the three spatial directions. In the E direction, this error is 0.068 mm/year, while in the N direction, it increases to 0.093 mm/year. In the U direction, the error is 0.078 mm/year. On the other hand, the VMD-LSTM model demonstrates a notable improvement in accuracy, with an average absolute error of 0.031 mm/year in the E direction, 0.060 mm/year in the N direction, and 0.060 mm/year in the U direction.

Meanwhile, the DVMD-LSTM model outperforms both the LSTM and VMD-LSTM models, showcasing its remarkable predictive accuracy. Specifically, in the

Table 5.5 Velocity values obtained by each station under the optimal noise model

Site	ENU	Trend (mm/year)			
		TURE	LSTM	VMD-LSTM	DVMD-LSTM
ALBH	E	−0.041	0.020	0.055	−0.044
BURN		−0.108	−0.005	−0.051	−0.116
CEDA		−0.726	−0.528	−0.693	−0.736
FOOT		0.02	0.015	0.001	0.009
GOBS		0.659	0.656	0.672	0.682
RHCL		0.811	0.666	0.805	0.783
SEDR		0.354	0.341	0.378	0.313
SMEL		0.026	0.009	0.023	0.021
ALBH	N	0.327	0.245	0.276	0.295
BURN		0.124	0.080	0.116	0.130
CEDA		−0.065	−0.041	−0.227	−0.042
FOOT		0.009	0.029	−0.036	0.005
GOBS		0.063	0.078	0.029	−0.020
RHCL		1.253	0.743	1.132	1.071
SEDR		0.199	0.170	0.212	0.195
SMEL		0.020	−0.001	−0.025	0.017
ALBH	U	0.383	0.204	0.131	0.268
BURN		0.241	0.144	0.238	0.216
CEDA		0.016	0.159	0.074	0.137
FOOT		0.194	0.125	0.194	0.202
GOBS		0.301	0.278	0.283	0.262
RHCL		0.298	0.206	0.367	0.264
SEDR		0.017	0.022	0.082	0.04
SMEL		0.195	0.182	0.206	0.183

E direction, it achieves an average absolute error of 0.016 mm/year, followed by 0.042 mm/year in the N direction, and 0.047 mm/year in the U direction. This underscores the DVMD-LSTM model's superior performance in reducing velocity prediction errors.

Comparing these results with the LSTM model, the VMD-LSTM model displays an average enhancement of 37.67% in velocity prediction accuracy, while the DVMD-LSTM model surpasses them all with a remarkable 56.80% average improvement. Furthermore, when compared to the VMD-LSTM model, the DVMD-LSTM model achieves an additional average increase of 33.02% in velocity prediction accuracy. In summary, both the VMD-LSTM and DVMD-LSTM models demonstrate notable improvements in velocity prediction accuracy compared to the baseline

LSTM model. However, the DVMD-LSTM model stands out with its remarkable enhancements, reaffirming its exceptional predictive capabilities.

5.5 Conclusion

In response to the limitations observed in the VMD-LSTM model, namely lower prediction accuracy and insufficient consideration of noise characteristics in time series forecasting, this study introduces a novel high-precision GNSS time series prediction approach based on DVMD and LSTM. The proposed method undergoes comprehensive validation and testing using daily time series data from eight North American regional GNSS stations, spanning from 2000 to 2022, and includes observations in the E, N, and U directions. The experimental results yield the following key findings:

(1) The VMD-LSTM model demonstrates commendable prediction accuracy for each IMF component following VMD decomposition but struggles when it comes to forecasting the residual component. The DVMD-LSTM model, on the other hand, leverages VMD decomposition to capture the fluctuation characteristics within the residual component, resulting in a substantial enhancement of prediction accuracy for the residuals and an overall improvement in prediction precision.
(2) In comparison to the initial VMD-LSTM hybrid model, the DVMD-LSTM model showcases significant advancements in prediction accuracy. Specifically, it achieves an average reduction of 9.71% in RMSE values in the E direction, 8.84% in the N direction, and 11.02% in the U direction. Furthermore, the DVMD-LSTM model exhibits an average reduction of 9.17% in MAE values for the E direction, 8.55% for the N direction, and 10.61% for the U direction. Additionally, it consistently elevates the R^2 values, with an average increase of 20.68% in the E direction, 12.18% in the N direction, and 21.03% in the U direction across all measurement stations. These findings underscore the DVMD-LSTM model's superior predictive accuracy, adaptability, and robustness, outperforming the VMD-LSTM model consistently.
(3) Compared to the baseline LSTM model, the DVMD-LSTM model exhibits a substantial improvement of 36.50% in the accuracy of the average optimal noise model across all stations, attaining an impressive overall accuracy rate of 79.17%. This outcome affirms that the DVMD-LSTM model effectively incorporates the noise characteristics of the data during the prediction process, resulting in superior prediction outcomes. By evaluating the velocities computed based on the optimal noise models, the DVMD-LSTM model demonstrates an average increase of 33.02% in velocity prediction accuracy when contrasted with the VMD-LSTM model. These results further underline the DVMD-LSTM model's exceptional predictive performance.

In summary, this study introduces an innovative approach to address the shortcomings of the VMD-LSTM model in GNSS time series prediction. The DVMD-LSTM model not only enhances prediction accuracy but also adeptly considers noise characteristics, demonstrating its potential as a robust and precise forecasting tool for high-precision time series forecasting applications.

References

1. Ohta Y, Kobayashi T, Tsushima H et al (2012) Quasi real-time fault model estimation for near-field tsunami forecasting based on RTK-GPS analysis: application to the 2011 Tohoku-Oki earthquake (Mw 9.0). J J Geophys Res Solid Earth 117(B2)
2. Cina A, Piras M (2015) Performance of low-cost GNSS receiver for landslides monitoring: test and results. J Geomat Nat Haz Risk 6(5–7):497–514
3. Meng X, Roberts GW, Dodson AH et al (2004) Impact of GPS satellite and pseudolite geometry on structural deformation monitoring: analytical and empirical studies. J J Geodesy 77:809–822
4. Altamimi Z, Rebischung P, Métivier L et al (2016) ITRF2014: a new release of the international terrestrial reference frame modeling nonlinear station motions. J Geophys Res: Solid Earth 121(8):6109–6131
5. Blewitt G, Lavallée D (2002) Effect of annual signals on geodetic velocity. J Geophys Res Solid Earth 107(B7):ETG 9-1-ETG 9–11
6. Chen JH (2011) Petascale direct numerical simulation of turbulent combustion—fundamental insights towards predictive models. J P Combust Inst 33(1):99–123
7. Klos A, Olivares G, Teferle FN et al (2018) On the combined effect of periodic signals and colored noise on velocity uncertainties. J GPS Solut 22:1–13
8. Li Y (2022) Research and application of deep learning in image recognition. In: 2022 IEEE 2nd international conference on power, electronics and computer applications (ICPECA). IEEE, pp 994–999
9. Masini RP, Medeiros MC, Mendes EF (2023) Machine learning advances for time series forecasting. J Econ Surv 37(1):76–111
10. Bengio Y, Simard P, Frasconi P (1994) Learning long-term dependencies with gradient descent is difficult. J IEEE T Neural Networ 5(2):157–166
11. Kim HU, Bae TS (2019) Deep learning-based GNSS network-based real-time kinematic improvement for autonomous ground vehicle navigation. J Sens
12. Tao Y, Liu C, Chen T et al (2021) Real-time multipath mitigation in multi-GNSS short baseline positioning via CNN-LSTM method. J Math Probl Eng 2021:1–12
13. Xie P, Zhou A, Chai B (2019) The application of long short-term memory (LSTM) method on displacement prediction of multifactor-induced landslides. J IEEE Access 7:54305–54311
14. Zhao L, Li Z, Qu L et al (2023) A hybrid VMD-LSTM/GRU model to predict non-stationary and irregular waves on the east coast of China. J Ocean Eng 276:114136
15. Huang Y, Yan L, Cheng Y et al (2022) Coal thickness prediction method based on VMD and LSTM. J Electron 11(2):232
16. Zhang T, Fu C (2022) Application of improved VMD-LSTM model in sports artificial intelligence. J Comput Intel Neurosc
17. Han L, Zhang R, Wang X et al (2019) Multi-step wind power forecast based on VMD-LSTM. J IET Renew Power Gen 13(10):1690–1700
18. Xing Y, Yue J, Chen C et al (2019) Dynamic displacement forecasting of dashuitian landslide in China using variational mode decomposition and stack long short-term memory network. J Applied Sci 15:2951
19. Dragomiretskiy K, Zosso D (2013) Variational mode decomposition. IEEE Trans Signal Process 62(3):531–544

20. Carson JR (1992) Notes on the theory of modulation. Proc Inst Radio Eng 10(1):57–64
21. Malhotra P, Vig L, Shroff G et al (2015) Long short term memory networks for anomaly detection in time series. Esann 2015:89
22. Jin Y, Guo H, Wang J et al (2020) A hybrid system based on LSTM for short-term power load forecasting. J Energies 13(23):6241
23. He X, Bos MS, Montillet JP et al (2019) Investigation of the noise properties at low frequencies in long GNSS time series. J Geodesy 93(9):1271–1282
24. Tingley MP, Huybers P (2010) A Bayesian algorithm for reconstructing climate anomalies in space and time. Part II: comparison with the regularized expectation–maximization algorithm. J Climate 23(10):2782–2800
25. Mei L, Li S, Zhang C et al (2021) Adaptive signal enhancement based on improved VMD-SVD for leak location in water-supply pipeline. J IEEE Sens J 21(21):24601–24612

Chapter 6
Autonomous UAV Outdoors Navigation—A Machine-Learning Perspective

Ghada Afifi and Yasser Gadallah

Abstract Unmanned Aerial Vehicles (UAVs) are increasingly gaining traction due to their potential and major use case applications. The UAV is typically required to navigate autonomously in highly dynamic environments to deliver on its intended applications. Existing UAV positioning and navigation solutions face several challenges, particularly in dense outdoor settings. To this end, we present various technological approaches for autonomous UAV navigation outdoors. This chapter aims to provide an efficient real-time autonomous solution that enables the UAV to navigate through a dynamic urban or suburban environment. Particularly, we evaluate the performance of Machine Learning (ML)-based techniques in UAV navigation solutions. The computational complexity involved in standard optimization-based methods hinders its utilization for UAV navigation in dynamic environments. The use of ML-based approaches can potentially enable near-optimal UAV navigation, while providing a practical real-time calculation that is needed in such dynamic applications. We provide a comprehensive detailed analysis to evaluate the performance of each of the presented ML-based UAV navigation methods as compared to other existing navigation approaches that we also discussed in this chapter.

6.1 Introduction

UAVs are increasingly becoming integrated in many applications both in the civilian and military domains. The use cases of UAVs include goods delivery, surveillance and reconnaissance missions, traffic monitoring, combat missions, search and rescue missions, etc. In addition, their role in communication systems is proliferating at an accelerating pace [1]. It is expected that UAVs will assume a critical role in non-terrestrial network (NTN) arrangements in 6G and beyond.

G. Afifi (✉) · Y. Gadallah
The American University in Cairo, Cairo 11835, Egypt
e-mail: ghadasameh@aucegypt.edu

Y. Gadallah
e-mail: ygadallah@ieee.org

K. Yu (ed.), *Positioning and Navigation Using Machine Learning Methods*, Navigation: Science and Technology 14, https://doi.org/10.1007/978-981-97-6199-9_6

Several types of UAVs exist with different aerodynamic and robotic features. The four main types of UAVs include the single rotor, multirotor, fixed wing and hybrid vertical take-off and landing UAVs. The most widely used type of UAVs is the Multirotor type. This type can be further classified into different categories depending on the number of rotors on the UAV e.g., tri-rotor, quadrotor, etc.

The simple kinematic model of a UAV is given by

$$x(t) = vcos\theta, y(t) = vsin\theta, \tag{6.1}$$

where *(x(t), y(t))* corresponds to the initial 2D position of the UAV at time *t,* while *v* and θ represent the velocity and heading angle of the UAV [2].

To characterize the motion of a quadrotor UAV, we use two coordinate systems, namely, the space coordinate system and the body coordinate system [3]. The used body coordinate system's elements are the pitch, roll and yaw angles which describe the angle of rotation of the UAV around the x, y and z, respectively. The rotational axis model is therefore given by

$$R_x(\varnothing) = \begin{bmatrix} cos\varnothing & sin\varnothing & 0 \\ -sin\varnothing & cos\varnothing & 0 \\ 0 & 0 & 1 \end{bmatrix}, \tag{6.2a}$$

$$R_y(\theta) = \begin{bmatrix} 1 & 0 & 0 \\ 0 & cos\theta & sin\theta \\ 0 & -sin\theta & cos\theta \end{bmatrix}, \tag{6.2b}$$

$$R_z(\psi) = \begin{bmatrix} cos\psi & 0 & -sin\psi \\ 0 & 1 & 0 \\ sin\psi & 0 & cos\psi \end{bmatrix}, \tag{6.2c}$$

where $\varnothing$, θ and ψ correspond to the pitch, roll, and yaw angles, respectively.

The human-controlled UAVs are usually piloted remotely by an operator without the need of having a pilot onboard. Autonomous UAVs, on the other hand, are self-operated with no human intervention at all. This autonomy is achieved via an onboard autopilot system, computing systems, and a set of sensing systems in addition to the other onboard devices that are required for the mission at hand. It is therefore required to devise the systems, both hardware and software, that enable the autonomous UAV to conduct the aviation duties such as the self-localization and navigation, in addition to the other duties that constitute the mission for which the UAV was launched. It can be safely stated that the failure to properly localize and navigate the UAV accurately would result in compromising the mission or application for which it was launched. Therefore, ensuring the robustness and accuracy of the localization and navigation techniques used by autonomous UAVs is central to the success of the entire mission that they support.

Since there is a need for the autonomous UAV to self-localize and self-navigate, this implies the need to communicate with external resources to help in the localization and navigation process. For this purpose, there are many possibilities for this communication. One of the most widely used approaches is the one relying on the Global Navigation Satellite System (GNSS) in conjunction with the Inertial Measurement Unit (IMU) that is installed onboard the UAV. The GNSS provides autonomous geospatial localization services that utilize satellites.

The use of GNSS systems requires direct line-of-site (LoS) communication with the supporting satellites. Therefore, the use of such a system would only be possible in outdoor environments. Another candidate communication solution to use for the UAV localization and navigation is Wi-Fi. It can provide a proper solution in closed (indoor) areas and open areas with limited space such as malls and university campuses. The cellular communication system also presents a strong contender for such tasks. The wide prevalence of such systems in urban areas and the general robustness of its operations nowadays enable it to be a reliable alternative in outdoor and indoor environments, depending on its level of coverage in a given area where the UAV is required to operate. The localization and navigation techniques that can be used with autonomous UAVs generally need to have the following characteristics:

- Accuracy
- Real-time operation
- Efficiency
- High reliability
- Ability to respond to obstacles and threats.

There are many methodologies that have been followed in the literature and practically to devise the navigation techniques for autonomous UAVs, as will be discussed throughout this chapter. One of the most promising methodologies that are currently widely explored for this purpose is the one that depends on machine-learning (ML) techniques. ML techniques have the potential to provide real-time calculation with a close-to-optimal solution. They operate in many different configurations, depending on the type of used technique. In this chapter, we discuss the different aspects that relate to the development of navigation techniques for autonomous UAVs. We detail the challenges of the UAV navigation solutions and present some proposed solutions to address the existing limitations.

6.2 Approaches of UAV Outdoor Navigation

Various wireless technologies are currently utilized for enabling UAV navigation applications. The main enabling requirement for the UAV to deliver on its intended missions is its capability to determine its location at any point in time [4]. The UAV navigation solutions face many challenges in outdoor environments. The navigation techniques require a high level of accuracy to function correctly. Furthermore, the trajectory planning solutions need to satisfy a real-time calculation condition to adapt to the dynamic nature of the environment.

6.2.1 UAV Navigation Techniques: Technological Perspective

The UAV must be able to accurately determine its position with high accuracy. We present in the following a brief overview of some of the main wireless enabling technologies commonly used for UAV localization and navigation applications.

6.2.1.1 The Global Navigation Satellite System

Most commercial UAVs rely on the integration of a Global Navigation Satellite System (GNSS) and the Inertial Measurement Unit (IMU) for their navigation. The GNSS system has the advantage of global coverage in remote areas e.g., in the desert and rural areas. There are four core satellite navigation systems, namely, the Global Positioning System (GPS), the Russian Global Navigation Satellite System (GLONASS), Beidou and Galileo. The GPS, which is operated by the United States Space Force, is the most widely used GNSS.

The GNSS is composed of a constellation of satellites that circle the earth at an altitude of about 20,000 km. These satellites are able to provide the world with time, speed and position services. A GNSS is composed of a space segment, which is composed of the satellite constellation, a control segment which tracks the satellites and provides them with management and control information and a user segment which is the receiver of the GNSS's information, see Fig. 6.1. The number of possible users of this system has no specific limit, in general. It is important to note that each satellite within the constellation has a unique code. This pseudo random noise code is used by the receiver to determine the pseudo range from the satellite to the receiver.

This is calculated by multiplying the signal travel speed, which is the speed of light, by the time it took the signal to travel from the satellite to the receiver. This time is determined by the receiver through the correlation of the satellite's code to its own code. Then, it determines how much its code needs to be delayed matching that of the satellite. This delay is the travel time. Since there are errors in both satellite and receiver clocks, the calculated distances need to be adjusted to account for the internal clock errors. Specifically, a GNSS-enabled receiver estimates its pseudo range r_i from a given GNSS satellite i by

$$r_i = s \times [t_r(T_2) - t_s(T_1)], \tag{6.3}$$

where s corresponds to the speed of the GNSS signal (speed of light) and $[t_r(T_2) - t_s(T_1)]$ represents the signal propagation time to reach the receiver. The pseudo range estimated by the receiver incorporates the geometric distance d_i between the satellite and the receiver as well as synchronization clock errors and other error terms due to

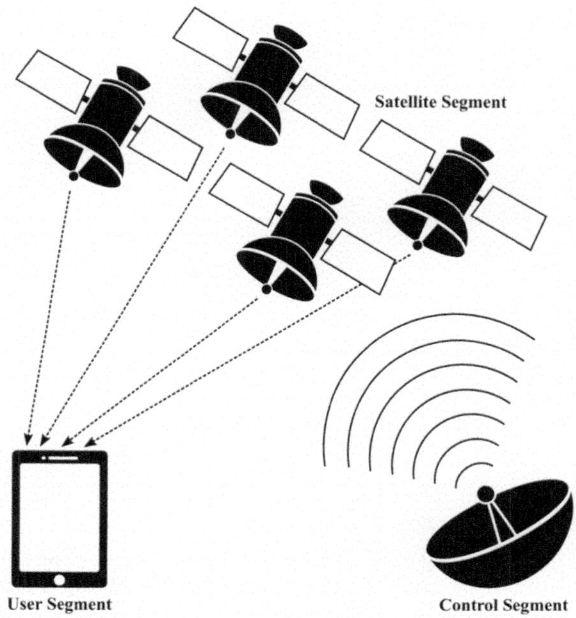

Fig. 6.1 The segments of the GNSS

the signal propagation through the atmosphere. More specifically, the pseudo range equation can be rewritten as such

$$r_i = d_i + s(\Delta t_r - \Delta t_s) + T + \alpha_f STEC + K_{p,r} - K_p^s + \mathcal{M}_p + \varepsilon_p, \tag{6.4}$$

where Δt_r and Δt_s correspond to the receiver and satellite clock offsets respectively, T is the tropospheric delay, and $\alpha_f STEC$ corresponds to a frequency dependent ionospheric delay term. $K_{p,r}$ and K_p^s represent receiver and satellite instrumental delay terms respectively, $\mathcal{M}_p$ corresponds to effect of multipath, and ε_p represents the receiver noise term. The signal propagation time can also be more accurately estimated through carrier phase measurements. Once the receiver determines the locations of 4 satellites along with its distance to each of them, it uses the triangulation principle [5, 6] to calculate its position on earth as shown in Fig. 6.2.

The GNSS can provide submeter localization accuracy in open outdoor environments. However, the GNSS does not perform well in dense urban and suburban environments due to signal blockage and reflections from high rise structures. As a result of this blockage, there may not be sufficient satellites to estimate the position of the UAV. Moreover, strong multipath signals would significantly degrade the positioning performance.

The global positioning system receivers suffer positioning errors due to some of the following potential sources of error:

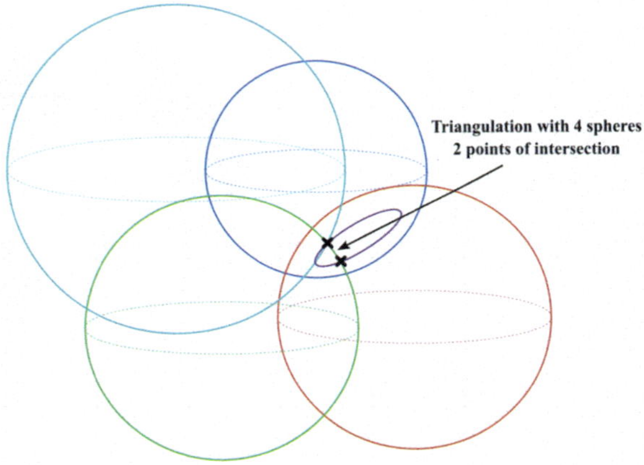

Fig. 6.2 Illustration of triangulation principle

- Receiver internal clock errors/inaccuracies: which lead to inaccurate determination of the distances to the different satellites. The receiver clock error is usually treated as an unknown parameter, while it is necessary to compensate for the satellite clock error to achieve precise positioning. Satellite clock errors are usually included in the navigation message or ephemeris.
- Satellite visibility: if the receiver cannot "see" at least 4 satellites necessary to determine its position due to building blockage or other barriers, it will not be able to correctly calculate its position.
- Interferences: there are several types of interferences that can affect the quality of the received signals such as multipath interference, ionospheric delay and tropospheric delay. These types of interference can affect the received satellite signals to varying degrees depending on the location and surrounding environmental conditions.
- Satellite distribution: in a good distribution, corresponding to a small geometric dilution of precision, satellites are spread in space, instead of in a line or crowded in a small space. This contrasts with the poor relative satellite positioning which occurs when the satellites are, for example, positioned in a line.

6.2.1.2 Imaging-Based UAV Navigation

This method relies on the use of camera images and videos for location determination. It determines poses from the camera images relative to a coordinate system of the surrounding environment, which may or may not be known in advance. Several image feature options are possible to use in the localization task, namely, points, lines, circles, etc. Of these feature options, points are the most widely used. For known environments, the determination of the camera pose location using a cloud of points is known as the problem of perspective-n-point (PnP) [7]. If $n \geq 6$ then the problem is linear. If $n = 3$, 4 or 5, then we have a non-linear problem. If $n = 2$, then the problem has no solution. When $n = 3$, the problem has at most 4 solutions.

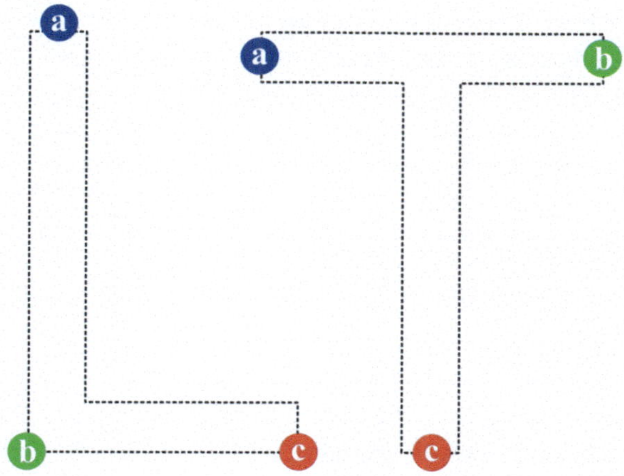

Fig. 6.3 Landmarks with feature points

When n = 4, the problem has a unique solution. As far as unknown environments are concerned, it is possible to reconstruct the environment from videos.

There are methods to solve such an environment in real time and online. This family of methods is termed Simultaneous Localization and Mapping (SLAM). There are other methods that solve this type of problems without real-time and online environment mapping. This type of method is an intermediate procedure that is termed structure from motion (SFM) family of techniques.

As an alternative to considering points as the determination feature, landmarks can be used for this purpose. Specifically, the UAV is equipped with a pan tilt gimbal where a color camera or binoculars are mounted [8, 9]. Typically, the UAV detects landmarks along its flight path and accordingly estimates its location with respect to the landmark. Feature points with good geometric configurations are selected to facilitate the landmark detection process. Landmarks with an "L" and "T" shaped configurations are commonly used as feature points as shown in Fig. 6.3. The concept of binocular ranging is based on the calculation of the vertical distance from the camera and estimating the disparity value in the horizontal direction of the left and right imaging planes.

The UAV is able to determine its location by estimating its distance to each feature point. The solution of the image-based localization problems is generally cumbersome especially in case of dynamic and large problem spaces.

6.2.1.3 Cellular Network-Based UAV Positioning and Navigation

One of the important UAV trajectory planning solutions relies on the existing cellular infrastructure to navigate the UAV. Some of such solutions associate the UAV with the cellular network. Specifically, the cellular network is responsible for navigating the UAV from a source to a destination. The trajectory planning algorithms in this case are executed at the cellular base station side. Other UAV solutions rely on cellular

Table 6.1 Broadcast signals for the 5G technology

Synchronization Signal Block (SSB)	Description of subblock functionality
Primary synchronization sequence (PSS)	Provides synchronization timing estimate—one of 3 possible sequences
Secondary synchronization sequence (SSS)	Provides cellular id—one of 336 possible sequences linked to the PSS which sums up to a total of 1008 possible cellular ids
Physical broadcast channel (PBCH)	Contains the Master Information block (MIB)
PBCH demodulation reference signal (DMRS)	Includes basic information to decode the System Information Block (SIB)

signals to navigate the UAV without actually interacting with the cellular network. Particularly, in such solutions, the UAV detects the periodic broadcast signals transmitted by the existing cellular infrastructure. For illustration purposes, we present the periodic broadcast signals for the 5G technology as an example shown in Table 6.1. Specifically, the 5G cellular stations broadcast the Synchronization Signal Block (SSB) periodically for synchronization purposes and to enable communication with the mobile users. The UAV can detect and utilize such broadcast signals for localization and navigation applications. The mission objectives determine the requirements of the UAV navigation solution.

The navigation solution aims to optimize a specific trajectory planning cost metric. The formulation of the trajectory planning cost metric depends on the application requirements. Many trajectory planning solutions aim to determine the shortest path or minimize the mission duration time with collision avoidance [10–12]. Other UAV navigation solutions formulate the navigation metric as a composite joint objective cost metric incorporating multiple objectives as follows

$$J(n|\pi_{sd}) = \sum_i w_i J_i(n), \tag{6.5}$$

where w_i corresponds to the weights associated with each mission objective.

6.2.2 UAV Navigation Techniques: Remarks

As we discussed earlier, the GNSS-based localization systems face many challenges including accuracy limitations, particularly in dense urban environments [13]. The signals transmitted by the GNSS satellites are prone to signal attenuation due to LOS blockage, especially in city canyons. To minimize the effect of such limitations, some GNSS-based localization solutions resort to the integration with other wireless networks e.g., Assisted GNSS (A-GNSS). The A-GNSS allows GNSS receivers to obtain information from other wireless network resources to assist in determining the receiver location. However, such solutions require communication with the wireless network and involve the transmission of detectable signals which might comprise the success of certain missions.

The imaging-based localization and navigation solutions, on the other hand, require substantial training and knowledge of the environmental terrain. Furthermore, vision-based solutions cannot be used in unknown dynamic environments.

Cellular networks are widely available worldwide providing an attractive alternative to GNSS, and imaging-based solutions. Furthermore, cellular networks typically have geometrically favorable configurations suitable for UAV navigation solutions in outdoor urban environments. As such, we focus our attention in the remainder of this chapter on cellular-based UAV navigation solutions.

To solve the UAV navigation problem, optimization-based techniques including Graph Search Algorithms, Ant Colony Optimization (ACO) and Genetic Algorithms, are commonly used. However, such optimization-based methods are typically iterative in nature and involve a high computational complexity. For example, the computational complexity of the exact optimization-based methods, e.g., Exhaustive Search (ES), is in the order of $O(nA)$ where n represents the variable problem size corresponding to the number of steps to reach the destination and A represents the 3-D flight area of the UAV. Similarly, the complexity of the heuristic optimization-based methods is in the order of $O(nt)$ where t represents the number of iterations needed for the algorithm to converge to the optimum solution. As such, iterative optimization-based techniques do not satisfy the real-time calculation objective governed by the dynamics of the environment and are impractical for use given the limited computational power onboard the UAV. Machine Learning (ML)-based methods can alternatively be utilized to solve the UAV navigation problem with near optimal accuracy. ML-based methods provide an attractive alternative to optimization-based techniques given their potential to satisfy the real-time calculation requirement.

6.3 ML-Based Autonomous UAV Navigation Techniques

We now focus our discussion on cellular-based UAV positioning methods to navigate the UAV in an outdoor environment. Our objective is to present an intelligent real-time autonomous UAV navigation solution that enables the UAV to travel through a dynamic urban or suburban environment efficiently.

6.3.1 Operational Requirements

The UAV is typically required to navigate through a complex and dynamic environment. The UAV should have the ability to determine a possible path from a starting point to a destination by optimizing a given trajectory planning cost metric. For this purpose, the UAV navigation solution should consider the environmental conditions and the UAV dynamic constraints. This necessitates the development of a UAV navigation solution that is suitable for dynamic outdoor urban and suburban environments. The UAV positioning solutions should provide the ability to localize the UAV up to a decimeter 3D accuracy for security and control reasons. While GNSS and Imaging-based solutions can be utilized to this end, they face several challenges in urban environments. As detailed in Sect. 6.2, cellular signals provide a favorable alternative for UAV navigation solutions in such environments. Accordingly, we utilize cellular signals to navigate the UAV along its route given the practicality and geometrically convenient configurations of such signals in outdoor environments. The UAV navigation solution is responsible for determining the optimal path to navigate to the destination. Moreover, the UAV navigation solution needs to consider the limitations of the UAV itself. The UAV has limited power and computational capability hence the need for an efficient navigation technique. Typically, the UAV is required to find the shortest path to reach its destination. For cellular-based UAV navigation, the UAV is also required to maintain connectivity to the cellular network along its path. In addition, it is imperative that the UAV determines a collision free path for successful mission completion. The UAV navigation solution needs to be capable of detecting and reacting to dynamic obstacles and threats in real time. Specifically, the UAV navigation solution should aim to determine a trajectory from a given starting point to a destination in such a way that optimizes a navigation cost metric while observing the real-time calculation limits. The navigation cost metric incorporates several objectives including, but not limited to, minimizing the path length, avoiding collision with dynamic threats, and maintaining cellular connectivity.

6.3.2 The Operating Environment

We consider dense urban and suburban outdoor environments for our use case application. This environment generally contains static and dynamic objects/obstacles that will normally be faced by a flying UAV. It is expected that the UAV will be tasked to conduct a given mission where it needs to fly from an initial point to a destination and possibly return back. We assume the presence of an existing cellular infrastructure that can be accessed for use in UAV positioning and navigation. We assume the cellular base stations (gNBs) are randomly placed 200–1000 m apart, as shown in Fig. 6.4. While flying in this environment, the UAV is expected to be exposed to different environmental conditions that would normally affect the communication signals on which the UAV relies for its localization and navigation. The chosen path loss model will depend on the technique being used and the operational conditions of the UAV [14].

In case of UAVs flying at low altitude, we assume a probabilistic path loss model given by

$$\varnothing(h_i, r_i) = \varnothing_{LOS} \times P_{LOS} + \varnothing_{NLOS} \times P_{NLOS}, \tag{6.6}$$

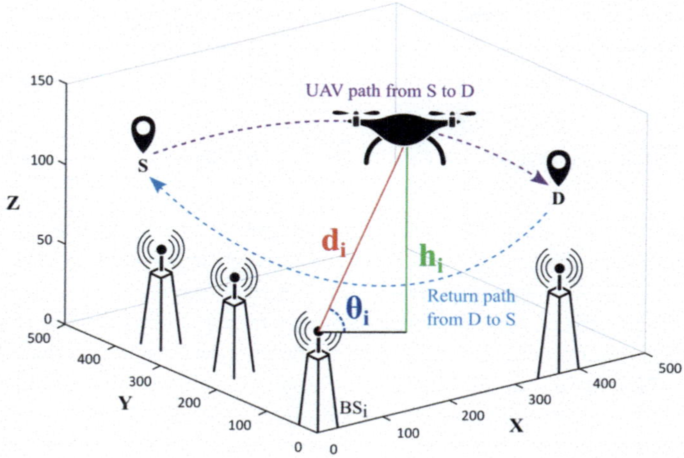

Fig. 6.4 The operating system model

where $\varnothing_{LOS}$ and $\varnothing_{NLOS}$ represent the mean path losses in case of line-of-sight (LoS) and non-line-of-sight (NLoS) situations, respectively. P_{LOS} is the probability of LoS situation with a given cellular base station, gNB_i, whereas P_{NLOS} is the probability of NLOS situation. $\varnothing_{LOS}$ and $\varnothing_{NLOS}$ are calculated as

$$\varnothing_{LOS} = 20\log\left(\frac{4\pi F d_i}{c}\right) + \eta_{LOS} \tag{6.7a}$$

$$\varnothing_{NLOS} = 20\log\left(\frac{4\pi F d_i}{c}\right) + \eta_{NLOS}, \tag{6.7b}$$

where F is the cellular carrier frequency, c is the speed of light and d_i is the Euclidian distance between the UAV and gNB_i. η_{LoS} and η_{NloS} represent the mean additional losses for the LoS and NLoS communications, respectively. P_{LOS} and P_{NLOS} are dependent on the environmental parameters, a and b, and are given by

$$P_{LOS} = \frac{1}{1 + aexp(-b(\frac{180}{\pi}\tan^{-1}\left(\frac{h_i}{r_i}\right) - a))}, \tag{6.8}$$

$$P_{NLOS} = 1 - P_{LOS}. \tag{6.9}$$

In case the UAV is flying at high altitude, we consider a Rician channel propagation model to represent the path loss between the cellular base stations and the UAV given by

$$\xi_i(dB) = 20\log_{10}(d_i) + 20\log_{10}(F) - 147.55, \tag{6.10}$$

where the index $1 \leq i \leq l$ corresponds to a given gNB_i and l corresponds to the number of gNBs along the route of the UAV.

6.3.3 Cellular Navigation Problem Formulation

We formulate the UAV navigation problem as an optimization problem to determine a feasible path between two given points. The navigation solution aims to optimize the path of the UAV by means of minimizing a given navigation cost metric. The navigation cost metric, C_π, typically incorporates multiple objectives as required by the application requirements e.g., minimizing the mission completion time, avoiding sharp and sudden turns, minimizing the positioning error along the path of the UAV, etc. [15–21]. Consequently, the UAV navigation problem is written as a constrained multi-objective optimization problem to determine the optimal trajectory π given by

$$\min_{\pi} C_\pi \tag{6.11}$$

subject to

$$\pi(0) = s, \tag{6.11a}$$

$$\pi(T) = d, \tag{6.11b}$$

$$v_{\min} < v < v_{\max}, \tag{6.11c}$$

where s and d correspond to the starting point and the destination, respectively, while $t = T$ represents the mission completion time. The navigation technique aims to optimize the formulated multi-objective optimization problem considering the limitations of the UAV and the environment. The constraints (6.11a) and (6.11b) allow the UAV to fly from a given starting point, s, to the destination, d, within time T. Finally, constraint (6.11c) limits the UAV velocity within the allowable range.

For example, we incorporate a multi-objective UAV navigation approach to optimize several objectives such as finding the shortest path, avoiding collisions with dynamic threats, maintaining connectivity with the cellular network as well as establishing a different return path for the UAV from the destination back to the start to conclude its mission. We therefore formulate the navigation cost metric C_π as a multi-objective metric composed of several constituents given by

$$C_\pi = \sum_i w_i C_i, \tag{6.12}$$

where the index $i = 1, 2, \ldots m$ corresponds to the objectives of the UAV navigation application. The cost C_i represents a given UAV navigation objective while w_i corresponds to the weight associated with a given navigation objective. Accordingly, we consider the first objective corresponding to determining the shortest path represented by C_1 to be given by

$$C_1 = a(t) + b(t), \tag{6.13}$$

where $a(t)$ corresponds to the distance from the starting point to the current position of the UAV while $b(t)$ is a heuristic corresponding to the Euclidean distance to the destination. We denote the collision avoidance objective by C_2. Particularly, C_2 penalizes the cost of the trajectory when the UAV is close to a possible collision according to

$$C_2 = \sum_{o \epsilon O_t} \begin{cases} 1, \mathit{if}\, |X - o| < d_{\min} \\ 0, \mathit{otherwise} \end{cases}, \tag{6.14}$$

where $o \epsilon O_t$ corresponds to the set of obstacles detected within the vicinity of the UAV at time t, while $d_{\min}$ represents the minimum distance to be maintained for collision avoidance. We let C_3 correspond to the objective of maintaining connectivity with the cellular network. We aim to maintain a certain signal-to-noise ratio (SNR) level with the nearest n base stations, denoted by $\mathrm{gNB_i}$ $\forall i = 1..n$. Consequently, we let C_3 be given by

$$C_3 = \sum_{i=1}^{n} \frac{\gamma_{th} - \gamma_i}{\delta_i}, \tag{6.15}$$

where γ_i represents the instantaneous SNR measured by the UAV from $\mathrm{gNB_i}$ while γ_{th} represents the desirable SNR threshold to be maintained. We let δ_i correspond to a normalization factor given by

$$\delta_i = \begin{cases} n(\gamma_{th} - \gamma_i), \mathit{if}\, (\gamma_{th} - \gamma_i) > 0 \\ \infty, \mathit{otherwise} \end{cases}. \tag{6.16}$$

Finally, we demonstrate an example of incorporating a different return path requirement within the UAV navigation cost metric, e.g., establishing a backup path, avoiding detection by following a different return path or satisfying the mission objectives. We let the fourth objective, denoted by C_4, correspond to the different return path requirement. For this purpose, we assign a binary value to penalize trajectories that do not satisfy the different return path objective. Particularly, C_4 is given by

$$C_4 = \begin{cases} 1\ if\ X \in \pi_{sd} \\ 0\ otherwise \end{cases}, \tag{6.17}$$

where π_{sd} represents the initial trajectory to be avoided.

6.3.4 Cellular ML-Based UAV Navigation

Our objective is to solve the formulated UAV navigation problem to determine the optimal path the UAV should follow to reach its destination within practical real-time boundaries. Opportunely, machine learning-based methods can potentially provide a near-optimal solution with practical real-time calculation that is needed in such dynamic applications.

ML-based navigation techniques can adopt an offline or online learning approach depending on the UAV computational capability and the dynamics of the environmental state. Particularly, the deep supervised learning and reinforcement learning-based methods are commonly applied to solve the presented UAV navigation problem. To this end, we present a deep supervised learning-based UAV navigation algorithm and a reinforcement learning-based algorithm to solve the formulated UAV navigation problem in a real-time fashion with near optimal accuracy. Furthermore, we analyze the efficiency of each of the presented approaches under a set of various operational conditions.

6.3.4.1 Deep Supervised Learning-Based UAV Navigation

In case the deep supervised learning-based approach is utilized, a deep neural network learns the environmental mapping of the suburban outdoor environment. Particularly, the UAV navigation problem is formulated as a black-box mapping between the inputs and the outputs. The inputs of deep neural network correspond to the state of the environmental trajectory planning whereas the output of the algorithm corresponds to the optimized trajectory the UAV should follow. This mapping is used to navigate the UAV to its destination, as shown in Fig. 6.5.

This approach requires prior knowledge of environmental maps as well as training data to learn the mapping process. The deep neural network architecture is composed of n-hidden layers each comprised of h nodes per layer. The deep neural network hyperparameter settings including the weights and dimensions, and the learning parameters, e.g., the performance goal e, the batch size $\mathscr{B}$, and the control parameter MU, are updated during the offline training phase as summarized in Algorithm 6.1.

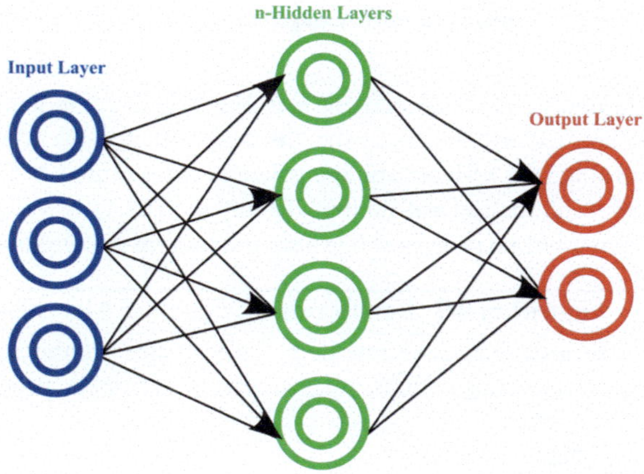

Fig. 6.5 Forward neural network architecture

Algorithm 6.1: Deep Supervised Learning Algorithm

Input: Coordinates of the UAV at time *t*, coordinates of the UAV at time *t -1*, coordinates of the surrounding cellular infrastructure, and the destination.

Output: Optimal UAV trajectory.

Neural network hyperparameter settings: Back propagation training FCN, Initial Mu ϵ [0,1], Mu increase and decrease factors, max. and min. Mu, max. training epochs E, Performance goal e, Batch size $\mathcal{B}$.

```
for epochs ← 1 to E
while (performance> e) do
Sample random batch from the training data set.
Evaluate network performance for the test data set.
Implement back propagation training.
Update hidden layer weights and biases.
if (gradient<threshold) then
break          // end Epoch
if (max.MU < MU < min.MU) then
break          // end Epoch
end
end
end
end
```

6.3.4.2 Reinforcement Learning-Based UAV Navigation

Alternatively, reinforcement learning-based techniques, including Q-learning, may be applied to determine the optimal path for the UAV [22–24]. Considering Q-learning, the Q-learning agent learns the optimal UAV navigation policy through direct interaction with the environment. The Q-learning agent can be trained in an offline, online or hybrid learning manner, as shown in Algorithm 6.2.

Algorithm 6.2: Reinforcement Learning Algorithm

Input: State-Action pair: Coordinates of the UAV, coordinates of the surrounding cellular infrastructure, the destination, and the selected UAV trajectory.
Output: Reward.

for episodes $i \leftarrow 1$ to j
  Detect obstacles and threats at time t
  Observe current system state s.
  while (destination not reached)
    Select action a to maximize Q-value.
    Evaluate system reward.
    Update observed Q-value in experience buffer $\mathbb{E}$.
    while *True* **do**
      Sample experience replay batch $\mathscr{B}$ from $\mathbb{E}$.
      Perform back propagation FCN
      Update network weights and biases.
      if (gradient<threshold) || (Learning rate > $\alpha_{\max}$) || (Performance< e) **then**
      break // end Episode
  end
  end
  end

Specifically, the Q-learning agent aims to maximize the Q-value of the observed system state as follows

$$Q(s, a) \leftarrow (1 - \alpha)Q(s, a) + \alpha\left[r(s) + \gamma \max_{a\prime} Q(s\prime, a\prime)\right], \tag{6.18}$$

where s is the system state including information pertaining to the position of the UAV and the surrounding environment. The Q-learning agent aims to minimize the cost of the trajectory. Hence, the reward, r, corresponds to the negative of the formulated trajectory cost metric.

To enforce the UAV system and dynamic constraints, the penalty method may be utilized. The penalty method is a method commonly used to reformulate a constrained objective function to an unconstrained optimization problem to facilitate calculation. As such, the penalty method can be utilized to enforce the constraints given by

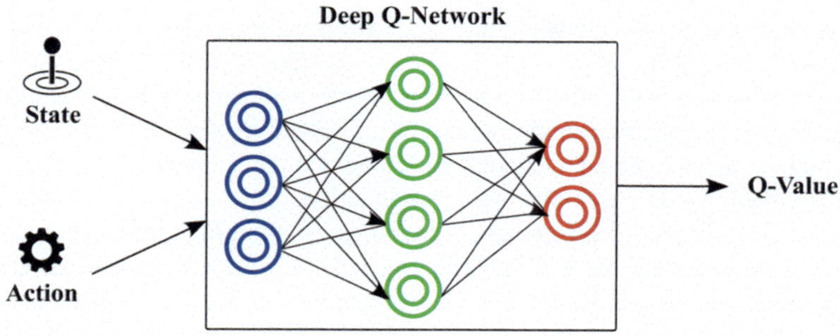

Fig. 6.6 Deep Q-network architecture

Eqs. (6.11a) to (6.11c) as follows.

$$r = -(\sum_i w_i C_i + \sum_i p_i), \tag{6.19}$$

where p_i is a penalty assigned for each constraint violation. The action, a, represents the trajectory that the UAV navigates to its destination while α is the learning rate of the reinforcement learning agent. The discount factor, γ, is parameter for adjusting the weight of anticipated future rewards in the learning process.

The Q-learning agent utilizes a deep Q-network (DQN) to predict the Q-value for a given action based on the system state. As shown in Fig. 6.6, the input to the DQN is composed of a state and action pair while the output is a scalar value proportional to the expected system reward. The trajectory that maximizes the cumulative reward is selected as the optimal trajectory.

6.4 Experimental Validation

We present experimental results and simulations to analyze the performance of cellular-based techniques and machine learning-based methods in UAV positioning and navigation solutions.

6.4.1 Field Test Experimentation

Several field tests have been conducted to verify the performance of cellular networks and machine learning-based methods in UAV positioning and navigation solutions. For example, a field test was conducted to demonstrate that cellular signals can improve the UAV's position root-mean squared error and the maximum position error by 30.69% and 58.86% respectively as compared to GNSS-based UAV localization

[11]. Another filed experiment was conducted to evaluate the effectiveness of cellular-based UAV localization using deep learning. Specifically, cellular field data were collected, and an augmentation process is proposed to train a deep neural network to localize the UAV in an Urban environment [28]. The authors aim to enhance the accuracy of fingerprint map-based localization methods through generating synthetic data that reflects the typical pattern of wireless localization information. The experimental results of this study provided promising outcomes. Another filed test is conducted to measure the performance of a UAV positioning and navigation solution in a 3-D setup utilizing carrier phase measurements while assuming limited GNSS presence. The authors proposed to leverage the relative stability of cellular base transceiver station (BTS) clocks to enable precise navigation with cellular carrier phase measurements. According to the conducted experimentations, this technique realizes a location estimation Root Mean Square Error (RMSE) of 0.8 m using 7 CDMA BSs and 0.36 m using measurements from 9 CDMA BSs [26]. Another experiment is conducted to analyze the performance of mobile device localization using cellular networks. The authors utilize multiple features and metrics of the LTE networks to generate a fingerprint grid map to localize the mobile device. The authors utilize a one-to-many augmenter to generate synthetic data to improve the performance of the proposed localization solution. According to the field experiments, the proposed technique achieves a localization accuracy of 13.7 m in an outdoor environment [25].

6.4.2 Simulation Results and Analysis

To validate the performance of the proposed ML approaches, we first find the optimal bound of the solutions. Then we compare the performance of the ML solutions against this optimal bound. In addition, we also present a GNSS-based technique from the literature and compare the proposed ML based techniques' performance against it.

The Optimal Bound

In order to assess the quality of the proposed ML solutions from the perspective of the optimality of the solution, we present a benchmark optimization-based technique. The optimal solution of the UAV navigation problem formulated in (6.11) can be found by utilizing classic optimization methods. However, exact optimization-based methods cannot be used as the formulated UAV navigation problem becomes intractable in high dimensional spaces [26]. Therefore, we resort to heuristic optimization-based techniques to determine the optimal bound of the solution. Particularly, we utilize the Ant Colony Optimization (ACO) method as a heuristic optimization technique [27]. The ACO technique iteratively solves for the optimal bound utilizing several ants which search the action space in parallel, as summarized in Algorithm 6.3. The ant movements through the action space are governed by a state transition probability which is proportional to the concentration of the ant pheromone levels. The state transition probability is thus given by

$$P_{ij}^{(k)}(t) = \frac{\phi_{ij}^{\alpha}(t)\mu_{ij}^{\beta}(t)}{\sum_{j \in allowed_k} \phi_{ij}^{\alpha}(t)\mu_{ij}^{\beta}(t)}, \tag{6.20}$$

where α and β correspond to the pheromone and expected heuristic factors, respectively. The quantity $\mu_{ij}(t)$ denotes the heuristic value corresponding to the Euclidean distance between node i and node j while $\phi_{ij}(t)$ represents the pheromone concentration at time t and is given by

$$\phi_{ij}(t+1) = (1-\rho)\phi_{ij}(t) + \Delta\phi_{ij}(t), \tag{6.21}$$

where $\rho < 1$ is the global pheromone volatility coefficient and $\Delta\phi_{ij}(t)$ is the pheromone increment amount.

Algorithm 6.3: The optimal bound

Input: ACO parameters including the number of ants, max. and current algorithm iterations, etc.

Output: The optimal UAV trajectory.

Initialization of pheromone levels and state transition probabilities.
for ACO iterations ← 1 to m
for number of ants ← 1 to j
  Evaluate state transition probability for each ant and update pheromone concentrations.
  if *destination reached* , **then**
   break
   else if max. no. of algorithm iterations reached **then** go to 1;
  end
 end
end

A Representative GPS-Based Technique

We also compare the performance of the presented cellular ML-based navigation solutions to a GPS-based technique. Specifically, we present the technique proposed in [10]. This GPS-based technique satisfies an in-time calculation according to the NASA Langley Research Center experimentations. The authors utilize NURBS curves [29] to generate UAV trajectories composed of $n+1$ waypoint locations given by

$$\mathrm{x(u)} = \frac{\sum_{i=0}^{n} h_i N_{i,k}(u) x_i}{\sum_{i=0}^{n} h_i N_{i,k}(u)}, 0 \le u \le n-k+2, \tag{6.22}$$

where $N_{i,k}(u)$ is the NURBS basis function while k represents the degree of the curve. The weight of each waypoint comprising the NURBS curve is given by h_i.

The technique then utilizes a Bayesian filtering algorithm to further improve the localization accuracy of the incoming GPS signals.

Performance Tuning of the ML-Based Positioning Techniques

We present the effect of various technique-specific hyper parameter settings on the performance of the presented ML-based UAV navigation techniques. For this purpose, we conduct MATLAB experiments to simulate the navigation of the UAV in a $500 \times 500 \times 300\,\text{m}^3$ outdoor suburban environment.

(1) Performance of the deep supervised learning-based technique

We determine the most suitable technique-specific hyper parameter settings of the presented deep supervised learning-based approach to deliver best results. As shown in Fig. 6.7, we demonstrate the effect of various neural network architectures on the performance of the deep learning-based technique [28]. Specifically, we vary the number of nodes per layer and the number of layers of the deep neural network. The performance of the presented deep supervised learning-based technique improves with increasing the deep neural network dimensions. This is expected as the neural network learns the environmental mapping more effectively with increasing the number of layers. Furthermore, the performance improves with increasing the number of nodes per layer allowing the network to accurately capture the non-linear relationships between the inputs and the outputs. The results demonstrate that a deep neural network architecture with 6 layers composed of 120 nodes per layer delivers optimized results.

(2) Performance of the reinforcement learning-based technique

We now investigate the effect of the various technique specific hyper parameter settings on the performance of the reinforcement learning-based UAV navigation solution. We demonstrate the effect of varying the DQN architecture to determine

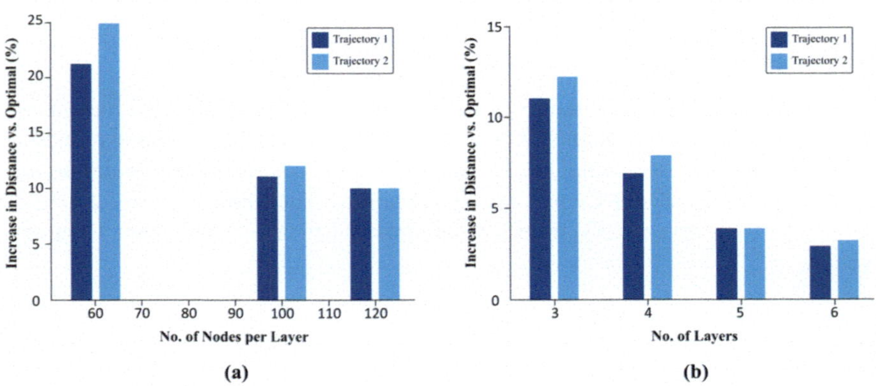

Fig. 6.7 Deep Learning-based results **a** Trajectory length vs. no. nodes per layer. **b** Trajectory length vs. no. of hidden layers

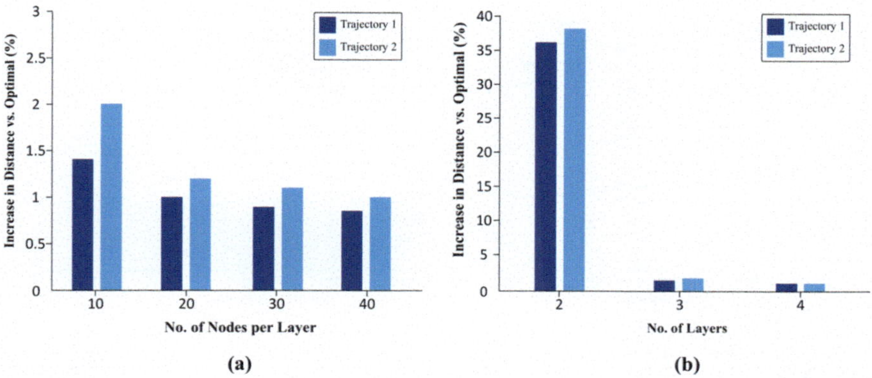

Fig. 6.8 Reinforcement Learning-based results **a** Trajectory length vs. no. nodes per layer. **b** Trajectory length vs. no. network layers

the suitable number of layers and nodes per layer needed to deliver the best results. The results demonstrate that the performance improves with increasing the number of nodes per layer of the DQN network, as shown in Fig. 6.8. This is expected as the DQN agent is able to more accurately map the Q-value function to each state-action pair as it learns through direct interaction with the environment. However, the performance of the proposed solution declines as the number of layers of the DQN increases beyond 3 layers as it becomes prone to overfitting.

6.4.3 Overall Results and Analysis

We demonstrate the performance of the proposed ML-based UAV navigation techniques as compared to the benchmark optimization-based techniques as well as the representative GPS-based solution. We assess the efficiency of each of the proposed trajectory planning approaches to fulfill the mission objectives under various operational conditions. Specifically, our objective is to find the shortest trajectory distance, maintain connectivity with the cellular network, collision avoidance and determining a different return path as presented in Sect. 6.3.

(1) Results of mission objective metrics

We utilize the components of the composite UAV navigation metric to assess the efficiency of the presented navigation techniques. As shown in Fig. 6.9, we measure the distance travelled by the UAV to reach various destinations. The trajectory norm represents the Euclidean distance to travel from a given source to a destination.

The presented ML- based methods perform closely to the optimal bound and the GPS-based solution in terms of traversed trajectory length, as demonstrated in Fig. 6.9a. The reinforcement learning-based technique yields shorter trajectories as compared to the deep supervised learning-based method as well as the GPS-based

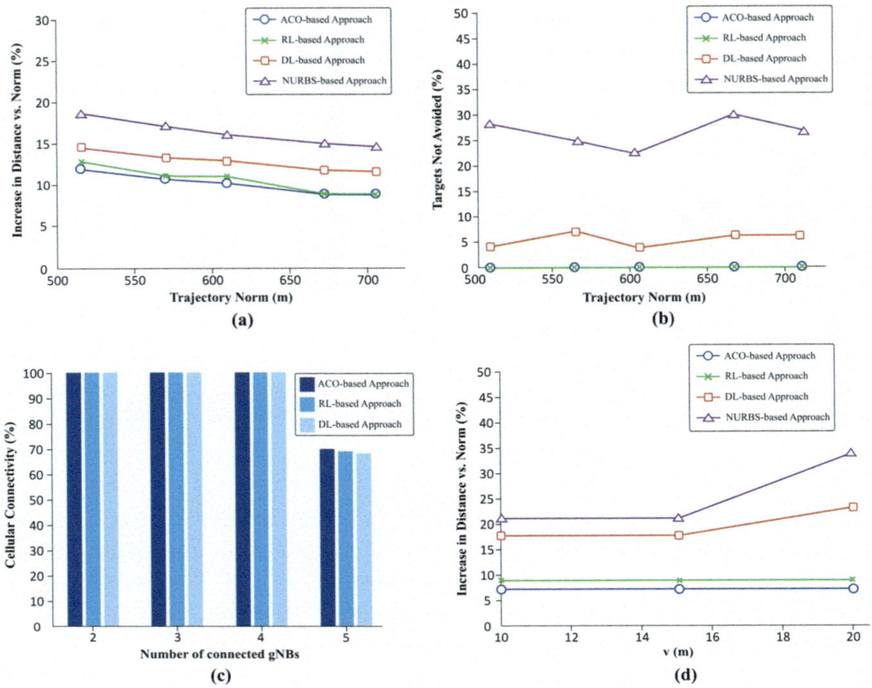

Fig. 6.9 Overall results **a** Traversed distance vs. trajectory norm **b** Collisions versus trajectory norm **c** Cellular connectivity versus. gNB concentration **d** Return path length over initial trajectory versus vicinity

technique. This is attributed to the ability of the reinforcement learning agent to evaluate all possible trajectories and select the path that maximizes the Q-value.

We also assess the efficiency of the ML-based methods in fulfilling the collision avoidance objective. As shown in Fig. 6.9b, the ACO-based algorithm determines a collision free trajectory from the source to the destination. The reinforcement learning-based solution effectively avoids collision by means of eliminating obstacles from its action space as compared to the deep supervised learning and the GPS-based methods.

As shown in Fig. 6.9c, both the deep supervised learning and the reinforcement learning-based solutions are effective in maintaining connectivity with the cellular network. The simulation results show that the presented ML-based solutions perform closely to the optimal bound achieved by applying the ACO-based technique.

We also evaluate the performance of the navigation algorithm in fulfilling the objective of finding a different return path on its trip back to the source. We vary the size of the vicinity, *v*, corresponding to the area surrounding the initial trajectory to be avoided. The reinforcement learning-based solution performs closely to the ACO-based technique in meeting this objective. However, the presented GPS-based solution does not provide an effective mechanism for establishing a different return route. Figure 6.9d shows the results.

(2) Computational complexity evaluation

We analyze the computational complexity of the presented ML-based UAV navigation solutions. We illustrate the computational complexity of each of the presented solutions in the form of the big-O notation. The big-O notation is a mathematical illustration that is commonly used to describe the limiting behavior of a function when the argument tends towards a particular value or infinity. Heuristic optimization-based methods, including the ACO-based solution, are iterative in nature.

The computational complexity of the ACO-based solution is proportional to the number of iterations needed to determine the optimal bound. The Big-O complexity of the ACO-based solution is given by

$$BigO_{ACO} = O(njlog(m)), \tag{6.23}$$

where n is the variable problem size, m is the number of ants and j corresponds to the number of needed iterations. Hence, this optimization-based method cannot be used in large problem size dimensions in real time.

The online computational complexity of deep supervised learning-based approach is proportional to the deep neural network architectural size and dimensions. The Big-O complexity of this technique is given by

$$BigO_{DL} = O(Cn), \tag{6.24}$$

where n is the variable problem size corresponding to the length of the trajectory and $C > 0$ is a constant corresponding to the network dimensions and is given by

$$C = n_{inputs} \times n_{Layer1} \times \cdots \times n_{Layern} \times n_{outputs}. \tag{6.25}$$

The computational complexity of the reinforcement learning-based approach is proportional to the size of the DQN used to estimate the reward as well as the degrees of freedom representing the directions the UAV is allowed to fly in. Accordingly, the Big-O computational complexity of the reinforcement learning-based algorithm is given by

$$BigO_{RL} = O(\partial n) + O(nlog(A + 2)), \tag{6.26}$$

where ∂ corresponds to the dimensions of the Q-network and A represents the action space proportional to the degrees of freedom of the UAV. Consequently, the ML-based solutions may potentially satisfy the real-time calculation objective as they involve lower computational complexity as compared to the optimization-based methods.

Simulations results demonstrate that the presented machine-learning based UAV navigation solutions perform closely to the optimal bound while providing much

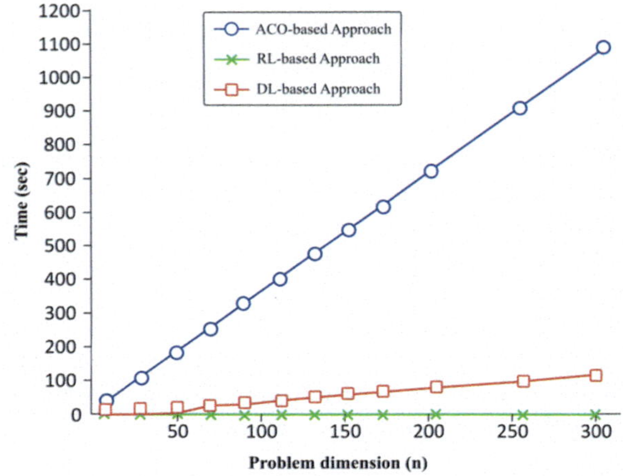

Fig. 6.10 Computational complexity versus variable problem dimension

faster computational results as shown in Fig. 6.10. The UAV can therefore determine a feasible path to navigate to its destination efficiently within real-time bounds.

(3) Remarks on the comparative performance between the deep learning and the reinforcement learning based solutions

According to our results, the ML-based techniques can be practically used for UAV navigation applications. Both the deep supervised learning and reinforcement learning-based solutions perform closely to the optimal bound in terms of traversed trajectory length while maintaining connectivity with the cellular network. The ML-based solutions also managed to fulfill the collision avoidance and different return path objectives of the multi-objective UAV navigation problem.

Based on our analysis, the computational complexity of the deep supervised learning-based solution is proportional to the dimensions of the deep neural network needed to learn the environmental map whereas the complexity of the reinforcement learning agent is governed by degrees of freedom of the UAV. As such, the deep supervised learning-based technique is effective to navigate through a small complex terrain with a high degree of freedom. However, the reinforcement learning-based solution is recommended for use over the deep supervised learning-based technique in large navigation areas where the computational complexity of the deep learning-based technique to learn the environmental mapping becomes significantly large.

6.4.4 Discussing the Experimental Versus Simulation Results

Field test experimentation results show that the use of cellular networks to support the UAV navigation produces promising results, particularly in outdoor urban environments. Several experimental studies have shown that the utilization of ML-based techniques for UAV localization and trajectory planning can actually be used in real life applications. Specifically, the ML-based algorithms have the potential to enable the UAV to adapt to the dynamics of the environment in real-time fashion. Based on the simulation results that we also presented in this chapter, it is clear that there is similarity in the findings with the experimental results that were obtained from field experiments. Specifically, the simulations conducted also demonstrate the effectiveness of utilizing cellular signals for UAV navigation. This is attributed to the fact that we have modelled the performance of ML-based and optimization-based UAV positioning and navigation algorithms in an urban environment with actual cellular infrastructure parameters. The cellular signals from the surrounding infrastructure can be utilized to position and navigate the UAV to its destination effectively. Both experimental and simulation results show that ML-based solutions can provide near optimal results with lower computational complexity, as compared to standard optimization-based methods. We can therefore safely conclude from these observations that the use of simulation can be a reasonable alternative to resorting to field tests when these field trials cannot be conducted for one reason or the other. However, it is imperative that the final stage of testing a given technique prior to enabling it for field use should be done in real life field environments.

6.5 Conclusions

The autonomous navigation of UAVs through complex and dynamic environments has major civilian and military use case applications. The objectives of the UAV navigation mission are dictated by the application requirements. The UAV navigation problem can be cast as a constrained multi-objective optimization problem to determine a feasible trajectory to reach the destination. The UAVs navigate through dynamic environments with the need for a real-time trajectory determination to react to obstacles and threats. Furthermore, the UAV has limited power and computational capability. Traditional optimization-based methods are rendered unsuitable for use given their large computational complexity. Hence, there is a need for effective UAV navigation solutions capable of providing near optimal paths in a computationally efficient manner. Machine learning-based techniques have the potential for fulfilling near-optimal solutions with real-time performance. Deep supervised learning and reinforcement learning-based methods have been applied to solve the UAV navigation problem in real-time through the use of existing cellular infrastructure. The presented ML-based solutions have demonstrated their efficiency to determine feasible paths fulfilling the objectives of the UAV navigation application. The reinforcement learning-based agent can adapt in a hybrid online and offline manner to the surrounding environmental state yielding improved results. We have performed a detailed experimental validation through simulations to determine the most suitable ML-based approach for use under various operating conditions.

References

1. Alsuhli G, Fahim A, Gadallah Y (2022) A survey on the role of UAVs in the communication process: a technological perspective. Comput Commun 194:86–123. https://doi.org/10.1016/j.comcom.2022.07.021
2. Li B, Li Q, Zeng Y, Rong Y, Zhang Y (2022) 3D trajectory optimization for energy-efficient UAV communication: a control design perspective. IEEE Trans Wirel Commun 21(6):4579–4593. https://doi.org/10.1109/TWC.2021.3131384
3. Zhi Z, Liu L, Liu D (2020) Enhancing the reliability of the quadrotor by formulating the control system model. In: International conference on sensing, measurement & data analytics in the era of artificial intelligence (ICSMD), Xi'an, China 242–246. https://doi.org/10.1109/ICSMD50554.2020.9261660
4. Afifi G, Gadallah Y (2021) Autonomous 3-D UAV localization using cellular networks: deep supervised learning versus reinforcement learning approaches. IEEE Access 9:155234–155248. https://doi.org/10.1109/ACCESS.2021.3126775
5. Tippenhauer NO, Pöpper C, Rasmussen KB, Capkun S (2011) On the requirements for successful GPS spoofing attacks. In: Proceedings of the 18th ACM conference on computer and communications security, vol 18, pp 75–86
6. Patil V, Atrey PK (2020) GeoSecure-R: secure computation of geographical distance using region-anonymized GPS data. In: IEEE sixth international conference on multimedia big data (BigMM), vol 6, 28–36. https://doi.org/10.1109/BigMM50055.2020.00015
7. Fischler MA, Bolles RC (1981) Random sample consensus: a paradigm for model fitting with applications to image analysis and automated cartography. Commun ACM 24:381–395
8. Wang J, Gu D, Liu F (2019) Research on autonomous positioning method of UAV based on binocular vision. In: Chinese automation congress (CAC), Hangzhou, China 3588–3593. https://doi.org/10.1109/CAC48633.2019.8996413
9. Zhou Y, Tang D, Zhou H, Xiang X, Hu T (2019) Vision-based online localization and trajectory smoothing for fixed-wing UAV tracking a moving target. In: IEEE/CVF international conference on computer vision workshop (ICCVW), Seoul, Korea. https://doi.org/10.1109/ICCVW.2019.00024
10. Banerjee P, Corbetta M (2020) In-time UAV flight-trajectory estimation and tracking using Bayesian filters. In: IEEE aerospace conference, pp 1–9. https://doi.org/10.1109/AERO47225.2020.9172610
11. Ragothaman S, Maaref M, Kassas ZM (2019) Multipath-optimal UAV trajectory planning for urban UAV navigation with cellular signals. In: IEEE 90th vehicular technology conference (VTC2019-Fall), Honolulu, HI, USA, pp 1–6. https://doi.org/10.1109/VTCFall.2019.8891218
12. Ge J, Liu L, Dong X, Tian W (2020) Trajectory planning of fixed-wing UAV using Kinodynamic RRT* algorithm. In: International conference on information science and technology (ICIST), Bath, London, and Plymouth, United Kingdom, vol. 10, pp 44–49. https://doi.org/10.1109/ICIST49303.2020.9202213
13. Lachow I (1995) The GPS dilemma: balancing military risks and economic benefits. Int Secur 20(1):126–148. JSTOR, www.jstor.org/stable/2539220. Accessed 5 June 2021
14. Afifi A, Gadallah Y (2022) Cellular network-supported machine learning techniques for autonomous UAV trajectory planning. IEEE Access 10:131996–132011. https://doi.org/10.1109/ACCESS.2022.3229171
15. Susarla P et al (2020) Learning-based trajectory optimization for 5G mmWave uplink UAVs. In: IEEE international conference on communications workshops (ICC workshops), Dublin, Ireland, pp 1–7. https://doi.org/10.1109/ICCWorkshops49005.2020.9145194
16. Zeng Y, Xu X (2019) Path design for cellular-connected UAV with reinforcement learning. arXiv:1905.03440
17. Bast SD, Vinogradov E, Pollin S (2019) Cellular coverage-aware path planning for UAVs. In: IEEE international workshop on signal processing advances in wireless communications (SPAWC), Cannes, France, vol 20, pp 1–5. https://doi.org/10.1109/SPAWC.2019.8815469

18. Zhang S, Zeng Y, Zhang Y (2019) Cellular-enabled UAV communication: a connectivity-constrained trajectory optimization perspective. IEEE Trans Commun 67(3):2580–2604. https://doi.org/10.1109/TCOMM.2018.2880468
19. Bulut E, Guevenc I (2018) Trajectory optimization for cellular-connected UAVs with disconnectivity constraint. In: Proceedings of the IEEE international workshop communication workshops (ICC workshops), pp 1–6
20. Chen Y, Huang D (2020) Trajectory optimization for cellular-enabled UAV with connectivity outage constraint. IEEE Access 8:29205–29218. https://doi.org/10.1109/ACCESS.2020.2971772
21. Zheng H, Guo J, Yan P (2018) A hybrid trajectory planning algorithm for UAVs in cluttered environments. In: The 9th international conference on mechanical and aerospace engineering (ICMAE), Budapest, pp 389–393. https://doi.org/10.1109/ICMAE.2018.8467706
22. Challita U, Saad W, Bettstetter C (2019) Interference management for cellular-connected UAVs: a deep reinforcement learning approach. IEEE Trans Wireless Commun. https://doi.org/10.1109/TWC.2019.2900035
23. Liu Q, Shi L, Sun L, Li J, Ding M, Shu F (2020) Path planning for UAV-mounted mobile edge computing with deep reinforcement learning. IEEE Trans Veh Technol 69(5). https://doi.org/10.1109/TVT.20202982508
24. Xie H, Yang D, Xiao L, Lyu J (2021) Connectivity-aware 3D UAV path design with deep reinforcement learning. IEEE TVT 70(12):13022–13034. https://doi.org/10.1109/TVT.2021.3121747
25. Mohamed A, Tharwat M, Magdy M, Abubakr T, Nasr O, Youssef M (2022) DeepFeat: robust large-scale multi-features outdoor localization in LTE networks using deep learning. IEEE Access 10:3400–3414. https://doi.org/10.1109/ACCESS.2022.3140292
26. Khalife J, Kassas ZM (2018) Precise UAV navigation with cellular carrier phase measurements. In: IEEE/ION position, location and navigation symposium (PLANS), pp 978–989. https://doi.org/10.1109/PLANS.2018.8373476
27. Li B, Qi X, Yu B, Liu L (2020) Trajectory planning for UAV based on improved ACO algorithm. IEEE Access 8:2995–3006. https://doi.org/10.1109/ACCESS.2019.2962340
28. Rizk H, Shokry A, Youssef M (2019) Effectiveness of data augmentation in cellular-based localization using deep learning. In: IEEE wireless communications and networking conference (WCNC), Marrakesh, Morocco, pp 1–6. https://doi.org/10.1109/WCNC.2019.8886005
29. Chen H, Guo C, Wang Z, Wen T, Zeng Z, Lin Z (2020) The trajectory planning system for spraying robot based on k-means clustering and NURBS curve optimization. In: The annual conference of the IEEE industrial electronics society, Singapore, vol. 46, 5356–5361.https://doi.org/10.1109/IECON43393.2020.9255172

Chapter 7
Magnetic Positioning Based on Evolutionary Algorithms

Meng Sun, Kegen Yu, and Jingxue Bi

Abstract The spatially discernible indoor magnetic field indicates locations through different magnetic readings at various positions. Therefore, magnetic positioning has garnered attention due to its promising localization accuracy and infrastructure-free nature, significantly reducing the investment in localization. Since the magnetic field covers all indoor environments, magnetic positioning holds the potential to create a ubiquitous indoor positioning system. This chapter investigates the stability of the magnetic field concerning factors such as devices, testers, materials, and dates. Compensation methods for different types of magnetic features are studied based on fluctuation patterns to achieve accurate positioning results. Evolutionary algorithm-based optimization strategies are proposed for online localization, tailored to the types of used magnetic features. Testing experiments validate the feasibility and efficiency of utilizing evolutionary algorithms to enhance magnetic positioning performance.

7.1 Introduction

The indoor geomagnetic field is a composite of the Earth's magnetic field and those of electromagnetic sources (e.g., power supplies) and/or ferromagnetic materials (e.g., iron furniture, central air conditioners) [1]. The uneven distribution of such a combined magnetic field has been exploited for indoor localization. Starting with robot localization in corridor environments, Suksakulchai et al. [2] used magnetic field disturbances as recognition signatures for localizing a robot. Experimental outcomes validated the feasibility of using sequential magnetic disturbance data

M. Sun (✉) · K. Yu
China University of Mining and Technology, No. 1 Daxue Road, Xuzhou 221116, China
e-mail: msun@cumt.edu.cn

K. Yu
e-mail: kegen.yu@cumt.edu.cn

J. Bi
Shandong Jianzhu University, Fengming Road No. 1000, Shandong Province, Jinan 250101, China
e-mail: bijingxue19@sdjzu.edu.cn

K. Yu (ed.), *Positioning and Navigation Using Machine Learning Methods*, Navigation: Science and Technology 14, https://doi.org/10.1007/978-981-97-6199-9_7

for positioning in corridor environments. Gozick et al. [3] investigated the stability and long-term variation of ambient indoor magnetic field, affirming the utility in distinguishing indoor landmarks, guideposts, and rooms. They confirmed that indoor magnetic maps can be developed for localization with only a mobile phone. Li et al. [4] observed that the characteristics of the magnetic field change with locations, making magnetic fingerprinting possible. They indicated the limited available information of magnetic data as a significant drawback of magnetic positioning. Accordingly, numerous efforts have focused on developing precise and stable magnetic positioning methods. Magnetic localization methods can be categorized into two main groups based on the utilization of magnetic features: single-point-based magnetic positioning (SPMP) and sequence-based magnetic positioning (SBMP) [5].

SPMP utilizes discrete features including 3-axis components, horizontal intensity, and total intensity of the magnetic field for positioning. As depicted in Fig. 7.1, SPMP requires an offline pre-built magnetic features database (also known as magnetic maps) for online matching. During the offline stage, the fingerprint associated with the ith position can be described as:

$$\Xi_i = \left\{B_x, B_y, B_z, B_h, B, x_i, y_i\right\} \tag{7.1}$$

where x_i and y_i denotes the coordinates of the ith indoor position, B_x, B_y, B_z, B_h and B are the triple-axis, horizontal, and total intensities of the magnetic field, respectively, which are also magnetic features. A discrete magnetic database is constructed by generating fingerprints at all the arranged indoor positions. The database can also be generated by crowdsourcing methods described in [6, 7], or the SLAM approach in [8, 9]. These methods aim at reducing labor-intensive measurement workload.

For online matching, the k-nearest neighbors (KNN) algorithm [10], and mean square deviation (MSD) [11] are commonly used. However, directly employing

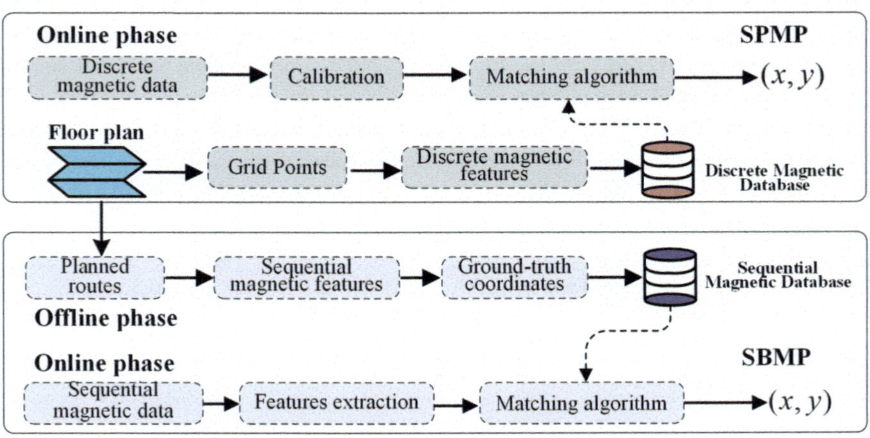

Fig. 7.1 The process of indoor magnetic positioning

KNN and MSD with low-dimensional magnetic feature data does not yield optimal positioning results. Researchers have tackled this issue by proposing optimized approaches or introducing supplementary information to magnetic features. For example, [12] proposes the multi-magnetic fingerprint fusion (MMFF) method, which combines coarse estimation with fine estimation for accurate positioning. In [13], the multi-parameter matching model of least magnetic distance (MPMD) is proposed to optimize the low accuracy problem of MSD and offer better noise immunity than MSD. Moreover, various optimization works are explored by using the particle filter, such as the integrated particle filter [14], genetic particle filter (GPF) [10], improved particle filters (IPF) [15, 16], sensitivity-based adaptive particle filter [17], and more. Generally, PF can well handle the problem of low accuracy caused by low-dimensional magnetic data, yet it always suffers from high computational complexity because lots of particles are required to perform filtering. To enhance the specificity of magnetic fingerprints, researchers have integrated phone attitudes with magnetic features to generate an augmented magnetic vector [18], aiming to reduce mismatching and improve localization accuracy. In [7], orientation-aided magnetic fingerprints are designed for processing crowdsourced magnetic data, improving fingerprint fidelity when massive data are used.

SBMP relies on sequential magnetic data to estimate indoor positions. As indicated in Fig. 7.1, SBMP needs an offline pre-built magnetic database for online matching. Generally, the sequential magnetic features contain the total intensity of the magnetic field along the designated routes. To construct a sequential magnetic database, the tester should hold a device while moving along the predefined routes to simultaneously measure total magnetic intensities and position information. The ith sequential fingerprint with n magnetic intensities can be expressed as:

$$\Xi_i = \left\{B_1, x_1, y_1, B_2, x_2, y_2, \ldots, B_j, x_j, y_j, \ldots B_n, x_n, y_n\right\} \tag{7.2}$$

where B_j is the jth total magnetic intensity of the magnetic sequence, $j \in \{1, \ldots, n\}$, n denotes the number of total magnetic intensities, respectively. During practical usage, n should be determined according to positioning accuracy. Generally, the higher the precision required, the longer the length of the magnetic sequence, which means that n has a greater number. For online matching, the length of the real-time measured consecutive magnetic data should be consistent with n. Therefore, online matching evaluates the similarity between the measured magnetic intensities and the fingerprint sequence, while the positions in the fingerprint defined by Eq. (7.2) will not be involved in the matching process. However, if one piece of sequential magnetic fingerprint is recognized, the $n - th$ position will be output as the positioning result.

Since measuring the magnetic sequence only requires walking along the planned routes, building the sequential magnetic database offers time advantages over generating a discrete magnetic database, which requires repeated measurement works at reference positions. However, the construction of a sequential magnetic database is restricted by users' walking patterns and speeds. To address this problem, Kuang et al. [19] and Asraf et al. [20] proposed to construct sequential magnetic data by connecting discrete magnetic features at reference points. Although this approach is free from moving along the planned roads, it diminishes the specificity of magnetic sequence and lacks adaptability to varying walking speeds.

Among online matching algorithms of SBMP, dynamic time warping (DTW) [21, 22], waveform-based DTW [23], and least squares method (LS) [11] are popular methods for evaluating sequence similarity. In [2], a sequential least-squares approximation method is used to match real-time measured magnetic data with stored signatures. DTW and LS are easy to implement but DTW faces high time costs with large datasets. Other efforts like the Gauss–Newton iterative (GNI) method [19], binary grid (BG) [24], leader–follower mechanism [25], and bags of words (BOW) [26] have been made. These methods try to find connections between magnetic sequences and positions by processing sequence data. The sequential magnetic data describes the data patterns of the specific locations or routes. Therefore, the specific patterns provide favorable conditions for machine learning. For instance, in [27], convolutional neural networks (CNN) is adopted to learn relationships between magnetic patterns and indoor positions, achieving localization accuracy better than 1.01 m in 75% of the cases. Moreover, recurrent neural networks (RNN) [28], deep recurrent neural networks (DRNN) [29], and probabilistic neural networks (PNN) [30] are also introduced for SBMP, yielding promising positioning results. Although these works demonstrate good performance using machine learning models for sequence matching, the problem of the high time cost should be always considered for practical application.

It is observed that classic distance-based matching in SPMP is inadequate for high-precision positioning due to limited magnetic information. Although the machine learning-based SBMP models achieve promising results, the high algorithm complexity would not be suited for implementation on mobile phones. To tackle these problems, this chapter will concentrate on the optimizations of the SBMP and SPMP methods. Additionally, the chapter will investigate the factors influencing magnetic data measurements and compensation methods (explained in Sect. 7.2). To address the limitations of distance-based SPMP and the high algorithm complexity of the machine learning-based SPMP, we introduce the mind evolutionary algorithm (a machine learning method, described in Sect. 7.3) for SPMP and the enhanced genetic algorithm-based extreme learning machine (EGA-ELM, discussed in Sect. 7.4) for SBMP, respectively.

7.2 Magnetic Signal Measurement and Processing Methods

7.2.1 Influence Factors Analysis of Magnetic Field Data

Analyzing the influence of factors on magnetic signal measurements can enhance the performance of magnetic positioning. Experiments were thus conducted to investigate the impacts of different kinds of factors on magnetic measurements, such as

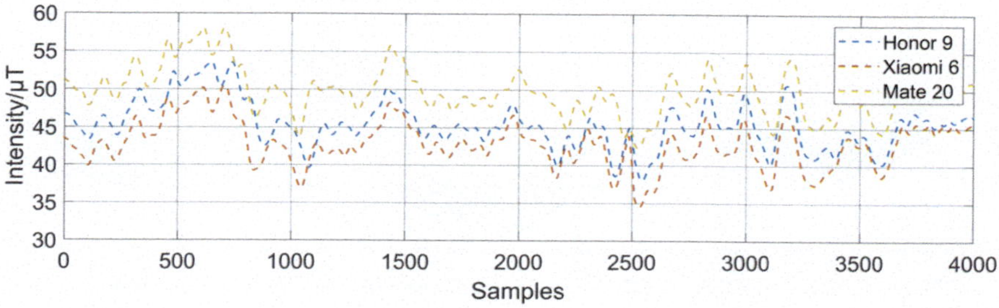

Fig. 7.2 Magnetic sequence data measured by different brands of phones

equipment, attitudes, dates, and diverse indoor media (iron, wood, electronic devices, etc.).

(1) Different mobile phones. The magnetic data measured by one type of mobile phone is usually different from that measured by another type of mobile phone. Even for different mobile phones of the same type, variations still exist due to production differences. Figure 7.2 shows the magnetic sequences measured by three mobile phones along the same indoor route. It can be seen that there are significant differences in the recorded geomagnetic curves. Taking the sequence measured by Honor 9 as the reference, the other two curves display a uniform upward or downward shift. If the magnetic sequence database is constructed based on the measurements of Honor 9, leveraging other mobile phones to perform magnetic positioning will introduce substantial errors. However, the three magnetic curves in Fig. 7.2 still reveal consistent changing tendencies, demonstrating that performing sequence matching using the changing trend of the magnetic sequence will be better than directly using discrete individual magnetic measurements.

(2) Different attitudes of devices. The 3-axis magnetic readings are recorded in the device frame, influenced by the manner in which individuals hold their devices. Within the same area, Fig. 7.3 presents the generated 3-axis magnetic maps corresponding to three distinct device attitudes (30°, 45°, and 60°, respectively). It can be found that identical geomagnetic component maps exhibit distinct patterns across varying device postures. Even in the same area, the same particular component showcases significant fluctuations due to different device orientations. This suggests that constructing a magnetic database for positioning based on a fixed device orientation can not accommodate the diversity of phone-holding styles. Therefore, device orientation raises a crucial challenge that demands primary consideration during both offline database construction and online positioning, which need magnetic field data independent from device attitudes.

(3) Long-time observations. Earth's magnetic field is susceptible to solar influence, which also in turn impacts the indoor magnetic field. This phenomenon results from the amalgamation of the earth's magnetic field and the magnetization generated by indoor magnetic materials. As illustrated in Fig. 7.4, within the same indoor area, the maps of the total magnetic intensity were generated with an 18 month interval. During this time interval, the structure of the testing area is fixed. Figure 7.4 reveals

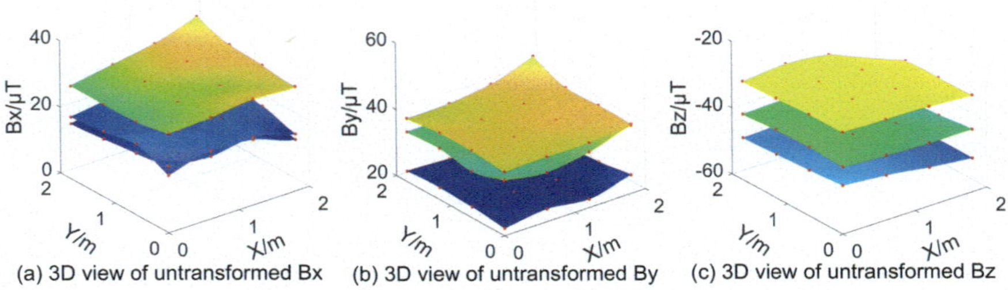

Fig. 7.3 The magnetic maps of the 3-axis components of the magnetic field under three attitudes within the same area

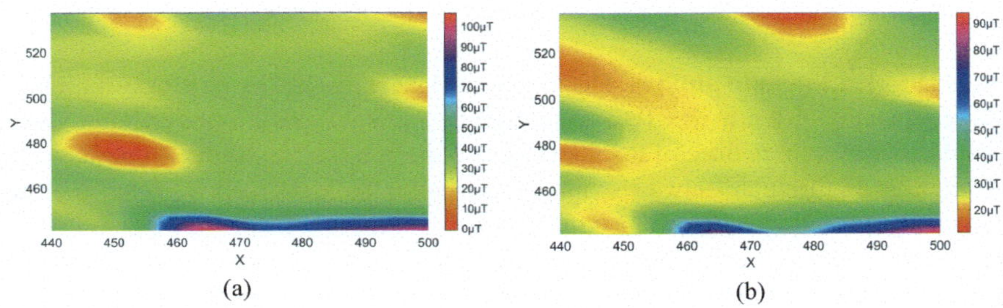

Fig. 7.4 The magnetic map of the same area with an 18-month interval. **a** map first generated; **b** map generated after 18 months. The units of the axes in the figure are decimeters

alterations in the magnetic field's strength after 18 months. Therefore, due to the fluctuation of the magnetic field's strength, it is necessary to regularly update the magnetic database for accurate localization.

(4) Magnetic sequences measured on different dates. To observe the temporal variations in magnetic sequences, the total intensities were measured on three different dates with one-month interval using the same device along the same indoor route. As depicted in Fig. 7.5, the measured magnetic intensity sequences exhibit similar distributions to those in Fig. 7.2. This observation indicates that although the magnetic sequence has better differentiation compared to the discrete magnetic features, the magnetic intensity sequences still dynamically change over time because the magnetic intensity readings have been changed, as observed in Fig. 7.4. This kind of sequential change exhibits periodic upward or downward shifts, yet the consistent changing trend of magnetic strength along the route remains for different devices (see Fig. 7.2) and dates.

(5) Magnetic sequences measured by different users. In real-life scenarios, users have different heights, leading to different heights of the device relative to the ground surface during data collection. Figure 7.6 shows the magnetic sequences collected by three users with heights of 1.92, 1.82, and 1.60 m, respectively. All three users walked along the same indoor route using the same device and very

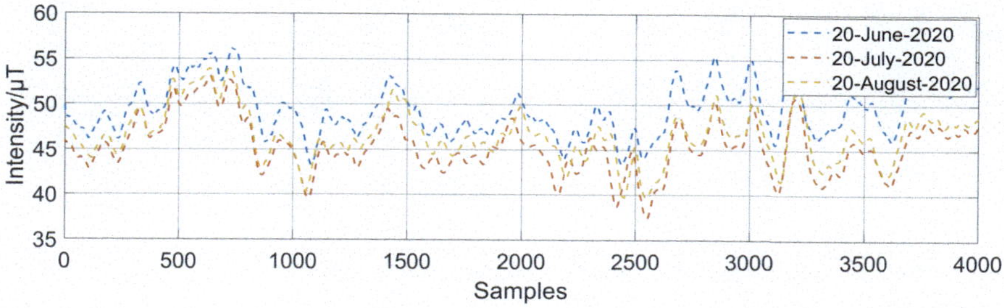

Fig. 7.5 Magnetic sequence data measured on different dates

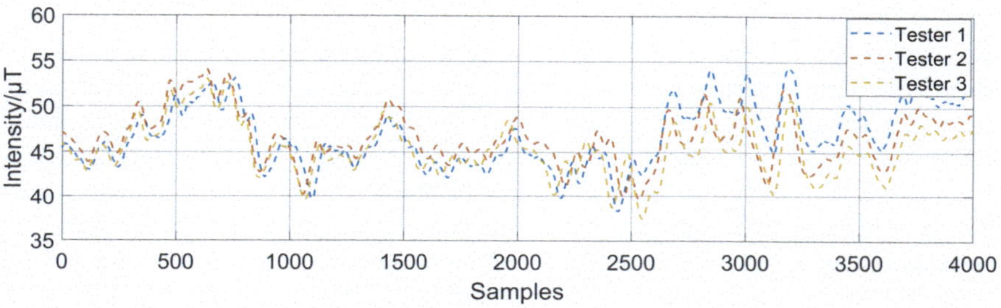

Fig. 7.6 Magnetic sequence data measured by different users

similar holding posture. It can be seen that the magnetic curves exhibit a consistent changing trend among users. However, similar to the patterns in Figs. 7.2 and 7.5, magnetic sequences obtained at different heights display very similar variations trends, exhibiting an overall upward or downward shift. Therefore, inferring from the changing trend of magnetic curves would yield better results than direct utilization of measured magnetic data. This inference applies across varying dates and devices.

(6) Different materials. An indoor scenario such as an office building has various devices or materials. Figure 7.7 illustrates the investigated materials including wood furniture, iron cabinets, fire hydrants, concrete walls, computers, and power supply equipment. An experiment was conducted to collect magnetic data near these materials. Initially, magnetic data were collected at a fixed distance from the materials (e.g., 1 m) for 100 s at a sampling rate of 50 Hz. Subsequently, the mobile phone is brought closer to the materials for another 100 s measurement, followed by returning to the original position for an additional 100 s of data collection. This approach aims to identify materials that impact the stability of the magnetic field, aiding in determining whether compensation or mitigation is necessary for positioning when the user is near any of these materials.

Figure 7.8a illustrates the magnetic intensity remains relatively stable when a person moves in close proximity to the phone. In Fig. 7.8b, as the phone approaches the wood material, there are no discernible fluctuations in the measurement sequence. This indicates that both wood material and the human body have negligible influence

(a) (b) (c) (d) (e) (f)

Fig. 7.7 The investigated materials. **a** wood furniture, **b** iron cabinet, **c** fire hydrant, **d** concrete wall, **e** computer, **f** power supply equipment

on the intensity of the magnetic field. However, the measured magnetic intensities decrease sharply when the mobile phone approaches iron cabinets, fire hydrants, and power supply devices within a certain distance, as shown in Fig. 7.8c, d, and g, respectively. Conversely, when the phone nears a concrete wall or computer hardware, the magnetic intensity shows a sudden increase, as shown in Fig. 7.8e and f, respectively. These experimental results reveal that common indoor materials have a certain range of influence on geomagnetic measurements. Once the mobile phone is within a certain range of some materials, the measured magnetic intensity undergoes significant change, but there is no effect when the phone is outside this range. This result emphasizes the necessity of considering intensity fluctuations when locating the user who is near such materials.

(7) Summary of influencing factors. The impact of several factors and patterns on indoor magnetic field intensity is summarized as follows:

① The indoor magnetic field intensity exhibits variability over time. These changes are related to Earth's inherent characteristics of geomagnetic field and ferromagnetic materials inside buildings. However, in structurally stable indoor buildings, the interrelations of magnetic field intensities among various points within the area remain consistent.

② Factors such as device height, dates, and hardware models can introduce overall deviations in magnetic field measurements. Variations in device attitudes lead to dynamic changes in the three-axis components. However, compared to the fluctuations in the three-axis components, the total intensity of the geomagnetic field is relatively stable.

③ Non-ferrous materials such as wood, and the human body do not generate influences on geomagnetic measurements. On the other hand, concrete walls, iron materials, electronic devices, and similar ferromagnetic elements can make impacts. However, this impact is local and momentary. It becomes negligible when moving away from the magnetic materials' impact range (e.g., beyond 1 m), having no bearing on the measurement of magnetic field data.

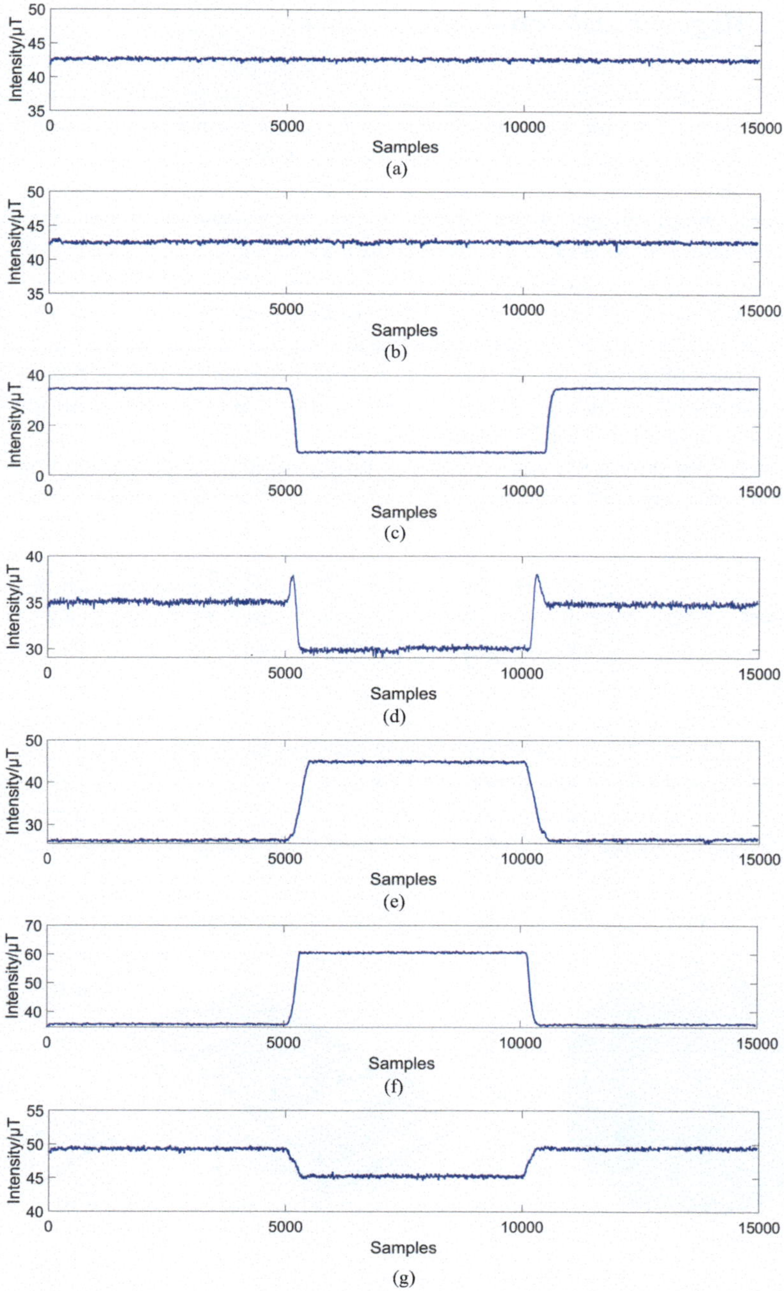

Fig. 7.8 The magnetic intensity readings influenced by different materials. **a** human **b** wood, **c** iron cabinet, **d** fire hydrant, **e** concrete wall, **f** computer, **g** power supply equipment

7.2.2 Magnetic Data Processing Methods

Section 7.2.1 introduced the dynamic changes of the magnetic field measurements. Considering the dynamic nature of the data, compensation is necessary before online positioning. This section will present methods for compensating different types of magnetic data.

Discrete magnetic data transformation. Figure 7.9 shows the geographic coordinate system (GCS) and the carrier coordinate system (CCS). Generally, the measured 3-axis components are in the CCS, which varies with the holding styles of phones. Since GCS is a system of latitude, longitude and altitude coordinates, the 3-axis components of the magnetic field in the GCS will be independent of the holding styles of phones. Therefore, using the magnetic data in the GCS will introduce fewer errors because the data is more stable than that in the CCS. This involves the transformation between GCS and CCS.

As Fig. 7.9b shows, the angles at which the smartphone rotates around the x, y, and z axes are called pitch, roll, and yaw, respectively. The rotation matrix from GCS to CCS is defined as:

$$\boldsymbol{C}_g^b = \boldsymbol{C}_n^3\,\boldsymbol{C}_n^2\boldsymbol{C}_n^1 = \begin{bmatrix} \cos\gamma & 0 & -\sin\gamma \\ 0 & 1 & 0 \\ \sin\gamma & 0 & \cos\gamma \end{bmatrix} \begin{bmatrix} 1 & 0 & 0 \\ 0 & \cos\theta & \sin\theta \\ 0 & -\sin\theta & \cos\theta \end{bmatrix} \begin{bmatrix} \cos\psi & -\sin\psi & 0 \\ \sin\psi & \cos\psi & 0 \\ 0 & 0 & 1 \end{bmatrix} \tag{7.3}$$

where $\boldsymbol{C}_n^1$, $\boldsymbol{C}_n^2$, and $\boldsymbol{C}_n^3$ represent the rotation matrices along the z, x, and y axes, $\boldsymbol{C}_g^b$ is an orthogonal matrix and denotes the rotation from GCS to CCS, respectively. Equation (7.3) reveals the rotation sequence as: $z \to x \to y$. To make the rotation, the three angles should be first obtained. The magnetic and gravity data are utilized

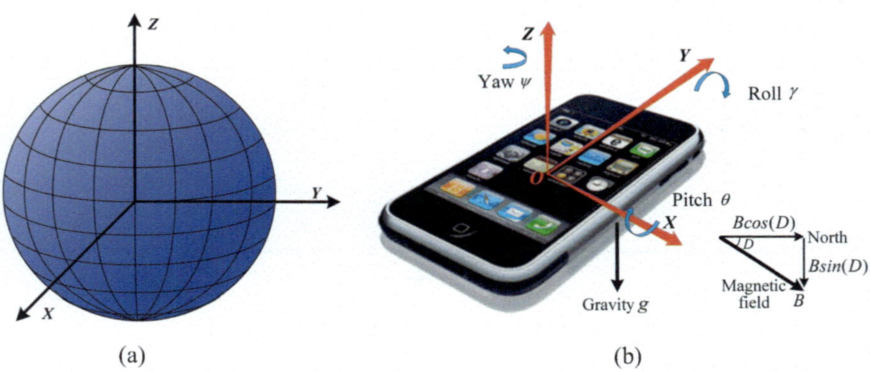

Fig. 7.9 The geographic coordinate system and carrier coordinate system. **a** geographic coordinate system; **b** carrier coordinate system

for the angle calculation, that is, the pitch and roll are computed using gravity data and the yaw is derived from magnetic data.

Since gravity always points to the center of the earth, the gravity readings can be expressed as $\boldsymbol{G} = [0, 0, g]^T$ and when the phone is placed horizontally, $g = 9.8m/s^2$. If the gravity is measured in the CCS, the gravity vector is denoted as $\boldsymbol{G}_b = [g_{bx}, g_{by}, g_{bz}]^T$. The relationship between $\boldsymbol{G}$ and $\boldsymbol{G}_b$ is as follows:

$$\begin{bmatrix} g_{bx} \\ g_{by} \\ g_{bz} \end{bmatrix} = \begin{bmatrix} \cos\gamma\cos\psi + \sin\gamma\sin\psi\sin\theta & -\cos\gamma\sin\psi + \sin\gamma\cos\psi\sin\theta & -\sin\gamma\cos\theta \\ \sin\psi\cos\theta & \cos\psi\cos\theta & \sin\theta \\ \sin\gamma\cos\psi - \cos\gamma\sin\psi\sin\theta & -\sin\gamma\sin\psi - \cos\gamma\cos\psi\sin\theta & \cos\gamma\cos\theta \end{bmatrix} \begin{bmatrix} 0 \\ 0 \\ g \end{bmatrix} \tag{7.4}$$

According to Eq. (7.4), pitch, roll, and gravity have the following relationship:

$$\begin{bmatrix} g_{bx} \\ g_{by} \\ g_{bz} \end{bmatrix} = \begin{bmatrix} -\sin\gamma\cos\theta \\ \sin\theta \\ \cos\gamma\cos\theta \end{bmatrix} g \tag{7.5}$$

Therefore, pitch and roll can be calculated by using gravity data as follows:

$$\begin{cases} \gamma = atan(-g_{bx}/g_{bz})\gamma \in [-\pi, \pi] \\ \theta = arcsin(g_{by}/g)\theta \in [-\pi/2, \pi/2] \end{cases} \tag{7.6}$$

Assuming the 3-axis magnetic data in the CCS and GCS are $\boldsymbol{B} = [b_x, b_y, b_z]^T$ and $\boldsymbol{B}_g = [b_{gx}, b_{gy}, b_{gz}]^T$, the yaw is computed as:

$$\psi = -arctan\left(\frac{sin\gamma sin\theta b_x + cosb_y - cos\gamma sin\theta b_z}{cos\gamma b_x + sin\gamma b_z}\right)\psi \in [0, 2\pi] \tag{7.7}$$

Based on (7.6), (7.7), magnetic data, and gravity data, the rotation angles and $\boldsymbol{C}_g^b$ can be obtained. The relationship between $\boldsymbol{B}$ and $\boldsymbol{B}_g$ is defined as:

$$\boldsymbol{B} = \boldsymbol{C}_g^b \boldsymbol{B}_g \tag{7.8}$$

which describes the transformation of the magnetic data from GCS to CCS. However, the required transformation for magnetic positioning is from CCS to GCS. According to the rules of the matrix operations, the following equation can be obtained:

$$(\boldsymbol{C}_g^b)^{-1}\boldsymbol{B} = (\boldsymbol{C}_g^b)^{-1}\boldsymbol{C}_g^b\boldsymbol{B}_g = \boldsymbol{B}_g \tag{7.9}$$

where $(\boldsymbol{C}_g^b)^{-1}$ denotes the inverse matrix of $\boldsymbol{C}_g^b$. Since $\boldsymbol{C}_g^b$ is an orthogonal matrix, we obtain the rotation matrix from CCS to GCS as follows:

$$\boldsymbol{C}_b^g = (\boldsymbol{C}_g^b)^{-1} = (\boldsymbol{C}_g^b)^T \tag{7.10}$$

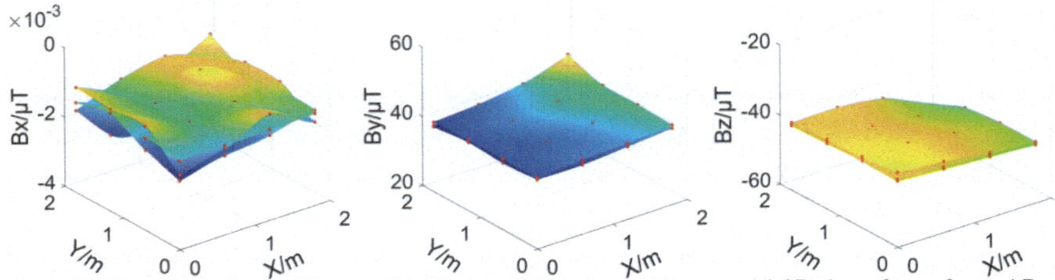

(a) 3D view of transformed Bx (b) 3D view of transformed By (c) 3D view of transformed Bz

Fig. 7.10 The 3-axis magnetic components transformed into the geographic coordinate system

Based on Eqs. (7.3) and (7.10), the transformation is made as:

$$\boldsymbol{B}_g = (\boldsymbol{C}_g^b)^T \boldsymbol{B} = \begin{bmatrix} \cos\gamma\cos\psi + \sin\gamma\sin\psi\sin\theta & \sin\psi\cos\theta & \sin\gamma\cos\psi - \cos\gamma\sin\psi\sin\theta \\ -\cos\gamma\sin\psi + \sin\gamma\cos\psi\sin\theta & \cos\psi\cos\theta & -\sin\gamma\sin\psi - \cos\gamma\cos\psi\sin\theta \\ -\sin\gamma\cos\theta & \sin\theta & \cos\gamma\cos\theta \end{bmatrix} \begin{bmatrix} b_x \\ b_y \\ b_z \end{bmatrix} \quad (7.11)$$

Figure 7.10 presents the transformed maps of the 3-axis components in Fig. 7.3. It can be seen that the intensity of the x component is nearly zero ($\times 10^{-3}$ μT), and the maps of y and z components illustrate the same overlapping maps, which are different from the separate maps in Fig. 7.3. Therefore, utilizing the above process for discrete magnetic data compensation can produce more stable 3-axis data and standard magnetic maps for online positioning.

Deviation mitigation of magnetic sequence data. As stated in Sect. 7.2.1, the sequential magnetic data has an overall deviation caused by factors like different mobile phones, dates, testers, etc. For a planned indoor route, although the measured magnetic patterns have differences, the same changing trend independent of the influence factors exists. Therefore, extracting this trend for online positioning is important for high-precision localization. We propose to use the wave-like features of the original magnetic sequences. If a magnetic sequence $\boldsymbol{S}$ has n samples, we construct a new data sequence, termed slope data $\boldsymbol{S}'$, by:

$$\boldsymbol{S}'(i) = \begin{cases} (S(i+m) - S(i))/m, i+m \leq n \\ (S(n) - S(i))/(n-i), i+m > n \& i < n \end{cases} \quad (7.12)$$

where m is a constant, recommended to be set at one-fifth of the sampling frequency. Figure 7.11 shows the extracted slope data from magnetic sequences measured on different dates or using different phones. Compared to Fig. 7.2 and Fig. 7.5, it is evident that the deviations observed in the original magnetic patterns have been effectively reduced by extracting the slope data. Therefore, before conducting sequence-based magnetic positioning, deviation mitigation using Eq. (7.12) is suggested.

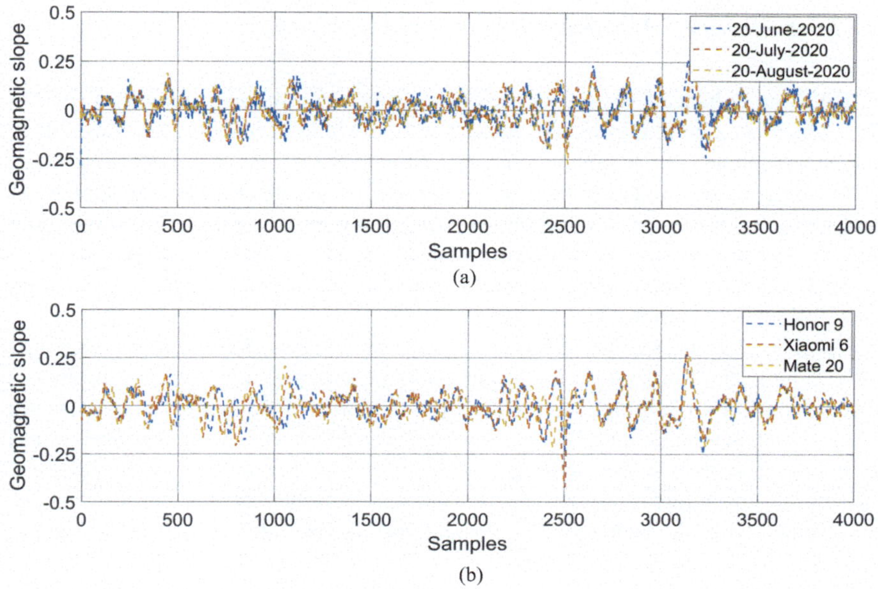

Fig. 7.11 The extracted magnetic slope curves. **a** different dates; **b** different mobile phones

7.3 SPMP Based on an Enhanced Mind Evolutionary Algorithm

The SPMP methods (e.g., KNN, MSD, etc.) face challenges in accurately evaluating similarities due to similar magnetic readings being captured at different indoor locations. To address this issue, we propose utilizing the enhanced mind evolutionary algorithm (EMEA) for location estimation. Unlike traditional distance-based methods, EMEA employs a global search strategy to find optimal positioning results rather than directly assessing the Euclidean distance between the real-time measured magnetic data and the magnetic fingerprint. Based on the high sampling rate of the magnetometer, a large amount of geomagnetic data can be gathered within one second, providing favorable conditions for applying learning algorithms. Theoretically, within this data, there should always be a piece of collected magnetic data that closely matches the ground-truth position. Following this principle, EMEA is employed to search for the optimal position using all the collected magnetic data.

The following parts will introduce the theory behind EMEA and define the EMEA-based SPMP. Experimental tests are presented to show the advantages of EMEA-based SPMP against distance-based methods.

7.3.1 Enhanced Mind Evolutionary Algorithm

EMEA is an enhanced version of the classic mind evolutionary algorithm (MEA) [31], which is a popular optimization method in the field of machine learning. MEA has the advantages of fast convergence and better-optimizing capability. It emulates the learning modes and activities of the human mind and utilizes a groupization strategy to find the optimal solution. MEA needs a population to perform searching and its learning process is realized by similartaxis and dissimilation operators, which divide the population into several superior and temporary subgroups, allowing competition and evolution among them.

Figure 7.12 shows that MEA has keywords: *similartaxis*, *dissimilation*, *local billboard*, *global billboard*, *superior group*, and *temporary group*. The search process begins with initializing and evaluating a population with N individuals, assigning scores based on their adaptability within the search space. A scoring function should be defined to deliver scores. The individuals with higher scores become centers of superior groups, while those with lower scores are regarded as the centers of temporary groups. After the centers are confirmed, the superior and temporary groups are generated. Then, the evolution starts.

Similartaxis. During this step, local competition is first executed within subgroups. The best individual posts its score on the local billboard. As competition proceeds, the best individual of one subgroup also dynamically changes. One subgroup is mature if the following condition is satisfied:

$$|S_{\max} - S_{\min}| \leq \varsigma \tag{7.13}$$

where $S_{\max}$ and $S_{\min}$ denote the maximal and minimal scores of the subgroup, respectively, and ς is a threshold. If one subgroup is mature, the highest score will be posted on the global billboard. The local competitions within all the subgroups are executed until all the subgroups mature.

Dissimilation. This step is also called global competition. It involves comparing scores on the global billboard to determine which subgroups to discard or regenerate. The superior subgroup with the lowest score will be replaced by a temporary subgroup with a better score. A new temporary subgroup will be regenerated in the search space.

Similartaxis and dissimilation are operated independently, with the global billboard capturing evolutionary information from each generation, steering the process

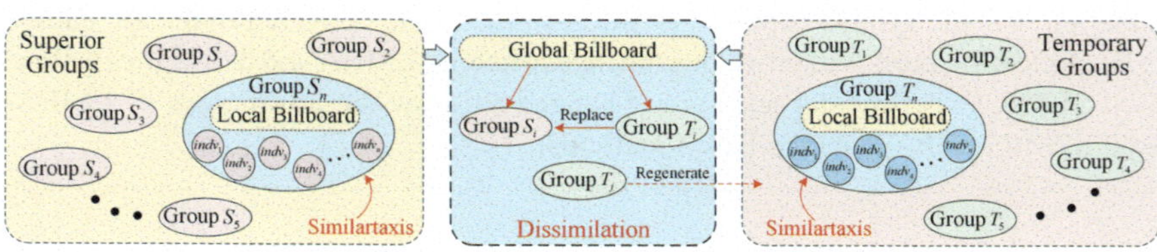

Fig. 7.12 Mind evolutionary algorithm

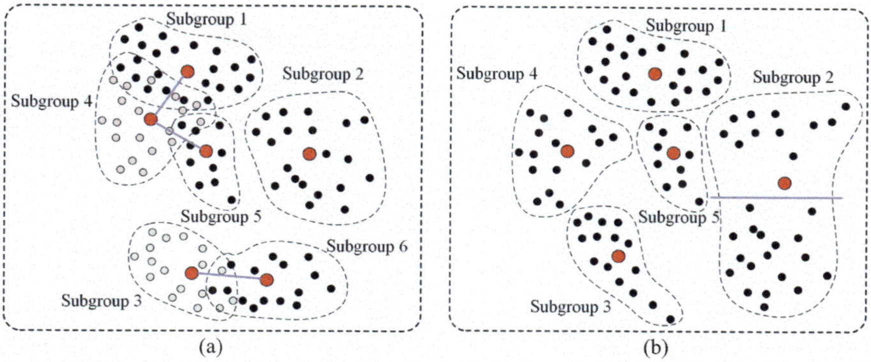

Fig. 7.13 The possible distributions of initial generated subgroups of MEA

toward an optimal direction. The evolutionary process finishes if the number of iterations reaches the predefined number or no superior subgroup is replaced. Under this condition, the best individual of the best superior subgroup is output as the searching result.

The performance of MEA depends on the number of subgroups, which is defined before the evolution starts. However, after the subgroups are generated, the subgroups may intersect (Fig. 7.13a) or the individuals of one subgroup are relatively scattered (Fig. 7.13b). In these cases, fixed numbers of superior or temporary subgroups will not cover all search space, which will be detrimental to the optimal search. Therefore, the enhanced MEA is proposed to address these shortcomings.

EMEA dynamically assigns the numbers of superiors or temporary subgroups using variance calculation and center control. The definitions are as follows:

Variance calculation. After generating the subgroups, the variance of individuals within one subgroup should be evaluated. If a subgroup is expressed as $\{z_i, i = 1, 2, 3, \ldots, n\}$, the variance is calculated as:

$$var = \frac{1}{n}\sum_{i=1}^{n}(z_i - \overline{z})^2 \tag{7.14}$$

where $\overline{z}$ is the mean value of $\{z_i\}$, which are the characteristics of the individuals. A large variance suggests distributions similar to those in Fig. 7.13b, indicating a need for subdivision of that subgroup.

Center control. Once the subgroup centers are confirmed, the distance between each pair of centers should be calculated. This helps determine whether a pair of subgroups are too close or intersect, as shown in Fig. 7.13a. In this case, the algorithm combines two subgroups and generates a new group in the search space.

To perform MEA optimization, both variance calculation and center control need to define thresholds empirically. For example, in the case of magnetic positioning, coordinates can be considered as the characteristic of individuals. Therefore, the

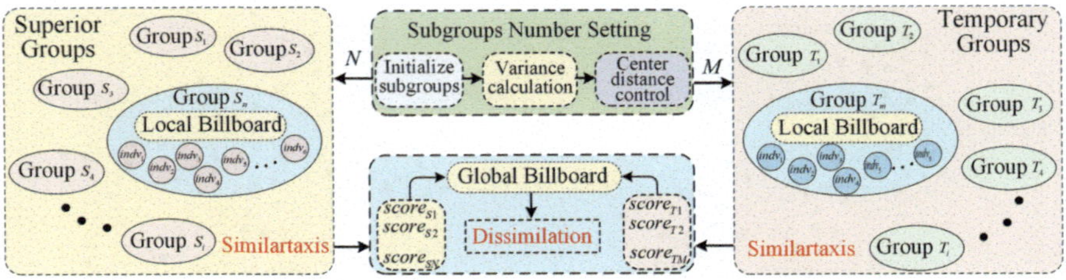

Fig. 7.14 Enhanced mind evolutionary algorithm

optimization work is executed by calculating the variance of individuals' coordinates and the plane distance of the centers of two subgroups. The calculated variance and plane distance of a pair of subgroups' centers should be compared with the predefined thresholds to decide whether to execute division or merging operations. This process continues until no further subdivision or combination of subgroups occurs. Finally, the numbers of superior or temporary subgroups are assigned as follows:

$$T = M + N \tag{7.15}$$

$$\begin{cases} N = N + 1, s_{\text{imax}} < \delta \\ M = M + 1, s_{\text{imax}} > \delta \end{cases} \tag{7.16}$$

where T, M, N denote the numbers of total, temporary, and superior subgroups, s_{imax} is the maximal score of the ith group, δ is a predefined threshold, respectively.

As illustrated in Fig. 7.14, EMEA independently determines the numbers of superior and temporary subgroups through the above strategies, then followed by executing similartaxis and dissimilation for evolution, ultimately leading to the final optimal result.

7.3.2 Definition of the Localization Model

To establish an EMEA-based magnetic positioning model, aligning the evolutionary process of EMEA with the positioning process is crucial. As previously mentioned, the high sampling rate of the magnetometer (e.g., 100 Hz) generates a consistent frequency of geomagnetic data. Leveraging this data, numerous temporary magnetic positions are obtained, serving as the initial population for EMEA. Consequently, the first step of the localization model, "initialization," is finished.

As seen from Fig. 7.15, in the context of the localization problem at the time k, with a sampling rate of n, there exist n temporary magnetic positions expressed as:

$$\boldsymbol{G}(k) = \{g_k(x_1, y_1), \ldots, g_k(x_i, y_i), \ldots, g_k(x_n, y_n)\} \tag{7.17}$$

Fig. 7.15 Schematic diagram of the EMEA-based SPMP within two consecutive moments

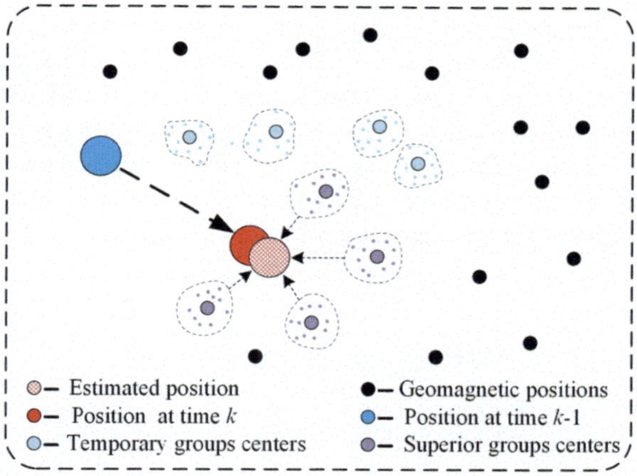

where $\boldsymbol{G}(k)$ represents the population of EMEA, $g_k(x_i, y_i)$ is the ith individual of $\boldsymbol{G}(k)$, x_i and y_i are the coordinates of the ith individual, $i \in \{1, 2, \ldots, n\}$. The following step is to score individuals. For single magnetic localization, scoring is related to the previous true magnetic position $\boldsymbol{M}(k-1)$, as shown in Fig. 7.15. The ith individual's score is calculated as:

$$s\{k, i\} = \frac{1}{\sqrt{(X_{k-1} - x_i)^2 + (Y_{k-1} - y_i)^2}} \tag{7.18}$$

where (X_{k-1}, Y_{k-1}) denotes the coordinates of $\boldsymbol{M}(k-1)$. As defined in Sect. 7.3.1, the individuals with the highest scores are selected as the centers of subgroups. Then, the variance calculation is made by calculating the coordinates variance of the subgroups:

$$var = \frac{1}{n}\sum_{i=1}^{n}\left[(x_i - \overline{x})^2 + (y_i - \overline{y})^2\right] \tag{7.19}$$

where $\overline{x}$ and $\overline{y}$ represent the mean coordinates of the individuals. If the variance surpasses a predefined threshold, a subgroup is divided. During our test, the threshold of variance is 0.5. The next step is center control. With the coordinates of the selected centers, the distance between two subgroups is computed as:

$$d = \sqrt{(x_{ci} - x_{cj})^2 + (y_{ci} - y_{cj})^2} \tag{7.20}$$

where (x_{ci}, y_{ci}) and $\left(x_{cj}, y_{cj}\right)$ represent the positions of the ith and jth centers, respectively. Two close subgroups will be combined if their center distance is below a certain threshold, which is set as 1 m.

As mentioned in Sect. 7.3.1, variance calculation and center control iteratively proceed until no further subgroup divisions or combinations occur. After that, as shown in Fig. 7.14, EMEA utilizes Eq. (7.16) to assign the numbers of superior and temporary subgroups. Then, the evolution starts by performing similartaxis and dissimilation until no superior subgroup is replaced or reaches the maximal iteration number. The best individual with the highest score of the superior subgroup is output as the estimated magnetic position.

7.3.3 SPMP Experimental Results

To evaluate EMEA's feasibility for magnetic positioning, testing was made at China University of Mining and Technology (CUMT), Xuzhou. The magnetic database is constructed in the form of Eq. (7.1). The 3-axis components of magnetic field data are transformed to GCS through coordinate transformation as described in Sect. 7.2.2. Comparative evaluations included state-of-the-art approaches such as KNN [10], multi-magnetic fingerprint fusion (MMFF) [12], and mean square differences (MSD) [11], alongside the MEA and EMEA-based magnetic positioning methods. The localization results are shown in Fig. 7.16.

The red lines in the error boxes reveal the median errors of different approaches. The median errors of distanced-based algorithms (KNN, MSD, and MMFF) are greater than 2 m. Generally, distance-based methods take the mean value of the measured magnetic data for positioning calculation. Although this operation can reduce the influence of random errors and measurement noise, it also loses the diversity of data. Therefore, mismatching always occurs, leading to poor positioning accuracy. This can be validated in Fig. 7.16, where large errors are always generated using the three distance-based methods.

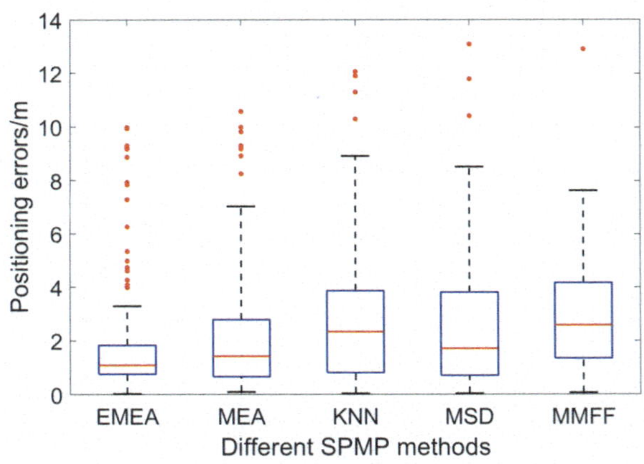

Fig. 7.16 Geomagnetic positioning performance comparison of different SPMP methods

Conversely, both MEA-based SPMP and EMEA-based SPMP obtain a median error of within 2 m, outperforming distance-based approaches. Moreover, the EMEA-based SPMP achieves better results than that of the MEA-based method. By utilizing all magnetic data for localization without averaging, EMEA-based SPMP minimized large positioning errors, affirming the efficacy of evolutionary algorithms for magnetic positioning optimization. The experimental results demonstrate the feasibility and efficiency of employing evolutionary algorithms for superior magnetic positioning performance compared to traditional distance-based methods.

7.4 SBMP Using an Enhanced Genetic Algorithm-Based Extreme Learning Machine

Sequence-based magnetic positioning (SBMP) often outperforms SPMP due to the higher-dimensional and more discriminative nature of magnetic sequence features. SBMP involves pattern identification by evaluating similarity rather than the individual magnetic signal strength. For example, DTW method [22, 32] estimates locations by selecting the position corresponding to the most similar magnetic pattern. Utilizing magnetic patterns for localization allows for the application of machine learning models that establish connections between magnetic sequences and positions. However, compared to state-of-the-art models (like CNN [27], RNN [29], etc.), these learning models always suffer from high computational complexity, so they may not be suited for some practical applications. To address this complexity issue, this chapter proposes using an extreme learning machine (ELM) for SPMP, aiming to mitigate algorithmic complexity issues.

The following parts introduce the theory of ELM and the optimization method based on an enhanced genetic algorithm (EGA). Experimental tests are presented to show the advantages of EGA-based ELM over state-of-the-art models.

7.4.1 Extreme Learning Machine

As shown in Fig. 7.17, an extreme learning machine (ELM) operates with a simplified structure, featuring only an input layer, a single hidden layer, and an output layer. Unlike the complex structures of CNN or RNN models, ELM utilizes a single layer for learning, ensuring low time cost. Constructing an ELM involves the definition of three key parameters: $\boldsymbol{\omega}$, b, $\boldsymbol{\beta}$. The connection between the input layer and the hidden layer is determined by the weight vector $\boldsymbol{\omega}$ and the activation threshold b. The hidden and output layers are linked via the weight vector $\boldsymbol{\beta}$. Given the condition that an ELM model has n input nodes and l hidden nodes, with input data $\boldsymbol{X} = \left[\boldsymbol{x}_1, \boldsymbol{x}_2, \ldots, \boldsymbol{x}_Q\right]^T$, $\boldsymbol{x}_{\rm j} = \left[x_{1j}, x_{2j}, \ldots, x_{nj}\right]^T \in \boldsymbol{R}^n$, and output data $\boldsymbol{Y}=\left[\boldsymbol{y}_1, \boldsymbol{y}_2, \ldots, \boldsymbol{y}_Q\right]^T$, $\boldsymbol{y}_{\rm j} = \left[y_{1j}, y_{2j}, \ldots, y_{\rm mj}\right]^T \in \boldsymbol{R}^m$, the model is expressed by the

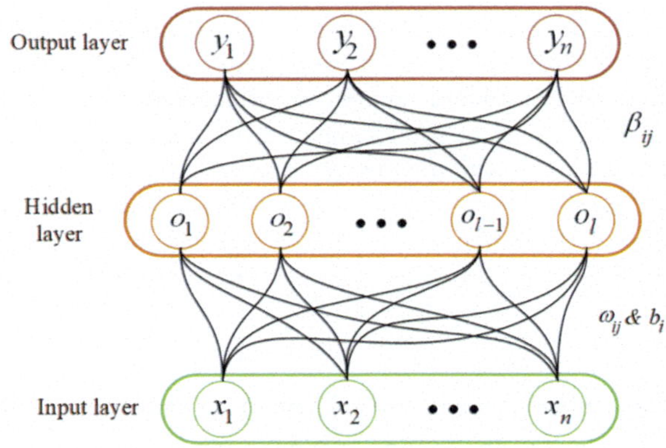

Fig. 7.17 The structure of extreme learning machine

following equation:

$$y_j = \sum_{i=1}^{l} \boldsymbol{\beta}_{ij} G_i(\boldsymbol{\omega}_i, b_i, \boldsymbol{X}), j = 1, 2, \ldots Q \tag{7.21}$$

where $\boldsymbol{\beta}_i = [\beta_{i1}, \beta_{i2}, \ldots \beta_{il}]^T$, $\boldsymbol{\omega}_i = [\omega_{1i}, \omega_{2i}, \ldots, \omega_{ni}]^T$, and $G_i(\boldsymbol{\omega}_i, b_i, \boldsymbol{X})$ is the output of the $i-th$ hidden node, respectively. $G_i(\boldsymbol{\omega}_i, b_i, \boldsymbol{X})$ is an entry of the hidden layer's output matrix $\boldsymbol{G}$, which is described as:

$$\begin{aligned} &\boldsymbol{G}(\boldsymbol{\omega}_1, \boldsymbol{\omega}_2, \ldots, \boldsymbol{\omega}_l, b_1, b_2, \ldots, b_l, \boldsymbol{x}_1, \boldsymbol{x}_2, \ldots, \boldsymbol{x}_Q) \\ &= \begin{bmatrix} f(\boldsymbol{\omega}_1 \bullet \boldsymbol{x}_1 + b_1) f(\boldsymbol{\omega}_2 \bullet \boldsymbol{x}_1 + b_2) \cdots f(\boldsymbol{\omega}_l \bullet \boldsymbol{x}_1 + b_l) \\ f(\boldsymbol{\omega}_1 \bullet \boldsymbol{x}_2 + b_1) f(\boldsymbol{\omega}_2 \bullet \boldsymbol{x}_2 + b_2) \cdots f(\boldsymbol{\omega}_l \bullet \boldsymbol{x}_2 + b_l) \\ \vdots \\ f(\boldsymbol{\omega}_1 \bullet \boldsymbol{x}_Q + b_1) f(\boldsymbol{\omega}_2 \bullet \boldsymbol{x}_Q + b_2) \cdots f(\boldsymbol{\omega}_l \bullet \boldsymbol{x}_Q + b_l) \end{bmatrix}_{Q \times l} \end{aligned} \tag{7.22}$$

where $f(\bullet)$ represents the activation function.

Assuming the target labels are $\boldsymbol{T}$, where $\boldsymbol{T} = [\boldsymbol{t}_1, \boldsymbol{t}_2, \ldots, \boldsymbol{t}_Q]_{m \times Q}$, $\boldsymbol{t}_j = [t_{j1}, t_{j2}, \ldots, t_{jm}]^T \in \boldsymbol{R}^m, j = 1, 2, \ldots, Q$, the relationship between $\boldsymbol{G}$ and $\boldsymbol{T}$ is:

$$\boldsymbol{G\beta} = \boldsymbol{T}' \tag{7.23}$$

where $\boldsymbol{T}'$ is the transpose matrix of $\boldsymbol{T}$. In theory, an ELM model with Q input training data samples can approximate any desired training error ε under the condition that $f(\bullet)$ is infinitely differentiable within any interval. Therefore, the training error ε is expressed as:

$$\left\| \boldsymbol{G}_{Q \times l} \boldsymbol{\beta}_{l \times m} - \boldsymbol{T}' \right\| < \varepsilon \tag{7.24}$$

which indicates that an ELM network is defined when parameters $\boldsymbol{\omega}$, b and $\boldsymbol{\beta}$ are known. Moreover, $\boldsymbol{\beta}$ can be calculated using $\boldsymbol{\omega}$, b and $\boldsymbol{T}$ as follows:

$$\boldsymbol{\beta}=\boldsymbol{H}^{\dagger}\boldsymbol{T} \tag{7.25}$$

where $\boldsymbol{H}^{\dagger}$ is further given by:

$$\boldsymbol{H}^{\dagger}=(\boldsymbol{G}^{T}\boldsymbol{G})^{-1}\boldsymbol{G}^{T} \tag{7.26}$$

The output matrix of the hidden layer is related to $\boldsymbol{\omega}$ and b, which are involved in the calculation of the weight vector $\boldsymbol{\beta}$. Therefore, an ELM network is uniquely determined by the assigned $\boldsymbol{\omega}$ and b. However, ensuring optimal parameters remains a challenge. Better initial parameters contribute to ELM approaching the training target with smaller errors. The relationship between initial parameters and training errors can be represented as a nonlinear function $g(\bullet)$ as follows:

$$\begin{cases} g(\boldsymbol{\omega}_{11}, \boldsymbol{\omega}_{12}, \ldots, \boldsymbol{\omega}_{1n}, b_{11}, b_{12}, \ldots, b_{1n}) \rightarrow e_1 \\ g(\boldsymbol{\omega}_{21}, \boldsymbol{\omega}_{22}, \ldots, \boldsymbol{\omega}_{2n}, b_{21}, b_{22}, \ldots, b_{2n}) \rightarrow e_2 \\ \qquad \vdots \\ g(\boldsymbol{\omega}_{n1}, \boldsymbol{\omega}_{n2}, \ldots, \boldsymbol{\omega}_{nn}, b_{n1}, b_{n2}, \ldots, b_{nn}) \rightarrow e_n \end{cases} \tag{7.27}$$

Equation (7.27) denotes that $g(\bullet)$ is a nonlinear function of the independent variables $\boldsymbol{\omega}$ and b. Therefore, the problem of determining the optimal parameters is transformed into solving the extreme value of the nonlinear function.

7.4.2 Enhanced Genetic Algorithm

Genetic algorithm (GA) is a popular method for solving nonlinear problems. A classic GA includes genetic operations of *selection*, *crossover*, and *mutation*, as well as a population with N individuals. Every individual carries its chromosome that can describe its characteristics. The chromosome is made up of genes, which can be formed by methods like binary encoding [33], value encoding [34], etc. The population evolves using the three operators until reaching convergence. The speed of convergence depends on whether the optimal individual can be found quickly. As shown in Fig. 7.18, to accelerate convergence, an enhanced genetic algorithm (EGA) is introduced by simultaneously optimizing the three genetic operators.

Roulette-wheel based selection strategy. One individual is selected depending on its fitness for the search space. The probability of being selected is defined as:

$$p_{si} = \frac{g_i}{\sum_{j=1}^{N} g_j}, i, j \in \{1, 2, \ldots, N\} \tag{7.28}$$

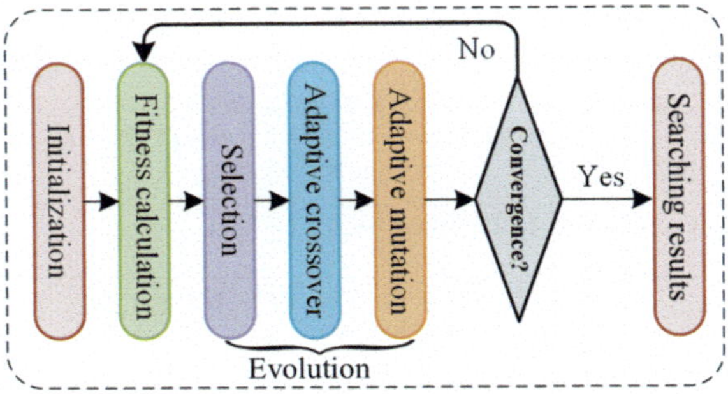

Fig. 7.18 The flowchart of the enhanced genetic algorithm

where p_{si} is the probability, g_i is the fitness of the ith individual, respectively. For every round of selection, the roulette-wheel method is performed to select several candidates. The best individual is then extracted from candidates based on their fitness values. Better individuals have a larger probability of being selected. Since a population has N individuals, the above process will be executed N times.

Adaptive crossover. This operation is made according to a crossover probability p_c, which remains constant in classic GA. Crossover exchanges genes between two individuals, aiming to produce improved offspring during evolution. To facilitate the passage of beneficial genes to the next generation, a dynamic crossover probability is calculated as follows:

$$\mathrm{p}_c = p_{\mathrm{cmax}} - \frac{n \times (p_{\mathrm{cmax}} - p_{\mathrm{cmin}}) \times \overline{\boldsymbol{F}_i} - m}{m \times \overline{\boldsymbol{F}_i}} \tag{7.29}$$

where p_{cmax}, p_{cmin}, and $\overline{\boldsymbol{F}_i}$ represent the maximal and minimal crossover probability, and the average fitness of the ith generation of the population; and m and n denotes the maximal and current numbers of evolutions, respectively. Equation (7.29) links the current crossover probability with the mean fitness of the population and the number of evolution. Given the following crossover condition:

$$p_c \le r, p_c, r \in [0, 1] \tag{7.30}$$

where r is a random number, which is used to control whether crossover should be executed. If Eq. (7.30) is satisfied, the crossover is operated. Based on (7.29) and (7.30), it can be found that crossover will be easier to make as evolution progresses because p_c gradually decreases.

Adaptive mutation. This operation is to produce new genes according to a mutation probability p_m, which also remains constant in classic GA. Initially, mutation generates new genes that can help GA to search in more directions. However, gene mutation can make bad genes, slowing down convergence or causing divergence. At the later stage of evolution, GA tends to find optimal search directions and better individuals are within the population. Under such conditions, reducing the frequency of mutation is necessary to ensure quick convergence of the algorithm. Similar to the adaptive crossover probability, the adaptive mutation probability is defined as:

$$\mathrm{p}_m = p_{\mathrm{mmax}} + \frac{n \times (p_{\mathrm{mmax}} - p_{\mathrm{mmin}}) \times \overline{\boldsymbol{F}_i} - m}{m \times \overline{\boldsymbol{F}_i}} \tag{7.31}$$

where p_{mmax}, p_{mmin}, and $\overline{\boldsymbol{F}_i}$ represent the maximal and minimal mutation probability, and the average fitness of the ith generation of the population, and m and n denotes the maximal and current numbers of evolutions, respectively. Given the mutation condition:

$$p_m \leq t, p_m, t \in [0, 1] \tag{7.32}$$

where t is randomly assigned and determines whether mutation should be performed. As evolution proceeds, Eq. (7.31) reveals that p_m gradually increases, it will be more difficult to satisfy (7.32). Therefore, the probability of mutation is reduced at the later stage of evolution.

7.4.3 Localization Model Definition Using EGA-Based ELM

To develop the localization model, the most important thing involves employing the EGA to estimate the optimal initial parameters $\boldsymbol{\omega}$ and b for the extreme learning machine (ELM). This problem has been stated in Sect. 7.4.1. Following the EGA process, the EGA-based ELM can be constructed as follows:

Chromosome formation. Every individual of EGA represents a potential solution for optimal parameter estimation. $\boldsymbol{\omega}$ and b are used for the formation of chromosomes. The genes of chromosomes are formed by using the value encoding method [34]. If the ELM has N input nodes and l hidden nodes, the number of elements of $\boldsymbol{\omega}$ is $N \times l$. The length of an individual's chromosome is calculated as:

$$L = N \times l + l \tag{7.33}$$

Therefore, the chromosome of the individual can be expressed as:

$$\mathrm{ch}rom = \left[\omega_{11}, \omega_{12}, \ldots, \omega_{ij}, \ldots, \omega_{nl}, b_1, \ldots b_j, \ldots, b_l\right] \tag{7.34}$$

where ω_{ij} represents the weight connecting the ith input node and the jth hidden node, and b_j is the jth activation value.

Fitness function definition. To evaluate an individual's adaptability, EGA needs a fitness function to assign scores. As defined in (7.27), the individuals are scored by $g(\bullet)$. If the training data has Q samples, the training error is calculated by:

$$F = 1 \Big/ \sum_{i=1}^{Q} |\boldsymbol{y}_i - \boldsymbol{t}_i| \tag{7.35}$$

where $\boldsymbol{F}$ denotes the fitness of the individual, $\boldsymbol{y}_i$ and $\boldsymbol{t}_i$, $i = \{1, 2, \ldots, Q\}$, represent the ELM prediction and the target labels, respectively.

Convergence definition. The convergence condition is defined by specifying the maximum number of iterations, which means that EGA stops evolution if the number of iterations reaches the predefined number. If setting a training goal, the condition can be defined by using the mean fitness of the population as follows:

$$\overline{\boldsymbol{F}} = \sum_{j=1}^{N} \boldsymbol{F}_j \tag{7.36}$$

where $\boldsymbol{F}_j$ is the fitness of the jth sample, $j = \{1, 2, \ldots, N\}$. In theory, if most samples of the population contain better parameters, Eq. (7.36) gives a smaller value. Convergence is considered achieved if $\overline{\boldsymbol{F}}$ falls below a certain threshold.

EGA searching. Before EGA evolution, the chromosome length must be determined based on the numbers of input and hidden nodes. Then, according to Eqs. (7.33) and (7.34), EGA generates a population with N samples by randomly assigning values to $\boldsymbol{\omega}$ and b. Based on Q pieces of training data, Q training errors are obtained and the individual's fitness is determined using (7.35). As Fig. 7.19 shows, once the EGA reaches convergence condition, the chromosome of the best sample is decoded and the optimal initial parameters for the ELM model are obtained.

The outlined steps are the construction process of the EGA-ELM model. The general principle is to utilize EGA and training data to estimate the optimal parameters of ELM. With the optimal parameters, the ELM model finishes the learning work of the links of geomagnetic sequences and positions.

Localization model. Based on the EGA-ELM model, the key problem is how to use the model to train the dataset. It is important to segment the geomagnetic sequences and make them location distinguishable, the length of the magnetic

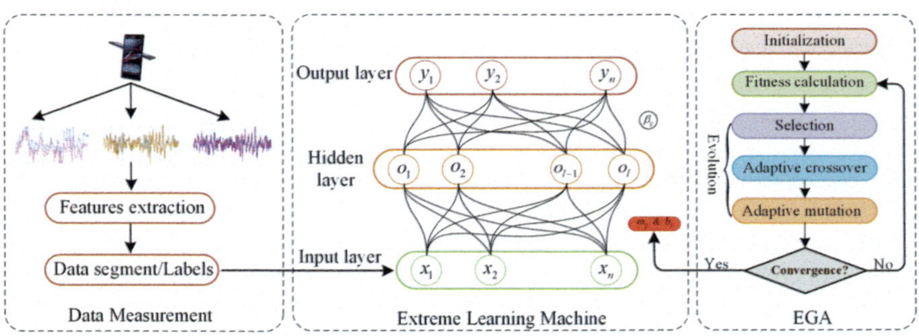

Fig. 7.19 The architecture of the EGA-ELM-based SBMP

segments depends on the changing trend (see Figs. 7.2, 7.5 and 7.6) of the magnetic field, and should be consistent with the number of input nodes of the ELM model. One piece of magnetic sequence should be labeled by only one indoor position.

As Fig. 7.19 shows, with these definitions, testers should collect magnetic data and make partitioning according to segment methods, which can be achieved by using a sliding window. After labeling magnetic segments, inputting data into the EGA model obtains optimal initial parameters. Then, the ELM-based positioning model is built using the optimal initial parameters.

7.4.4 SBMP Experimental Results

To evaluate the performance of the EGA-ELM-based magnetic positioning model, testing was made at CUMT, Xuzhou. The magnetic sequences are collected by different testers using different phones on different dates. Table 7.1 summarizes the configuration details of the EGA-ELM and provides insights into the training and testing datasets. Additionally, Table 7.2 compiles information about the testers and the phones used in the experiment. For comparison, the back-propagation neural network (BP), convolutional neural network (CNN) [27], recurrent neural network (RNN) [28], radial basis function neural network (RBF), and learning vector quantization neural network (LVQ) are implemented. The detailed configuration and testing results are available in Table 7.3 and Fig. 7.20.

Figure 7.20a reveals that EGA-ELM can obtain a mean positioning error of about 1 m, which is better than the popular learning models like BP, RNN, LVQ, and RBF. Despite the CNN model displaying superior accuracy, its mean model construction time is 12 times longer than that of the EGA-ELM model, as indicated in Fig. 7.20b. The model construction time of RNN, BP, and LVQ is about 6 times, 4 times, and 3

Table 7.1 The detailed information of the EGA-ELM model

Parameter	Value
Number of testers	5
Number of mobile phones	5
Number of testing datasets	4
Number of testing days	7
Population size of GA	50
Iterations of GA	50
Sampling rate of magnetometer	25 Hz
Number of training data segments	5372
Number of testing data segments	2686
Path length for data collection	210 m
Dimension of training data segment and sliding window	100

Table 7.2 The detailed information of testers and phones

Tester no	Height (cm)	Gender	Phones	Magnetometer
1	182	Male	HONOR 9	AKM09911
2	188	Male	Xiaomi 6	AK09916C
3	160	Female	MATE 20	AKM09918
4	183	Male	OPPO Reno	MMC5603X
5	175	Male	HONOR Magic	AKM09918

Table 7.3 Configuration of different learning models

Learning models	Description	Considered values
EGA-ELM BP, RNN RBF & LVQ	Input dimension	100
	Hidden layer nodes	200
	Learning rate	0.01
	Training goal	0.0001
CNN	Input dimension	10×10
	Number of convolutional layers	5
	Number of pooling layers	5
	Number of fully-connected layers	1

times longer than that of the EGA-ELM model, respectively. Although RBF model construction is fast, its mean positioning accuracy is about 1.5 m, which is worse than that of the EGA-ELM. It can be concluded that the proposed EGA-ELM method can achieve high-precision localization with low time cost. This feature is significant for real-time updating of the sequence-based magnetic positioning models.

7.5 Summary

This chapter provides extensive studies for magnetic positioning using evolutionary algorithms. The factors that affect the magnetic measurements are investigated. Deviation mitigation methods for different types of magnetic features are presented. To improve magnetic positioning, an enhanced MEA and an EGA-ELM are proposed for performing single-point-based magnetic positioning (SPMP) and sequence-based magnetic positioning (SBMP), respectively. The testing results demonstrate that the EMEA-based SPMP outperforms the classic distance-based methods such as KNN, MSD, etc., and the EGA-ELM-based SBMP performs better than popular machine learning models (such as BP, CNN, etc.) in terms of positioning accuracy and time cost of models. The experimental results verify the feasibility of using evolutionary algorithms for enhanced magnetic positioning. Future research will focus

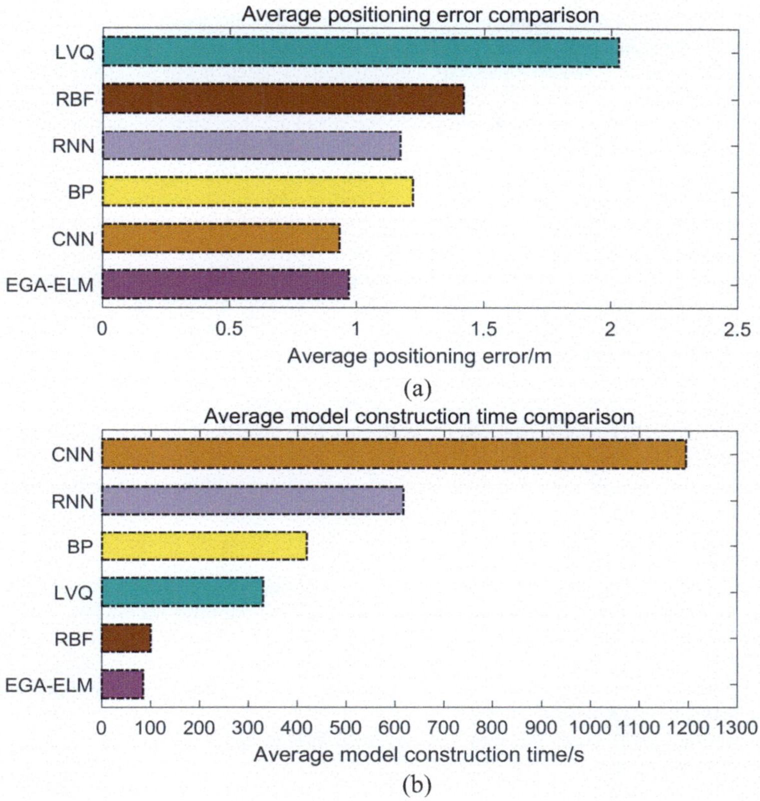

Fig. 7.20 Comparisons of positioning errors and model construction time of different models. **a** average positioning error comparison; **b** average model construction time comparison

on extending and integrating the proposed methods with other indoor localization methods.

References

1. He S, Shin KG (2018) Geomagnetism for smartphone-based indoor localization: challenges, advances, and comparisons. ACM Comput Surv 50(6):97
2. Suksakulchai S et al (2000) Mobile robot localization using an electronic compass for corridor environment. In: IEEE international conference on systems, man and cybernetics, vol 5, pp 3354–3359
3. Gozick B et al (2011) Magnetic maps for indoor navigation. IEEE Trans Instrum Meas 60(12):3883–3891
4. Li B et al (2012) How feasible is the use of magnetic field alone for indoor positioning? In: 2012 international conference on indoor positioning and indoor navigation, pp 1–9

5. Sun M et al (2021) Indoor geomagnetic positioning using the enhanced genetic algorithm-based extreme learning machine. IEEE Trans Instrum Meas 70:1–11
6. Chen L, Wu J, Yang C (2020) MeshMap: a magnetic field-based indoor navigation system with crowdsourcing support. IEEE Access 8:39959–39970
7. Hou L et al (2019) Orientation-aided stochastic magnetic matching for indoor localization. IEEE Sens J 20(2):1003–1010
8. Solin A et al (2018) Modeling and Interpolation of the ambient magnetic field by gaussian processes. IEEE Trans Robot 34(4):1112–1127. https://doi.org/10.1109/tro.2018.2830326
9. Viset F, Helmons R, Kok M (2022) An extended Kalman filter for magnetic field SLAM using gaussian process regression. Sensors 22(8):2833
10. Sun M et al (2020) Indoor positioning integrating PDR/geomagnetic positioning based on the genetic-particle filter. Appl Sci-Basel 10(2):668
11. Kang R, Cao L (2017) Smartphone indoor positioning system based on geomagnetic field. In: 2017 Chinese automation congress (CAC), pp 1826–1830
12. Liu GX et al (2020) Focusing matching localization method based on indoor magnetic map. IEEE Sens J 20(7):10012–10020. https://doi.org/10.1109/JSEN.2020.2991087
13. Wang J et al (2019) Performance test of MPMD matching algorithm for geomagnetic and RFID combined underground positioning. IEEE Access 7:129789–129801
14. Shi LF et al (2023) Pedestrian indoor localization method based on integrated particle filter. IEEE Trans Instrum Meas 72:1–10
15. Huang H et al (2018) An improved particle filter algorithm for geomagnetic indoor positioning. J Sens 2018. https://doi.org/10.1155/2018/5989678
16. Zhang M et al (2017) Indoor positioning tracking with magnetic field and improved particle filter. Int J Distrib Sens Netw 13(11). https://doi.org/10.1177/15501477177418
17. Zheng M et al (2017) Sensitivity-based adaptive particle filter for geomagnetic indoor localization. In: 2017 international conference on communications in China (ICCC), pp 1–6
18. Lee S, Chae S, Han D (2020) ILoA: indoor localization using augmented vector of geomagnetic field. IEEE Access 8:184242–184255
19. Kuang J et al (2018) Indoor positioning based on pedestrian dead reckoning and magnetic field matching for smartphones. Sensors 18(12):21. https://doi.org/10.3390/s18124142
20. Ashraf I et al (2019) GUIDE: smartphone sensors-based pedestrian indoor localization with heterogeneous devices. Int J Commun Syst 32(15):e4062. https://doi.org/10.1002/dac.4062
21. Subbu KP, Gozick B, Dantu R (2011) Indoor localization through dynamic time warping. In: 2011 IEEE international conference on systems, man, and cybernetics, pp 1639–1644
22. Qiu K et al (2018) Indoor geomagnetic positioning based on a joint algorithm of particle filter and dynamic time warp. In: 2018 ubiquitous positioning, indoor navigation and location-based services (UPINLBS), pp 1–7
23. Hui L et al (2014) TACO: a traceback algorithm based on ant colony optimization for geomagnetic positioning. China conference on wireless sensor networks. Springer, Berlin Heidelberg, pp 208–222
24. Ashraf I, Hur S, Park Y (2018) MPILOT-magnetic field strength based pedestrian indoor localization. Sensors 18(7):22. https://doi.org/10.3390/s18072283
25. Stein P et al (2014) Leader following: a study on classification and selection. Robot Auton Syst 75:79–95
26. Montoliu R, Torres-Sospedra J, Belmonte, O (2016) Magnetic field based indoor positioning using the bag of words paradigm. In: 2016 international conference on indoor positioning and indoor navigation, pp 1–7
27. Ashraf I et al (2020) MINLOC: magnetic field patterns-based indoor localization using convolutional neural networks. IEEE Access 8:66213–66227
28. Bae HJ, Choi L (2019) Large-scale indoor positioning using geomagnetic field with deep neural networks. In: IEEE international conference on communications (ICC), pp 1–6
29. Bhattarai B et al (2019) Geomagnetic field based indoor landmark classification using deep learning. IEEE Access 7:33943–33956

30. Chen Z et al (2023) Geomagnetic vector pattern recognition navigation method based on probabilistic neural network. IEEE Trans Geosci Remote Sens. https://doi.org/10.1109/TGRS.2023.3273552
31. Chengyi S, Yan S, Keming X (2000) Mind-evolution-based machine learning and applications. In the 3rd World Congress on Intelligent Control and Automation, Vol 1, pp 112–117
32. Subbu KP, Gozick B, Dantu R (2013) LocateMe: magnetic-fields-based indoor localization using smartphones. ACM Trans Intell Syst Technol 4(4):73
33. Bajpai P, Kumar M (2010) Genetic algorithm–an approach to solve global optimization problems. Indian J Comput Sci Eng 1(3):199–206
34. Kumar A (2013) Encoding schemes in genetic algorithm. Int J Adv Res IT Eng 2(3):1–7

Chapter 8
Indoor Acoustic Localization

Zhi Wang, Naizheng Jia, Can Xue, and Wei Liang

Abstract Until now, in the satellite-denied environment, there is no mature and stable universal solution for high-precision Location Based System (LBS) similar to the Global Navigation Satellite System, making it still an open field. Near-ultrasonic positioning, as an emerging medium-range positioning technology, has natural advantages such as low synchronization costs, strong compatibility with smart devices, independence from image acquisition, and signals that are not easily obstructed by rooms, making it an optimal solution for low-cost, high-precision safety positioning. This chapter focuses on positioning in the practical and complex satellite-denied environments, based on portable smart terminal platforms, leveraging the complementary advantages of near-ultrasonic and inertial navigation technologies. It combines positioning data with theoretical and practical requirements for training privacy-secure solutions, aiming to provide novel and practical robust positioning schemes. By diverging from conventional electromagnetic wave-based methods for indoor positioning, it offers innovative perspectives and approaches.

Keywords Acoustic · Indoor positioning system · Near-ultrasonic localization · NLOS · TDOA

8.1 Introduction

In environments where traditional GPS and RF-based localization systems prove ineffective, near-ultrasonic acoustic positioning presents itself as a formidable alternative. This chapter delves into the utilization of high-frequency mechanical waves, specifically those exceeding 18 kHz, to facilitate precise localization within indoor settings. Distinguished from electromagnetic waves, near-ultrasonic waves are not

Z. Wang (✉) · N. Jia · C. Xue
College of Control Science and Engineering, Zhejiang University, Hangzhou, China
e-mail: zjuwangzhi@zju.edu.cn

W. Liang
State Grid Jiangsu Electric Power Co., Ltd., Electric Power Research Institute, Nanjing, China

K. Yu (ed.), *Positioning and Navigation Using Machine Learning Methods*, Navigation: Science and Technology 14, https://doi.org/10.1007/978-981-97-6199-9_8

susceptible to electromagnetic interference and possess long wavelengths essential for the efficient detection of weak signals via sophisticated signal processing techniques including filtering and generalized cross-correlation [13].

Acoustic localization systems harness both building materials and the ambient air as transmission mediums, enabling precise navigation in environments that are structurally intricate and fraught with electronic noise. These systems find applications across various sectors, particularly in industrial settings where they counteract the disruption caused by lower frequency mechanical and electrical noises. Additionally, the inherent inaudibility of near-ultrasonic waves enhances their utility in noise-sensitive areas.

Contemporary implementations of near-ultrasonic positioning systems have demonstrated significant capabilities, notably in tracking pedestrian posture and position with accuracies reaching up to 10 cm. For instance, Liu et al. have showcased this capability [16], and systems like Guo Guo's employ signals with pseudo-random codes above 15 kHz for positioning, achieving similar levels of precision [15]. However, these systems are not devoid of limitations-they are typically bound by the maximum feasible distance for effective localization and necessitate meticulous manual positioning of base stations within highly controlled noise environments.

This chapter will articulate the fundamentals of acoustic signal ranging and elaborate on node auto-calibration techniques and robust algorithms aimed at bolstering the accuracy and reliability of indoor acoustic localization. Special emphasis will be placed on addressing the challenges posed by non-line-of-sight (NLOS) conditions and multipath interference, which are common in indoor spaces. Through a blend of theoretical analysis and empirical validation, this discourse seeks to furnish a thorough overview of the present capabilities and future prospects of indoor acoustic localization technologies, advocating their advancement as a reliable solution for environments devoid of satellite coverage.

The chapter on indoor acoustic localization is organized to provide a detailed exploration of this cutting-edge technology. The overall structure is shown as followed: It begins with an introduction that establishes the context and significance of using near-ultrasonic acoustic positioning as a reliable alternative in environments where traditional GPS and RF-based systems are ineffective. This section highlights the technological underpinnings and the unique attributes of ultrasonic waves that make them suitable for complex indoor environments.

Following the introduction, the chapter proceeds to a detailed examination of robust acoustic signal ranging. Here, the focus is on the specific methodologies employed in enhancing signal detection and processing, such as filtering and generalized cross-correlation, which are critical for navigating through electronically noisy and structurally complex spaces.

The subsequent section on node auto-calibration delves into the calibration techniques that ensure the accuracy and reliability of positioning nodes, which are integral components of any acoustic localization system. This part discusses both manual and self-calibration methods, emphasizing advancements that minimize human error and enhance operational efficiency.

The chapter then explores the core algorithms that drive indoor acoustic localization systems. It details the mathematical models and algorithmic strategies designed to address and mitigate common challenges like NLOS conditions and multipath interference, which are prevalent obstacles in indoor localization.

In the experimental results section, the chapter presents empirical data and case studies that validate the theoretical approaches discussed earlier. This part not only demonstrates the practical application and effectiveness of the technology but also highlights areas requiring further research and development.

The chapter concludes with a summary that synthesizes the key discussions, reiterates the potential of indoor acoustic localization technology, and outlines future directions for research. This final section solidifies the chapter's contribution to the field, advocating for continued advancement and application of this promising technology in satellite-denied environments.

8.2 Robust Acoustic Signal Ranging

In this section, the positioning base stations use Chirp signals. The base station is controlled by STM32 and comes with a built-in GNSS chip for timing. All localization algorithms are based on Time of Arrival (TOA) values for localization. At the same time, the Lora chip is used to synchronize and calculate the TOA value between the base station and the positioning terminal. Chirp signals are signals whose frequency increases linearly with time, also known as Linear Frequency Modulation (LFM) signals [11, 12]. They have good correlation properties and are widely used in the field of communications. The expression for Chirp signals is as follows:

$$s(t) = \cos 2\pi f_0 t + \pi k t^2 \tag{8.1}$$

In Eq. 8.1, f_0 represents the starting frequency, k denotes the frequency change slope of the Chirp signal, which controls the rate at which the frequency changes over time, and T signifies the duration of the signal. To ensure Chirp signals exhibit better correlation properties and higher signal recognition rates, this chapter selects a Chirp signal duration of 45ms, with a modulation frequency range from 16 to 19 KHz, and a signal frequency of 1 Hz.

When positioning devices emit near-ultrasonic high-frequency signals, audible noise may occur at the start and end phases of the sound, known as low-frequency leakage. To address this issue, this chapter introduces the use of Blackman window functions and rectangular window functions at the start and end phases of the signal. By attenuating the energy at these phases, the low-frequency leakage is reduced. At the same time, the energy in the signal's core area is maintained to ensure transmission distance. The signal after the window function transformation can be obtained as:

$$s(n) = s(n) * w(n) \tag{8.2}$$

Here, $*$ is the convolution function. And $w(n)$ is:

$$w(n) = \begin{cases} 1, & \frac{1}{4}N \le n \le \frac{3}{4}N, \\ 0.42 - 0.5\cos\left(\frac{\pi n}{N}\right) + 0.08\cos\left(\frac{2\pi n}{N}\right), & otherwise, \end{cases} \tag{8.3}$$

where N is the length of the signal.

Given the good correlation properties of Chirp signals, this chapter employs the Generalized Cross-Correlation (GCC) for detecting the arrival time of Chirp signals. Let $x(t)$ denote the received signal, and $s(t)$ represent the transmitted signal. The Generalized Cross-Correlation function can be defined as:

$$GCC_{x,s}(\tau) = \int_{-\infty}^{\infty} X(f)S^*(f)e^{j2\pi f\tau}\, df \tag{8.4}$$

where $X(f)$ and $S(f)$ are the Fourier transforms of $x(t)$ and $s(t)$ respectively, and $S^*(f)$ is the complex conjugate of $S(f)$. The variable τ represents the time delay between the received signal and the transmitted signal.

Using the waveform formed by the GCC, the arrival time of the signal can be inferred from the position of the peak as:

$$\tau_0 = \arg\max GCC_{x,s}. \tag{8.5}$$

Using the aforementioned algorithm, the GCC of Chirp signals with different signal-to-noise ratios and the original Chirp signals can clearly identify the position of the signal's maximum peak. Further, by estimating the signal propagation delay, the TOA values can be obtained, and the position coordinates can be solved. The TOA estimation is ensured by synchronize base stations and ranging tags.

However, in real environments, due to the influence of noise, multipath propagation, and NLOS, the peaks formed by GCC are not pronounced and it is difficult to directly calculate the signal's arrival time through the cross-correlation peak. To address this issue, a frame-by-frame normalized GCC detection method is proposed, which can get over the multipath effect in long range estimation as shown in Fig. 8.1.

The frame-by-frame normalized cross-correlation detection involves dividing the signal into frames, performing cross-correlation calculations for each frame, and normalizing the results of the cross-correlation signals. This process filters out the effects of multipath and NLOS on the original signal, thereby better identifying the peak values in the cross-correlation results. Figure 8.2 illustrates the frame-by-frame normalized GCC detection process.

The steps for frame-by-frame normalized cross-correlation are as follows [10]:

1. Set the signal sliding window frame, length L, and the sliding window step length is Δ, which should be greater than the length of the Chirp signal.
2. Calculate the cross-correlation coefficient for each frame according to the step length. Take the first half of each frame to combine, resulting in a new combined cross-correlation signal frame, with a frame length of $L/2$.

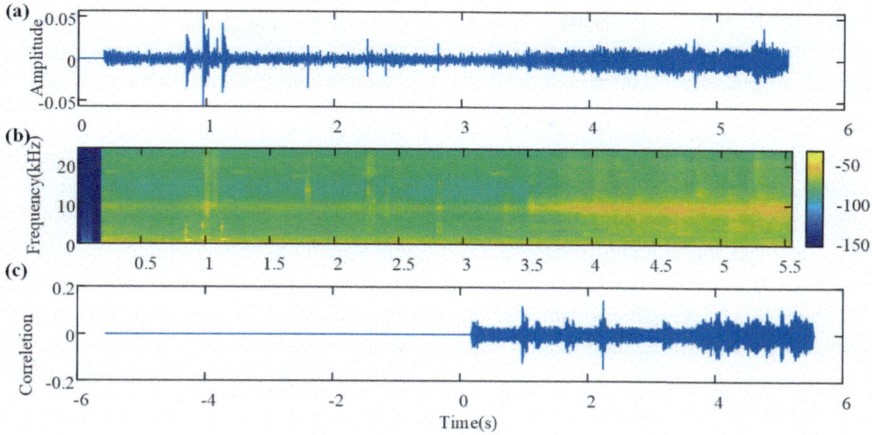

Fig. 8.1 Waveform (**a**), spectrum (**b**) and GCC diagram (**c**) of the received Chirps signal for the NLOS case

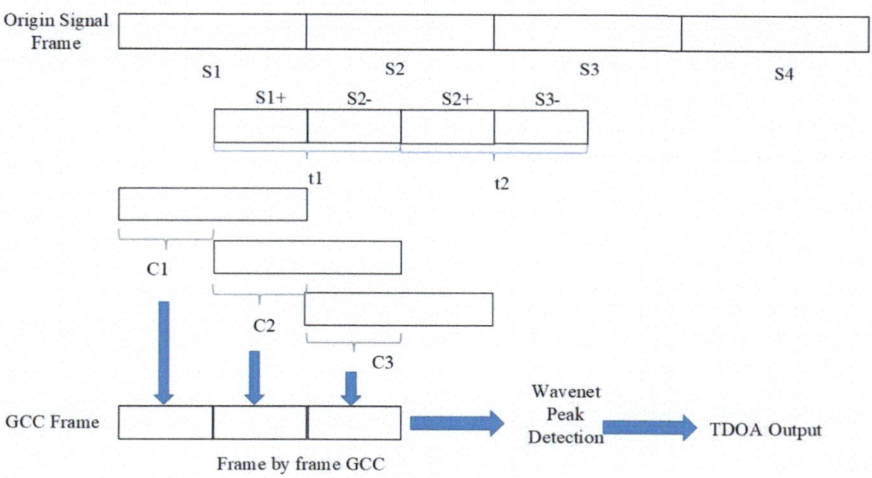

Fig. 8.2 The structure of frame-be-frame-GCC

3. Normalize the maximum amplitude of each frame signal, A_{max}, and the maximum value of the cross-correlation of each combined frame, C_{max}.
4. Use wavelet decomposition and the derivative method to find the position of the peak.

Wavelet transform can be represented as:

$$Wf(u, s) = s^n \frac{\mathrm{d}^n}{\mathrm{d}u^n} \left(f^* \bar{\theta}_s\right)(u) \tag{8.6}$$

where $\bar{\theta_s}$ is the rapidly decaying function of the wavelet signal's vanishing moments, and $s(t)$ is the original signal. It is evident that wavelet transform can remove noise at different frequencies. The algorithm for peak finding using wavelet decomposition is as follows:

1. Perform continuous wavelet transform on the output cross-correlation signal to obtain the wavelet transform coefficient spectra at different decomposition scales; the wavelet base signal is 'db2'.
2. Reconstruct the wavelet transform coefficient spectra at different decomposition scales.
3. Use the first derivative to find the extreme points for peak detection and merge the results of peak detection to form a sequence of peaks.
4. Determine the appropriate correlation peaks from the peak detection results using thresholding.

The results of the frame-by-frame normalized detection of the signal belonging to Fig. 8.1 are shown in Fig. 8.3. It can be seen that the original intricate and difficult to distinguish the waveforms of the inter-correlation diagram appeared six equally spaced, highly recognizable wave peaks, corresponding to the signal 1 Hz refresh rate. Therefore, based on the frame-by-frame normalized inter-correlation detection can well overcome the multipath effect brought by the bending space, and the arrival time of the signal can be calculated more accurately.

Based on the signal arrival time, the distance between the sending end and the receiving end can be calculated as:

$$d = t_0 c \tag{8.7}$$

where t_0 is the TOA value, c is the velocity of the sound in air.

Supposed there is n base station, and the coordinate of each base station is $(x_i, y_i, z_i), i = 1, 2, ..., n$. The localization target coordinate is (x_t, y_t, z_t), with

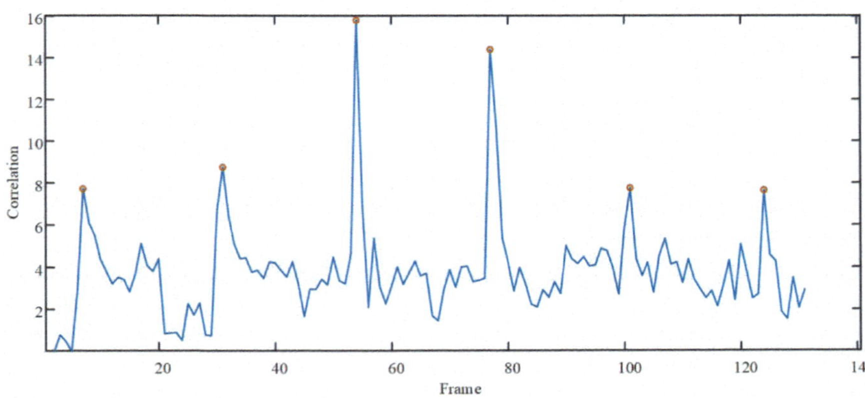

Fig. 8.3 GCC diagram of signals after frame-by-frame normalization of signals

Eq. 8.7, the distance matrix is shown as:

$$X = \begin{bmatrix} d_1 \\ d_2 \\ \vdots \\ d_n \end{bmatrix} = \begin{bmatrix} \sqrt{(x_1 - x_t)^2 + (y_1 - y_t)^2 + (z_1 - z_t)^2} \\ \sqrt{(x_2 - x_t)^2 + (y_2 - y_t)^2 + (z_2 - z_t)^2} \\ \vdots \\ \sqrt{(x_n - x_t)^2 + (y_n - y_t)^2 + (z_n - z_t)^2} \end{bmatrix} \quad (8.8)$$

Assume that the distance from the device to each base station calculated using Time Difference of Arrival (TDOA) is:

$$\hat{X} = \begin{bmatrix} \hat{d}_1 \\ \hat{d}_2 \\ \vdots \\ \hat{d}_n \end{bmatrix} = \begin{bmatrix} t_1 c \\ t_2 c \\ \vdots \\ t_n c \end{bmatrix} \quad (8.9)$$

The TDOA matrix is shown as:

$$X_{TDOA} = \hat{X} - \hat{X}(1) \quad (8.10)$$

The sections followed are based on Eq. 8.10.

8.3 Node Auto Calibration

Common TDOA positioning systems are typically composed of nodes and tags. After receiving signals from nodes, the tags calculate the difference in signal arrival time, further generating positioning results. This process relies on a premise: the distance estimations of the nodes are known. In satellite navigation systems, satellites can be considered as nodes, whose positions are provided by precise atomic clocks and ephemerides. However, indoor positioning systems do not have determined orbits or fixed placement positions, and thus, after system deployment, the nodes' positions need to be calibrated. Given that the TDOA algorithm requires the nodes' positions as known inputs, their accuracy directly impacts the overall precision of the positioning system. Therefore, accurately determining the nodes' positions is a core step for TDOA positioning. The calibration of nodes can be divided into manual calibration and self-calibration. Manual calibration requires the use of measuring tools such as tape measures, laser rangefinders, and total stations, which is not only time-consuming and labor-intensive but also prone to introducing calibration errors. When the area to be covered by the positioning system is large and requires a number of nodes, the difficulty and time required for precise calibration increase significantly. Adopting self-calibration technology based on signal interactions between nodes can avoid the errors introduced by manual calibration. Its accuracy depends on the signal interaction mode and the self-calibration algorithm, avoiding errors introduced by human factors, which is beneficial for improving the efficiency and precision of node calibration. It also overcomes difficulties faced with manual calibration in adverse environments. Therefore, the use of self-calibration technology based on signal interactions between nodes has significant importance in the practical application of positioning systems. In recent years, there have been a series of attempts to use Building Information Modeling (BIM) to improve the accuracy of self-calibration or positioning. These attempts have improved self-calibration accuracy in experiments but lack theoretical support and analysis. It is necessary to theoretically prove that map information constraints can reduce calibration errors caused by ranging errors. In the field of positioning, the CRLB is commonly used as the theoretical limit of positioning performance. The aim of this section is to quantitatively analyze the theoretical performance limits of self-calibration technology with BIM and, based on this, design a self-calibration algorithm that integrates BIM. The performance of the self-calibration algorithm will be analyzed and compared with CRLB, optimizing the self-calibration algorithm under the condition that the building information model remains unchanged.

For simplicity, consider a two-dimensional plane model. The main conclusions of this section can also be extended to three-dimensional situations, which will not be repeated here. The distance estimations of a series of unknown nodes are shown as Eq. 8.10.

Nodes with known positions (anchor nodes) are denoted as:

$$a = \{a_1, a_2, a_3, \ldots, a_m\}. \tag{8.11}$$

The ranging result between unknown nodes x_i and x_j is denoted as $d_{ij}(1 \leq i < j \leq n)$, and the ranging result between unknown node x_i and known node a_j is denoted as $d_{ij}(1 \leq i \leq n, 1 \leq j \leq m)$.

The self-calibration technique uses the known node distance estimations a, as well as all ranging results $d_{ij}(1 \leq i < j \leq n), d_{ij}(1 \leq i \leq n, 1 \leq j \leq m)$ to calculate all the position node distance estimations X. To simplify notation, consolidate all unknowns into a vector Z, which includes all unknown distance estimations in X. The total two-dimensional distance estimations of n unknown nodes amount to $2n$, that is, the dimension of vector Z.

$$Z = [z_1\ z_2\ z_3\ \ldots\ z_{2n}]^T. \tag{8.12}$$

To further simplify notation, consolidate all ranging results into a vector d, which contains a total of $N = \frac{n(n-1)}{2} + mn$ ranging values,

$$d = [d_{12}\ d_{13}\ \ldots\ d_{(n-1)n}\ d_{11}\ d_{12}\ \ldots\ d_{nm}]^T. \tag{8.13}$$

The cooperative self-calibration problem is to solve for all the unknown node distance estimations using the ranging vector d.

In typical indoor environments such as malls and museums, BIM can provide constraint models of the walls on which nodes are placed. When nodes are placed on walls, their two-dimensional distance estimations will inherently have constraints. The general model can be represented as follows:

$$f_i(z_{s_1}, z_{s_2}, z_{s_3}, \ldots, z_{s_p}) = 0, \quad i = 1, 2, \ldots, l. \tag{8.14}$$

$$S = \{s_1, s_2, s_3, \ldots, s_p\}, \ 1 \leq s_1 \leq s_2 \leq \ldots \leq s_p \leq 2n. \tag{8.15}$$

Here, $f_i(\cdot)$ represents the i-th constraint function, with a total of l constraint functions. S represents the set of all constrained unknowns.

The above provides a universal constraint model, which often degenerates into linear constraints in actual application scenarios. A single linear constraint can be represented as:

$$f = e_i{}^T Z = 0, \quad 1 \leq i \leq 2n. \tag{8.16}$$

Here, e_i is a $2n$-dimensional vector with the i-th element as 1 and the rest as 0. This vector represents the constraint produced when the i-th coordinate in the to-be-determined coordinate vector Z is known. In contrast, a full constraint appears as pairs of linear constraints, and a pair of full constraints can be represented as:

$$f = e_i^T e_{n+i}^T Z = 0, \quad 1 \leq i \leq n. \tag{8.17}$$

Suppose the unbiased estimator $\hat{Z}$ of the parameter Z satisfies k ($k < 2n$) continuous and differentiable constraints

$$f(\hat{Z}) = 0. \tag{8.18}$$

The $k \times 2n$ matrix $F(Z)$ is the gradient matrix of these constraints, satisfying the relationship:

$$F(Z) = \frac{\partial f(Z)}{\partial Z^T}. \tag{8.19}$$

Assuming the gradient matrix $F(Z)$ is row full rank (meaning k constraints are independent), then there exists a matrix $U \in R^{2n \times (2n-k)}$, whose columns form an orthogonal basis for the null space of $F(Z)$, implying the following relationship:

$$F(Z)U = 0. \tag{8.20}$$

$$E(\hat{Z} - Z)(\hat{Z} - Z)^T \geq U(U^T J U)^{-1} U^T. \tag{8.21}$$

$$U(U^T J U)^{-1} U^T = J^{-1} - J^{-1} F^T (F J^{-1} F^T)^{-1} F J^{-1}. \tag{8.22}$$

Thus, it follows that

$$cov(\hat{Z}) = E(\hat{Z} - Z)(\hat{Z} - Z)^T \geq J^{-1} - J^{-1}F^T(FJ^{-1}F^T)^{-1}FJ^{-1}. \quad (8.23)$$

Property 1 *$J^{-1}F^T(FJ^{-1}F^T)^{-1}FJ^{-1}$ is a non-negative definite matrix.*

Proof From the definition of the Fisher information matrix, it is easy to derive that J^{-1} is a non-negative definite matrix. Therefore, for $\forall x \in R^k$, by the property of non-negative definite matrices, we have

$$x^T(FJ^{-1}F^T)x = (F^Tx)^T J^{-1}(F^Tx) \geq 0. \quad (8.24)$$

From the above, it follows that $FJ^{-1}F^T$ is a non-negative definite matrix, and thus its inverse $(FJ^{-1}F^T)^{-1}$ is also a non-negative definite matrix. For $\forall x \in R^{3n}$, we have

$$x^T[J^{-1}F^T(FJ^{-1}F^T)^{-1}FJ^{-1}]x = (FJ^{-1}x)^T(FJ^{-1}F^T)^{-1}(FJ^{-1}x) \geq 0. \quad (8.25)$$

From the above, it is derived that $J^{-1}F^T(FJ^{-1}F^T)^{-1}FJ^{-1}$ is a non-negative definite matrix, proving Property 1. Hence,

$$CRLB(\hat{Z}) = J^{-1} - J^{-1}F^T(FJ^{-1}F^T)^{-1}FJ^{-1} \leq J^{-1}. \quad (8.26)$$

This proves that the Cramer-Rao matrix of the estimator $\hat{Z}$, subject to BIM constraints, is less than or equal to the unconstrained Cramer-Rao matrix in the positive definite sense. The diagonal elements of $CRLB(\hat{Z})$ are the lower bounds of the variance of $\hat{Z}$. This proves that BIMs are helpful in improving the performance of self-calibration.

In practical scenarios, another typical setup is for nodes to be fixedly placed at corners of walls. In this case, since the distance estimations of nodes placed at corners can be completely determined by BIM, it forms a "full constraint", equivalent to increasing the number of anchor nodes. Additionally, when the number of nodes placed at corners is sufficient, it is possible to eliminate the pre-arranged anchor nodes. A sufficient number of nodes at corners can serve the role of anchor nodes, considered as "self-constraint", implying that effective constraints can be formed solely with nodes, without the need for anchor nodes. This section's simulation experiment aims to compare the impact of BIM on CRLB under different constraint scenarios with the same number of constraints. The design of node and anchor node positions for this experiment is set in a 20 m × 15 m rectangular room. Stars numbered 1 to 8 represent nodes uniformly distributed on the four walls (considering the practical scenario where nodes are as evenly distributed as possible, with two nodes fixedly allocated to each wall); small squares represent known anchor nodes with positions at (5,5), (10,10), and (15,5) units; the rectangular outline represents walls. Simulation layouts and results show the comparison of the average variance of node coordinate errors under unconstrained, semi-constrained, fully constrained, and self-constrained

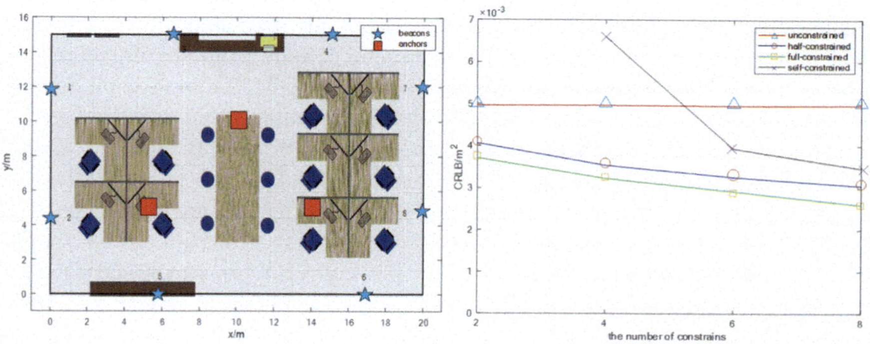

Fig. 8.4 Simulation results

scenarios (this metric is directly related to the CRLB, reflecting the theoretical lower limit of self-positioning errors), comparing 2, 4, 6, and 8 constraints respectively. For semi-constraints, the experiment sequentially adds groups of constraints: the x-coordinates of nodes 1 and 2, the y-coordinates of nodes 3 and 4, and so on. As a contrast, full constraints progressively turn the coordinates of nodes 1, 3, 5, and 7 into known constraints, similarly increasing the number of constraints from 2 to 8. After 10,000 Monte Carlo simulations, the average variance under unconstrained conditions is used as the baseline to compare the performance under semi-constrained and fully constrained conditions, with results shown in Fig. 8.4. The experimental results indicate that self-constraint, semi-constraint, and full constraint can effectively lower the theoretical error lower limit, with self-constraint being the least effective and full constraint being the most effective, but all significantly reduce the theoretical error lower limit. With two constraints, semi-constraint can reduce the error lower limit by 25.1%, and with eight constraints (under the most ideal semi-constraint condition, where every node has one coordinate determined), it can reduce the unknown coordinate error lower limit by 47.5%.

8.3.1 Other Method for Node Auto Calibration

Node self-calibration technology is a commonly used technique in the field of sensor networks, which can be divided into collaborative and non-collaborative types based on the interaction information [21]. In non-collaborative localization, calibration is carried out solely using the ranging information between unknown nodes and known nodes, which fundamentally has no significant difference from technologies such as TOA and Received Signal Strength Indication (RSSI), thus it is not separately discussed. Meanwhile, collaborative localization utilizes the ranging information between unknown nodes as well, theoretically providing higher positioning accuracy, making it a current research hotspot in self-calibration technology. In the data-rich collaborative self-calibration technology, the commonly used algorithms are mainly divided into two categories: methods based on semi definite programming relaxation (SDP) [2] and those based on multidimensional scaling Multidimensional Scaling (MDS) analysis [5]. Each of these methods has its advantages and disadvantages.

The node self-calibration technology based on MDS utilizes classical scaling analysis, where the core technique is the Singular Value Decomposition (SVD) of a matrix. Assuming the distance estimations of all unknown nodes are represented by a matrix X, with elements being the distance estimations in various dimensions (applicable for both two and three dimensions), we first convert it into a symmetric matrix $Q = X^T X$ and perform spectral decomposition on Q to obtain $Q = V D^2 V^T$. Thus, the distance estimations of the unknown nodes W can be calculated as $W = D_p V_p^T$, where p indicates the dimensionality of the nodes, with D_p and V_p^T representing submatrices/vectors respectively. This method allows for solving all unknown nodes' relative distance estimations in closed form, and absolute distance estimations can be obtained through appropriate transformations based on known distance estimations. This algorithm is termed the MDS-MAP algorithm [18].

While the MDS-MAP algorithm provides good localization in standard scenarios, it falters in real-world situations with low node connectivity since the fundamental approach of MDS, which projects distances in high-dimensional spaces to lower dimensions (two or three), aiming to maximize distance similarity across these spaces. This does not ensure high-precision localization. Therefore, building on MDS-MAP [17], Shang and Ruml proposed the MDS-MAP(P) algorithm, which starts with MDS-MAP's initial values and uses least squares to minimize the least squares error between measured distances and computational errors to further reduce inaccuracies. This approach significantly enhances self-calibration performance in scenarios of low connectivity while maintaining MDS-MAP's efficiency in common settings.

Due to their closed-form nature, MDS-based algorithms offer quick solutions but may not handle large-error scenarios well. Despite optimizations for low connectivity, their accuracy may severely decline. In such contexts, the Maximum Likelihood Estimation (MLE) algorithm would theoretically provide the highest accuracy. However, MLE for self-calibration poses a non-convex optimization challenge, difficult to solve directly due to potential trapping in local minima, whereas global optimization algorithms are too resource-intensive. To address this, methods based on SDP relaxation have been explored to simplify the optimization by converting non-convex quadratic distances into convex constraints [2, 8], thus transforming the self-calibration challenge into a semidefinite programming problem. Despite the complexity, the inherent advantage of convex optimization is that local optima are global optima, allowing for efficient resolution.

Nevertheless, while SDP relaxation algorithms show promise in low-connectivity scenarios compared to MDS-MAP(P), they underperform in noisy self-calibration scenarios. Expanding on the MDS-MAP(P) methodology, it adapted the basic distance geometry model to effectively utilize noise-influenced distance constraints. BISWAS et al. developed nonlinear optimization problems with lower and upper bounds or interval-based constraints, converting the challenge into a convex optimization problem through SDP relaxation. Employing maximum likelihood estimation ideas to minimize expected estimation errors, utilizing techniques like gradient descent, showed a distinct advantage in noise management over purely SDP-based self-calibration approaches [3, 4].

The Semidefinite Relaxation (SDR) algorithm has shown excellent performance in self-calibration problems, yet its computational speed struggles to meet practical application demands. To enhance the computational speed of SDR technology in self-calibration problems, an improved approach called Sub SDP (SSDP), which relaxes the original semidefinite matrix cone into a series of smaller-scale semidefinite sub-matrix cones, was proposed [26]. The subproblem decomposition algorithm significantly speeds up computation in semidefinite programming, a type of optimization problem known for its high computational complexity. Both theory and numerical simulations have shown that this approach not only achieves high localization performance similar to the SDR algorithm but also effectively reduces computation time. Moreover, the proposed scheme is not limited to self-calibration problems but can also be applied to other SDP issues.

In recent years, research focusing on node self-calibration under specific circumstances has emerged. For the local map fusion problem in scenarios of low node connectivity, a greedy optimization algorithm named SGO was introduced in [19]. This algorithm is more suitable for distributed optimization in sensor networks compared to the traditional nonlinear Gauss-Seidel algorithm. Further exploration of applying this algorithm to the SSDP and non-convex optimization has shown certain improvements. Similarly, inspired by the objective function similar to that in [4], a class of convex relaxation schemes based on the maximum likelihood estimation formula was proposed in [20]. On this basis, a computation-efficient edge-based algorithm was derived, enabling sensor nodes to solve these edge-based convex programs locally through communication only with their immediate neighbors. This algorithm relies on the Alternating Direction Method of Multipliers (ADMM), which converges to a centralized solution, can run asynchronously, and possesses computational fault tolerance. This distributed scheme was analyzed and numerically compared with other available methods, demonstrating the positive impact of ADMM on large-scale networks [20].

Zengfeng Wang and colleagues [27], capitalizing on the characteristics of RSS-based signals, first transformed the logarithmic attenuation model of signal strength into a multiplicative model, thus eliminating the exponential form in the maximum likelihood estimation expression. By squaring the optimization items in the maximum likelihood formula and then relaxing them, they made the form meet the basic requirements of convex optimization. From the transformation graph of the objective function, it is apparent that these two transformations primarily affect the steepness of the objective function, with minimal deviation on the extremes (indicated by white squares in Fig. 8.5). Based on this transformation, the SDP-LSRE self-calibration algorithm was proposed, demonstrating that, due to its better consideration of the RSS propagation model's characteristics, the schemes surpass the performance of state-of-the-art technologies in various network settings.

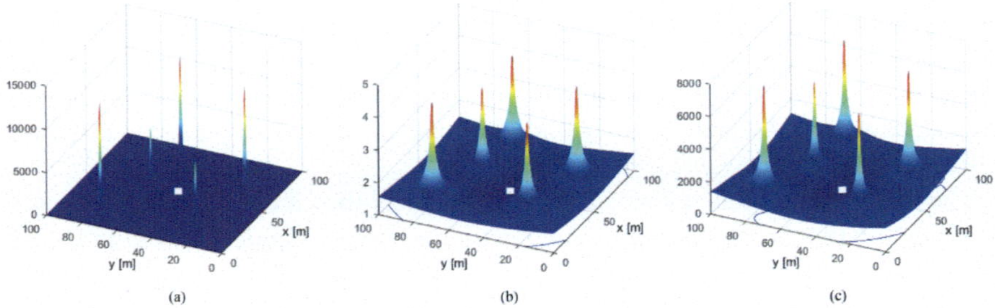

Fig. 8.5 Target function translation

Furthermore, in underwater autonomous vehicles, SDP-based optimization algorithms have also been successfully used to minimize the impact of unknown path loss exponents during underwater signal propagation [14]. Given the researchers' design of the PCAL algorithm, which is robust against signal shadowing effects, its performance in self-calibrating underwater autonomous vehicles under unknown and variable path loss exponents is superior to state-of-the-art technologies known for path loss exponents. Its effectiveness was proven by comparison with the Cramer-Rao Lower Bound (CRLB).

Recent studies have noted [9] that in the field of indoor positioning, positioning nodes are usually placed on walls. In such scenarios, constraints formed by walls can offer certain assistance to the self-calibration problem. However, this document did not propose a computationally efficient solution, instead utilizing computationally intensive algorithms like genetic algorithms and harmony search. Such schemes, being computationally demanding, are impractical for use, and currently, there lacks a self-calibration technology that effectively utilizes map information.

8.4 Acoustic Indoor Localization Algorithm

In the TDOA scenario we're focusing on here, the target that needs to be located sends out a signal, which is then picked up by multiple sensor nodes. These nodes work on the signals they receive, comparing them in pairs using something called GCC to figure out where the target is. The usual way of doing TDOA positioning picks one node as a reference, and subtracts the signal arrival times at the other nodes from it to get a set of TDOA values. When we know what the original signal's waveform looks like, we can cross-correlate the received signals with an ideal version of the signal to pinpoint the signal's arrival time. We can also cross-correlate the received signals with each other. However, when the original signal's waveform is unknown—like with sounds of explosions, underwater vessels, or bird calls, which are not artificially modified signals - then we only can cross-correlate the received signals with each other to calculate the differences in signal arrival times. But, the existing GCC techniques might not always work perfectly because signals can get distorted as they travel, and the cross-correlation values do not always meet the zero-sum condition. This is where the full set TDOA comes in handy, as it can significantly improve positioning accuracy in these situations. The research on the following techniques are diving into to further explore full set TDOA positioning [23–25].

In a passive localization system, the signal received by each node is denoted as u_t. This is the signal sent from the source, $s(t)$, which after traveling a period of d_i, is recorded at the node. The signal during its journey will be influenced by the channel conditions, introducing an error, η_i. Therefore, under ideal conditions, the received signal can be modeled as follows:

$$u_i(t) = s(t - d_i) + \eta \tag{8.27}$$

The error η_i should follow a Gaussian distribution when channel conditions are good. This error introduces some inaccuracies in the TDOA values when cross-correlating the received signals. Furthermore, depending on the GCC technique used, there can be an impact on the TDOA measurements. This impact relates to the signal's spectral intensity $S(w)$ and the noise's spectral intensity $N(w)$. The theoretical expression varies with the specific technique used, which we will not detail here. We denote the function that calculates the position of the maximum value after cross-correlation processing as $\Phi(\bullet)$, so the TDOA value d_{ij} obtained from cross-correlating signals u_i and u_j can be calculated as:

$$d_{ij} = \Phi(u_i, u_j). \tag{8.28}$$

d_{ij} includes the aforementioned signal error and generalized cross-correlation error. If both are in ideal conditions, the resulting d_{ij} can be considered the true TDOA value d_{ij}^o plus a smaller noise n_{ij}. The distribution of this noise can be approximated as a normal distribution. Hence, we can make the following assumptions about the noise in global TDOA [25]: 1. The noise in TDOA values originates from errors in signal propagation and errors introduced by cross-correlation techniques. 2. The distribution of this error can be approximated as normal.

Considering the errors between d_{ij} follows a multivariate normal distribution, under this condition, its CRLB takes a simpler form, which is beneficial for the analysis and design of the algorithm's optimality. It is noteworthy that traditional TDOA algorithms do not use all measurements d_{ij}, but rather perform a Gauss-Markov estimation before choosing a reference node for localization. This transformation process is particularly sensitive to NLOS noise, with NLOS signals introducing significant errors into each TDOA measurement after transformation, making it nearly impossible to eliminate NLOS errors in the final localization result. This is a primary motivation for exploring methods beyond the Gauss-Markov estimation for full-set TDOA localization.

In TDOA localization algorithms, several essential elements are required: the coordinates of multiple sensor nodes S, the TDOA measurement values d_{ij} or the corresponding distance difference measurements $r_{ij} = c \cdot d_{ij}$ (where c is the speed of signal propagation in the medium), and the noise cross-correlation matrix Q among the TDOA measurement values d_{ij}. The coordinates of sensor nodes are acquired through manual calibration or self-calibration techniques, and their accuracy will ultimately affect the precision of the full set TDOA; TDOA values are obtained through GCC between signals received by pairs of nodes. With n nodes, this method can yield up to $n(n-1)/2$ linearly independent TDOA values. Generally, these TDOA values contain line-of-sight errors introduced by signal distortion, cross-correlation algorithms, etc., which can be calculated after numerous measurements to obtain their cross-correlation matrix Q.

For uniform representation, subsequent parts will use the measurement r_{ij}, considering the signal propagation speed instead of TDOA measurements, with its true value denoted as r_{ij}^o, which can be defined as:

$$r_{ij}^o = \|u^o - s_i\| - \|u^o - s_j\|. \tag{8.29}$$

Here, u^o and s_i respectively represent the true position of the target to be located and the positions of the ith node. Squaring both sides of the above equation yields a linear expression:

$$r_{ij}^{o\,2} + 2r_j^o r_{ij}^o + 2(s_i - s_j)^T u^o + \|s_i\|^2 - \|s_j\|^2 = 0. \tag{8.30}$$

Incorporating the error term $n_{ij} = r_{ij} - r_{ij}^o$ existing in the real-world positioning scenario into the above formula and neglecting second-order noise yields a noisy linear expression:

$$2r_i^o n_{ij} \approx r_{ij}^2 + s_j^2 - s_i^2 + 2s_i - s_j^T u^o + 2r_{ij} r_j^o. \tag{8.31}$$

On this basis, a least squares expression is constructed:

$$Bn \approx h - G\theta, \tag{8.32}$$

where θ represents a vector containing the true coordinates u and auxiliary variables. B, G, and h are respectively matrices or vectors that can be calculated through node coordinates S and measurement values r_{ij}. Therefore, the unknown vector θ can be directly calculated using weighted least squares:

$$\theta = (G^T W G)^{-1} G^T W h, \tag{8.33}$$

where W is the known error weight matrix. This yields a closed-form weighted least squares solution, upon which further optimization can be performed to enhance its positioning performance, making it reach the CRLB performance. This solution can serve as an initial estimate for a more accurate solution.

Simulation results indicate in Fig. 8.6 that the SDP approach can achieve superior performance under low noise conditions and is among the closest to the MLE among various schemes. However, this outcome is only ideal when the channel conditions are perfect, and the collected TDOA values contain only LOS information. When the set

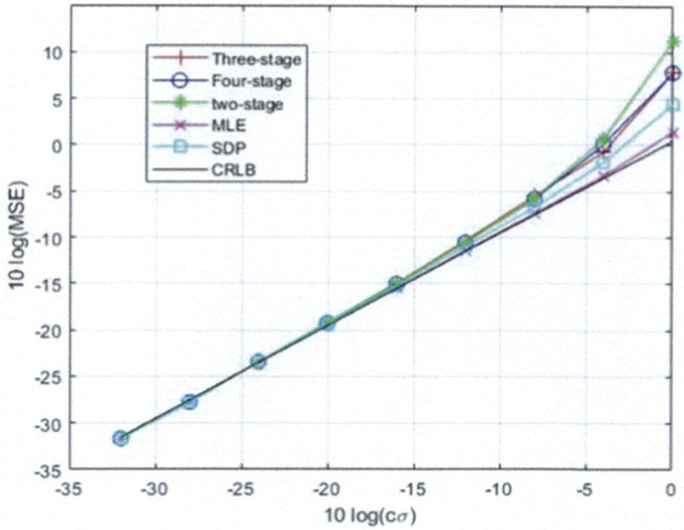

Fig. 8.6 Simulation in full set TDOA

includes NLOS information, relying solely on SDP techniques can result in significant errors. Moreover, in real-world scenarios, NLOS noise conditions, influenced by the channel and the signals used, and caused by reflection, refraction, diffraction, and other factors, vary. Thus, using a probability distribution to model NLOS noise is not a universally applicable solution. Further research will focus on developing a universal NLOS signal discrimination scheme to broaden the applicability of this approach.

8.5 Experimental Results

To validate the simulation conclusions of this chapter, a series of localization experiments using near-ultrasound and comparative trials with UWB (Ultra-Wideband) base stations were conducted inside the cable tunnels at the National Grid Experimental Center in Nanjing. The layout and the experimental base station equipment are illustrated in Fig. 8.7. The purpose of using a space with bends was to test the diffraction capabilities of ultrasound and UWB positioning systems in NLOS environments. Both UWB and ultrasound base stations were mounted on the same tripod, targeting the device to be located simultaneously. To prevent interference between different signals that could make it impossible to distinguish the arriving signals, Time Division Multiple Access (TDMA) was employed for signal separation. Given the narrowness of the cable tunnel, the environment could be approximated as a one-dimensional space for localization. The coordinates of the test positioning points and the measurements from different devices are shown in Table 8.1. Points A, B, and

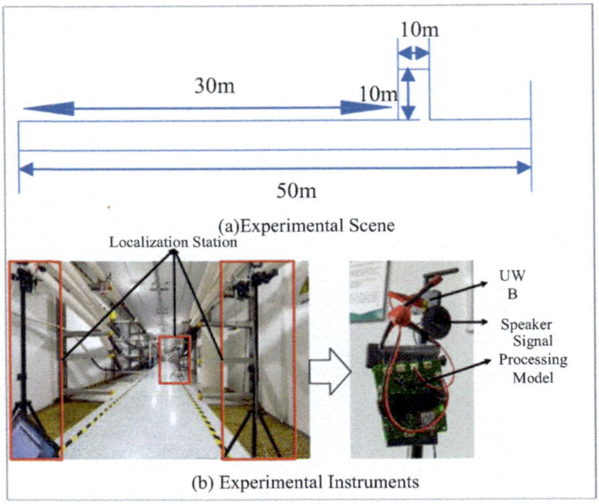

Fig. 8.7 Experimental scene and instrumentation

Table 8.1 Localization errors comparison

Pos.	Coord. (m)	Near-Ultrasonic error (m)	UWB error (m)
A	10	0.0901	0.3628
B	20	0.0621	0.2060
C	30	0.2162	0.3503
D	40 (NLOS)	0.6562	Unable to connect
D*	40	0.1601	Unable to connect
E	50	0.3295	0.5020

C represent linear one-dimensional localization measurements. Point D represents localization measurements in an NLOS bend space. Since the ranging values can be obtained through post-calibration means, the average error was calculated based on the consistency of signal arrival times (standard deviation).

It is noteworthy that in Table 8.1, at point D located 40 m away in a NLOS scenario, the UWB base station struggles to establish a signal connection between stations through diffraction, thus failing to measure accurate position coordinates. This demonstrates that ultrasound positioning is more suited for application in NLOS and bending environments than UWB positioning. In LOS conditions, the accuracy of ultrasound ranging can achieve very good results, comparable to that of UWB. In NLOS conditions, ultrasound positioning might fluctuate. Here, using the 2 σ of Gauss distribution rule to remove outliers, it's observed that the positioning accuracy remains high, meeting the requirements of the positioning system.

Figure 8.8 shows the error Cumulative Distribution Function (CDF) graph for near-ultrasound and UWB positioning at different points. Combined with the test point coordinates from Table 8.1 and the error comparison between near-ultrasound

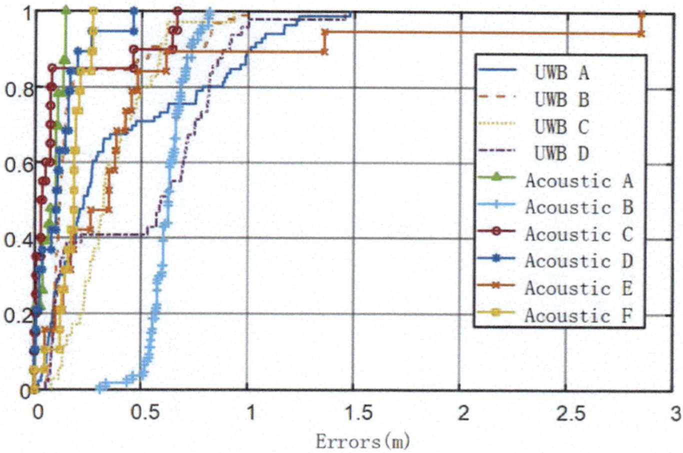

Fig. 8.8 Experimental results

and UWB positioning, as well as Figure 8.8, it can be concluded that in the narrow, metal-shielded, and bending space of cable tunnels, the testing performance of near-ultrasound base stations exceeds the existing UWB positioning technology in both accuracy and stability.

The evaluation study uses the root mean square error (RMSE) as a performance measure, and it is defined as $RMSE\sqrt{\sum_{i=1}^{C}\|\mathbf{u}^{(i)}-\mathbf{u}^{o}\|^2/C}$, where $\mathbf{u}^{(i)}$ is the estimate from the i-th run by the Monte Carlo (MC) method, and C is the number of MC experiments. If there is no extra explanation, the RMSEs of different methods are generated by 1000 MC trials.

We start with the source localization scenarios in 2D space (the dimension $K = 2$), where the source and M sensors are uniformly distributed in a square with a side length of 100 m. In each MC trial, the source broadcasts ranging signal $M - 1$ times, where M sensors can hear the signals from the source. In the $j-$th broadcast, sensor j is the reference sensor and the TDOA measurements $r_{j+1,j}, r_{j+2,j}, \ldots, r_{M,j}$ are recorded. Thus we can obtain a suitable full TDOA set, where the noise correlation matrix $\mathbf{R}$ consists of $M - 1$ diagonal blocks and the diagonal elements in each block are 1 and the other elements 0.5. If there is no additional explanation, the measurements in the full TDOA set have the same variances, so the covariance matrix $\mathbf{Q} = \sigma^2\mathbf{R}$, where σ^2 is the measurement noise power. The above scenario is just one way to obtain a full TDOA set in the simulations. In the real world the measurements are from the results of GCC, and the corresponding covariance matrix can be evaluated by real data.

A practical NLOS rejection method's performance should be guaranteed in different scenarios, including the LOS environments. Although localizing using LOS measurements is not a challenge for the current approaches, they may cause consid-

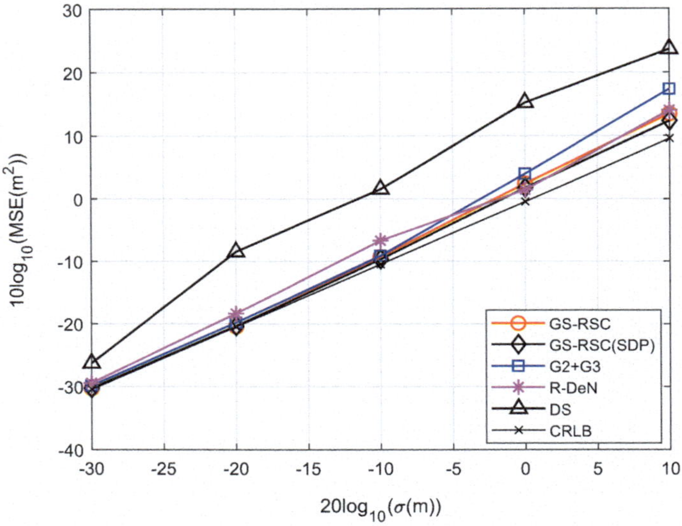

Fig. 8.9 The algorithms' MSEs versus the LOS noises' levels

erable performance loss since many LOS TDOAs are mistaken as NLOS. Therefore, it is essential to investigate whether the proposed GS-RSC scheme's performance is promising with LOS measurements.

LOS noises with known variance σ^2 are added to the true range difference measurements to simulate the noisy TDOA measurements. Besides the proposed method, we additionally introduce several state-of-the-art full-set algorithms. The G2+G3 [7], R-DeN [22] and data-selective (DS) [1] are included for comparison. Besides, we also perform the CRLB in Section as the criterion. The G2+G3and R-DeN are just NLOS measurement detection algorithms without localization capability. In order to give a fair comparison, we use a closed-form solution, Chan algorithm [6] or the CFS solution in this chapter to localize the source after removing the NLOS measurements.

The results of the compared full-set algorithms are shown in Fig. 8.9 versus multiple noise levels, which are generated by 1000 MC trials in a network of 7 sensors. Besides the CFS, we also provide the results generated by the SDP-FS solution, as the 'GS-RSC(SDP)' in Fig. 8.9. Due to their data-cleaning strategies, the DS and R-DeN methods have a little localization performance loss. The DS only chooses a stable number of measurements and discards the other useful ones that are from LOS, causing performance loss, especially when the LOS noise power is equal to or larger than 0.1 m^2. The R-DeN method needs to decide a stopping threshold when calculating, which limits its performance for a small LOS noise level. GS-RSC and G2+G3 perform similarly relative to the CRLB since their rejection modules nearly keep all the LOS signals, especially when the noise level is insignificant. GS-RSC (SDP) is more robust when σ^2 is bigger than 1 m^2, i.e., $20log_{10}(\sigma(m)) = 0$. Since the

Table 8.2 Algorithms' identification accuracy

Method	Accuracy of finding outliers				Accuracy of finding the
	n = 0 (%)	n = 2 (%)	n = 4 (%)	n = 6 (%)	NLOS measurements (%)
GS-RSC	100.0	98.6	97.3	96.0	100.0
G2+G3	100.0	96.8	94.2	84.1	1.6
R-DeN	100	99.0	85.8	53.2	0.0

GS-RSC (SDP) outperforms the other methods only when the LOS noise is significant, if its performance is close to the GS-RSC method, we do not reveal its results in the subsequent figures.

The localization scenarios with outliers have been examined in the former numerical examples. Besides, NLOS propagation is also a common phenomenon worthy of discussion. Specifically, if the path between a sensor, e.g., $\mathbf{s}_1$, to the source is occupied by obstacles, the corresponding range measurement will contain a positive bias. Therefore, the TDOA measurements related to $\mathbf{s}_1$ share the same NLOS bias. In the following numerical examples, we consider a randomly selected sensor $\mathbf{s}_i$ which is suffering from the NLOS propagation, and the corresponding range measurement r_i is the sum of its true value r_i^o and an NLOS bias η_i. The NLOS-path error η_i is uniformly distributed, which satisfies $\eta_i \sim (20, 40)\ m$. Due to the NLOS propagation, the full set contains six relevant NLOS measurements. Besides, we randomly chose 0 to 4 TDOAs to become outliers to simulate a more complex situation. First, we examine the full-set algorithms' ability to search outliers. Table 8.2 reveals the identification accuracy of different methods with the ability to classify when the variance of LOS noise is 0.1 m^2. We also show whether they can accurately remove the measurements polluted by the NLOS-path error. The results reveal that although the G2+G3 and R-DeN methods can still find the outliers under the interference of the NLOS path, the NLOS path is a latent harmful factor that they cannot detect, significantly limiting their localization performance.

Furthermore, in Fig. 8.10, we show the algorithms' localization performance versus the LOS noise's level. Since the G2+G3 and R-DeN methods cannot eliminate the influence carried by the NLOS path, their performance is dominated by the undetected NLOS measurements, which means that the NLOS path will significantly aggravate the localization error, even though they can deal with outliers. On the contrary, the GS-RSC and DS approaches can still maintain high accuracy and efficiently identify the NLOS measurements. Especially, the GS-RSC still achieves the CRLB performance in the presence of the NLOS path and outperforms the DS method in most cases.

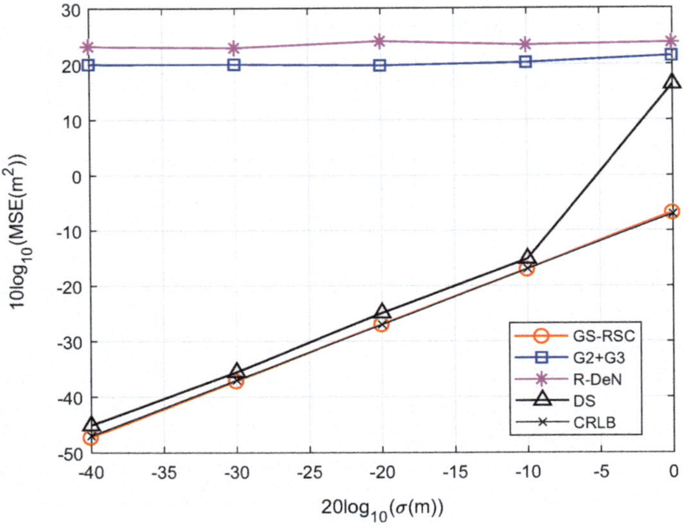

Fig. 8.10 The algorithms' MSEs versus the LOS noise levels with an NLOS path

8.6 Conclusion

In conclusion, this chapter offers a comprehensive exploration into leveraging near-ultrasonic signals for precise indoor positioning in environments where traditional GPS and RF-based methods falter. Through rigorous experimentation and innovative algorithmic approaches, the authors effectively demonstrate the viability and superiority of acoustic-based localization techniques, particularly in complex, satellite-denied indoor environments characterized by obstacles and NLOS challenges.

The research meticulously addresses the limitations of current indoor positioning systems by integrating acoustic signal processing with advanced mathematical modeling, such as SDP and TDOA techniques, to enhance accuracy and reliability. The experimental setup within the National Grid Experimental Center's cable tunnels provided real-world validation of the theoretical models, underscoring the potential of near-ultrasonic technology to outperform UWB systems in specific scenarios.

Key findings include the robust performance of SDP approaches under low noise conditions and the critical role of NLOS signal discrimination in maintaining high accuracy levels across a range of environmental conditions. The study's foray into node auto-calibration and the utilization of TDMA for signal distinction further illustrate the depth of technical innovation and practical applicability of the proposed solutions.

However, the authors also acknowledge the challenges that arise in highly obstructed or NLOS environments, where acoustic signal distortion can diminish the effectiveness of cross-correlation techniques used for TDOA estimation. This limitation underscores the need for ongoing research and development of more universally applicable NLOS discrimination strategies to broaden the scope of acoustic indoor localization technologies.

Overall, the chapter contributes significantly to the field of indoor positioning, presenting a viable acoustic-based framework that promises enhanced safety, precision, and cost-efficiency for Location-Based Services (LBS) in complex indoor spaces. Future work focusing on refining NLOS discrimination and exploring the integration of acoustic localization with other sensor technologies could further elevate the efficacy and application range of this promising approach.

References

1. Apolinário JA, Yazdanpanah H, Nascimento A, de Campos ML (2019) A data-selective ls solution to TDoA-based source localization. In: ICASSP 2019–2019 IEEE international conference on acoustics, speech and signal processing (ICASSP), IEEE, pp 4400–4404
2. Biswas P, Ye Y (2004) Semidefinite programming for ad hoc wireless sensor network localization. In: Proceedings of the 3rd international symposium on information processing in sensor networks, pp 46–54
3. Biswas P, Lian TC, Wang TC, Ye Y (2006) Semidefinite programming based algorithms for sensor network localization. ACM Trans Sens Netw (TOSN) 2(2):188–220
4. Biswas P, Liang TC, Toh KC, Ye Y, Wang TC (2006) Semidefinite programming approaches for sensor network localization with noisy distance measurements. IEEE Trans Autom Sci Eng 3(4):360–371
5. Buja A, Swayne DF, Littman ML, Dean N, Hofmann H, Chen L (2008) Data visualization with multidimensional scaling. J Comput Graph Stat 17(2):444–472
6. Chan YT, Ho KC (1994) A simple and efficient estimator for hyperbolic location. IEEE Trans Signal Proc 42(8):1905–1915
7. Compagnoni M, Pini A, Canclini A, Bestagini P, Antonacci F, Tubaro S, Sarti A (2017) A geometrical-statistical approach to outlier removal for TDoA measurements. IEEE Trans Signal Proc 65(15):3960–3975
8. Doherty L, El Ghaoui L et al (2001) Convex position estimation in wireless sensor networks. In: Proceedings IEEE INFOCOM 2001. Conference on computer communications. Twentieth Annual Joint conference of the IEEE computer and communications society (Cat No 01CH37213), vol 3. IEEE, pp 1655–1663
9. Gualda D, Ureña J, Alcalá J, Santos C (2019) Calibration of beacons for indoor environments based on a digital map and heuristic information. Sensors 19(3):670
10. Ji C, Chen M, Sun P, Shu H, Wang Z (2020) An encoding and decoding scheme for long-distance ultrasonic localization. In: 2020 39th Chinese control conference (CCC). IEEE, pp 5247–5252
11. Jia N, Shu H, Wang X, Xu B, Xi Y, Xue C, Liu Y, Wang Z (2022) Smartphone-based social distance detection technology with near-ultrasonic signal. Sensors 22(19):7345
12. Jia N, Cui W, Wang Y, Xue C, Liu G, Wang X, Cao Z, Wang Z (2023) Robust acoustic TOA estimation based on multipath extraction in frequency domain. In: 2023 13th international conference on indoor positioning and indoor navigation (IPIN). IEEE, pp 1–7
13. Lazik P, Rowe A (2012) Indoor pseudo-ranging of mobile devices using ultrasonic chirps. In: Proceedings of the 10th ACM conference on embedded network sensor systems, pp 99–112
14. Liu B, Zhu X, Jiang Y, Wei Z, Huang Y (2019) UAV and piecewise convex approximation assisted localization with unknown path loss exponents. IEEE Trans Veh Technol 68(12):12396–12400

15. Liu K, Liu X, Li X (2015) Guoguo: enabling fine-grained smartphone localization via acoustic anchors. IEEE Trans Mob Comput 15(5):1144–1156
16. Liu Y, Zhang W, Yang Y, Fang W, Qin F, Dai X (2019) RAMTEL: robust acoustic motion tracking using extreme learning machine for smart cities. IEEE Internet Things J 6(5):7555–7569
17. Shang Y, Ruml W (2004) Improved MDS-based localization. In: IEEE INFOCOM 2004, vol 4. IEEE, pp 2640–2651
18. Shang Y, Ruml W, Zhang Y, Fromherz MP (2003) Localization from mere connectivity. In: Proceedings of the 4th ACM international symposium on mobile ad HOC networking and computing, pp 201–212
19. Shi Q, He C, Chen H, Jiang L (2010) Distributed wireless sensor network localization via sequential greedy optimization algorithm. IEEE Trans Signal Process 58(6):3328–3340
20. Simonetto A, Leus G (2014) Distributed maximum likelihood sensor network localization. IEEE Trans Signal Process 62(6):1424–1437. https://doi.org/10.1109/TSP.2014.2302746
21. Tomic S, Beko M, Dinis R (2014) RSS-based localization in wireless sensor networks using convex relaxation: noncooperative and cooperative schemes. IEEE Trans Veh Technol 64(5):2037–2050
22. Velasco J, Pizarro D, Macias-Guarasa J, Asaei A (2016) TDoA matrices: algebraic properties and their application to robust denoising with missing data. IEEE Trans Signal Process 64(20):5242–5254
23. Wang Y, Ho K, Wang Z (2023a) Robust localization under NLOS environment in the presence of isolated outliers by full-set TDoA measurements. Signal Processing 212:109159
24. Wang Y, Sun P, Wang Z (2023b) Towards low-complexity state estimation for rigid bodies based on range difference measurements. Electron Lett 59(22):e13020
25. Wang Y, Sun P, Wang Z (2023c) Towards robust and accurate cooperative state estimation for multiple rigid bodies. IEEE Trans Veh Technol
26. Wang Z, Zheng S, Ye Y, Boyd S (2008) Further relaxations of the semidefinite programming approach to sensor network localization. SIAM J Optim 19(2):655–673
27. Wang Z, Zhang H, Lu T, Gulliver TA (2018) Cooperative RSS-based localization in wireless sensor networks using relative error estimation and semidefinite programming. IEEE Trans Veh Technol 68(1):483–497

Chapter 9
Scalable and Accurate Floor Identification via Crowdsourcing and Deep Learning

Fuqiang Gu, You Li, Yuan Zhuang, Jingbin Liu, and Qiuzhe Yu

Abstract Understanding the floor-level location of a user in a multi-storey building is crucial for various applications, including emergency response and shopping guides. Current floor identification systems face several challenges, such as low accuracy, the requirement for time-consuming site surveys, assumptions about user encounters, initial floor knowledge, and poor generalization. In this chapter, we present UnFI, a novel floor identification system that is both scalable and accurate, eliminating the need for site surveys, initial floor knowledge, and other assumptions. The system leverages widely-available smartphone sensors to determine a user's floor location. By automatically recognizing the ground floor and utilizing the stable pressure difference between floors, we avoid the need for cumbersome site surveys for fingerprint association. To ensure precise floor identification, we have developed deep learning-based methods for indoor/outdoor detection and floor identification. Experimental results demonstrate that UnFI outperforms existing systems and shows great potential for large-scale deployment.

9.1 Introduction

Floor identification is important for a variety of applications and services such as emergency response and rescue, shopping guide, and other location-based services [1–4]. For instance, the floor information of a fire scene will assist the first responders to quickly decide the rescue manner and hence save lives or property to maximum extent. Existing floor identification methods can be categorized as: cellular

F. Gu (✉)
College of Computer Science, Chongqing University, Chongqing, China
e-mail: gufq@cqu.edu.cn

Y. Li · Y. Zhuang · J. Liu
State Key Laboratory of Information Engineering in Surveying, Mapping and Remote Sensing, Wuhan University, Wuhan, China

Q. Yu
Meituan Co., Beijing, China

K. Yu (ed.), *Positioning and Navigation Using Machine Learning Methods*, Navigation: Science and Technology 14, https://doi.org/10.1007/978-981-97-6199-9_9

fingerprinting [5, 6], WiFi fingerprinting [7–10], and barometer-based methods [11, 12].

Cellular fingerprinting is similar to WiFi fingerprinting [13], as both involve a training phase and an identification phase, as illustrated in Fig. 9.1. During the training phase, the received signal strength (RSS) from visible cellular towers or WiFi access points (APs) is recorded along with the corresponding floor information. The vector of RSS collected at a specific location is termed a fingerprint. In the identification phase, the user sends a floor identification query with the measured RSS values to the server, and the floor level is inferred using machine learning methods like K-nearest neighbors (KNN).

The primary challenge for cellular/WiFi fingerprinting methods in floor identification is floor association, which involves associating collected cellular/WiFi RSS measurements with the correct floor levels. Most prior works depend on a site survey process to label the collected fingerprints, a task that is both time-consuming and labor-intensive. To reduce the labeling burden, some researchers have proposed barometer-based methods [11]. Instead of using wireless signal measurements, these methods use barometric readings to determine the user's floor. However, this approach requires all devices to be equipped with a barometer, limiting its applicability. Fusion methods combining WiFi RSS and barometer readings have been proposed to improve floor identification accuracy, but they also require devices to be equipped with a barometer [12, 14, 15].

The primary objective of this study is to eliminate the need for a site survey while expanding the applicability of floor identification methods. We assume that some devices are equipped with a barometer, and these barometric readings are used to help construct a WiFi radio map (fingerprint database). Using barometers during the creation of the radio map removes the necessity for a labor-intensive site survey. Once the WiFi radio map is established, it can be used for floor identification by all devices, including those without barometers. This significantly broadens the applicability of the floor identification method.

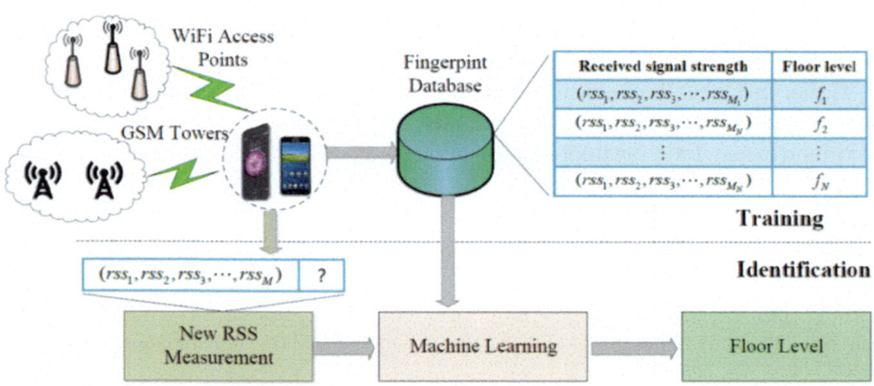

Fig. 9.1 Cellular/WiFi fingerprinting

Specifically, we introduce a novel identification method called UnFI, which uses WiFi fingerprints to determine a user's floor level. Compared to existing methods, UnFI is more scalable and accurate, and it does not require a site survey, initial floor knowledge, or other assumptions. Floor association is achieved by utilizing barometer readings when available and by automatically detecting the ground floor using GNSS and magnetometer measurements. This process is crowdsourced, requiring no effort from the user and eliminating the need for initial floor level input. To enhance floor identification accuracy, we propose a deep learning-based method. Our approach is evaluated through experiments conducted in a multi-storey office building and a shopping mall. The results demonstrate that our method achieves higher identification accuracy than existing methods and performs consistently across different phone models.

9.2 Related Work

Floor identification has attracted the attention of researchers in recent years. An early method for floor identification is SkyLoc [5], which uses widely available cellular signals to identify the current floor of a user in multi-floor buildings. SkyLoc can achieve a floor identification accuracy of about 73% by selecting features with high relevance for fingerprint matching. However, it requires to manually build a radio map consisting of cellular RSS and corresponding floor levels, which is time-consuming and labor-intensive. Also, the achieved accuracy is not quite satisfactory. Ai et al. [7] propose a method that uses WiFi signals to locate the floor and accelerometer and barometer readings to detect the floor change. While it reaches a high accuracy of 99%, it requires an intensive site survey process to construct a radio map. In [17], a deep learning-based AP-independent floor identification method is introduced, which leverages WiFi signals to generate images that are then fed to a convolutional neural network for floor identification. Zhang et al. [6] presents a floor identification method using cellular signals, which first uses denoising autoencoder for data noise reduction and feature extraction, and then utilize a Long Short-Term Memory (LSTM) network for floor identification. Qi et al. [9] introduce the confidence interval of WiFi signals, on which they further develop a fast floor identification method. However, such methods still require troublesome site survey.

To expedite the troublesome site survey process, researchers have made several efforts. Khaoampai et al. [8] propose a method called FloorLoc-SL, which collects WiFi fingerprints via a self-learning algorithm. While FloorLoc-SL does not require a site survey process, it achieves only an accuracy of 87% and asks the user to input the initial floor number when starting the system. FTrack [18] locates the floor by using smartphone accelerometer readings to detect the traveled floors. While FTrack reports an accuracy of 90%, it is not robust to varying device orientation and user's motion states, and requires the knowledge of initial floor level. F-Loc [19] improves the FTrack by considering both WiFi signals and accelerometer readings, and reports an accuracy of 95%. F-Loc constructs the WiFi radio map through crowdsourcing and smartphone sensing, but it relies only on accelerometer readings for detecting elevators and stairs, which is not robust if the user does not follow the assumed pattern. B-Loc [11] uses only barometer fingerprints to identify the floor level without requiring WiFi infrastructure. By constructing a barometer fingerprint map via crowdsourcing, B-Loc achieves an accuracy of 98%. However, B-Loc is only applicable for the barometer-equipped devices, which is still not pervasively available these days. BarFi [12] combines WiFi RSS with barometric pressure for floor localization. By utilizing a two-phase clustering method to train the RSS radio map, BarFi reaches an accuracy of about 96%. The main limitation of BarFi is that it requires to manually calibrate the barometric pressure among different devices and the user needs to give the initial floor information. A similar method is presented in [20], but the fusion of WiFi RSS and barometer readings is implemented through a Monte Carlo Bayesian algorithm and a Kalman filter. Different from the methods above, UnFI, a floor identification method proposed in this chapter, requires neither site survey and knowledge of floor plans, nor the initial floor information and manual calibration.

9.3 Proposed Method

Figure 9.2 provides an overview of the proposed UnFI method, which consists of a training phase and an identification phase. During the training phase, various sensor readings from smartphones are collected to create a fingerprint database. Specifically, GNSS, WiFi, and barometer data are gathered through a crowdsourcing approach, requiring no effort from the user, such as a site survey or manual input of the initial floor level. In the identification phase, the user's current floor number is determined by comparing the measured WiFi RSS with the data stored in the fingerprint database. The testing device can be a low-end phone equipped only with WiFi. To ensure high floor identification accuracy, we have developed a deep learning-based method, which will be detailed later.

UnFI consists of three key components: ground floor detection, floor association, and floor identification. The ground floor detection and floor association occur during the training phase, while floor identification takes place in the identification phase.

Ground Floor Detection: This component detects the ground floor by utilizing features extracted from GNSS and magnetometer measurements. To establish the ground floor's pressure value, the user must walk on the ground floor of the building at least once, recording sensor data (GNSS, WiFi, magnetometer, and barometer data) for a period of time, such as 5 min. The barometric pressure recorded on the ground floor serves as a reference value for determining different floors.

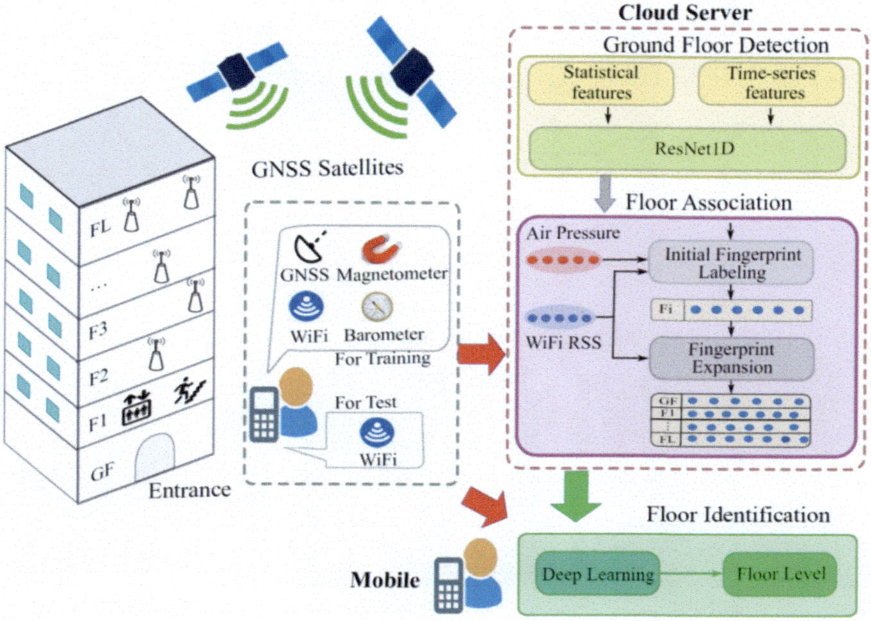

Fig. 9.2 Overview of UnFI

Floor Association: Using the reference barometric pressure value, the floor association component associates the collected WiFi RSS measurements with the corresponding floor levels. To expedite the construction of the fingerprint database, we employ semi-supervised learning with sensor data sequences that may not include ground floor data.

Floor Identification: After the fingerprint database is built, it is used to identify the current floor level of a user via a deep learning method.

The following subsections will elaborate on each of these three components.

9.3.1 *Ground Floor Detection*

We identify the ground floor by utilizing features extracted from GNSS and magnetometer measurements to detect the transition between indoor and outdoor environments. This ground floor information is crucial for associating WiFi RSS measurements with the corresponding floor levels. Although the light sensor has been effectively used for indoor/outdoor (IO) switch detection [21], it is influenced by the phone's orientation and weather conditions, requiring the user to hold the phone so

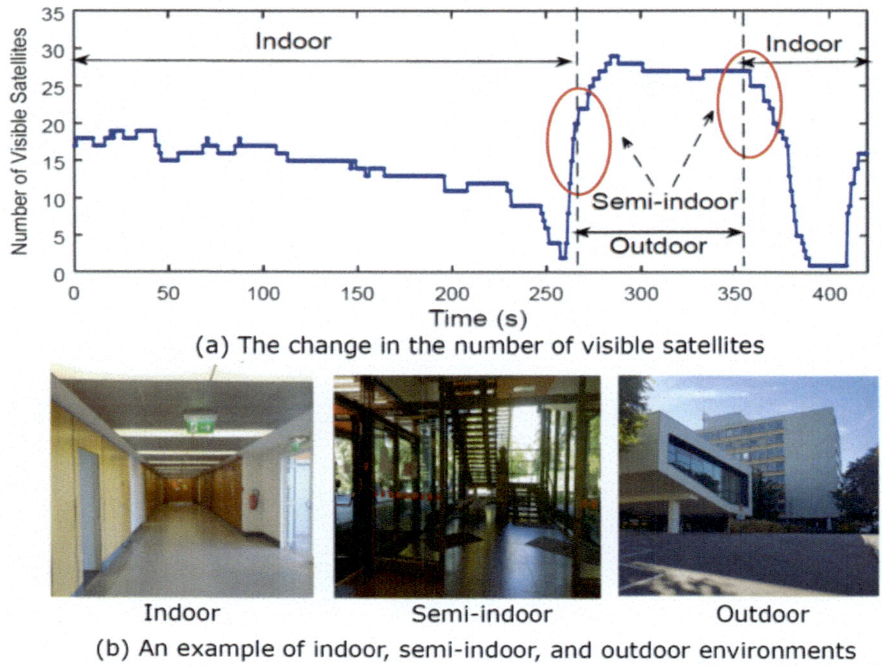

Fig. 9.3 An example of using the change in the number of visible satellites for IO switch detection. **a** The number of visible GNSS satellites changes as the user exits and enters a building. **b** Indoor, semi-indoor (entrance or exit areas), and outdoor scenarios

that the screen faces the ceiling or sky. To develop a more robust ground floor detection method, we rely on GNSS and magnetometer signals to detect the IO switch (Fig. 9.3).

Let Φ_{GNSS} denote the sequence of the GNSS measurements, namely

$$\Phi_{GNSS} = (g_0, \ldots, g_t, \ldots g_T) \tag{9.1}$$

where g_t is the GNSS measurements at time t, and T is the ending time of collecting this sequence. Similarly, the magnetometer measurement sequences and barometric pressure sequences are expressed as:

$$\Phi_{Mag} = (m_0, \ldots, m_t, \ldots m_T) \tag{9.2}$$

$$\Phi_{Pressure} = (p_0, \ldots, p_t, \ldots p_T) \tag{9.3}$$

where m_t and p_t represent the magnetic field and barometric pressure collected at time t, respectively. It should be noted that the length of the three sequences are usually different due to the varying sampling rates of different sensors. However, we can use time interpolation method to easily make the three sequences with the same length. Therefore, we use the same measurement index in Eqs. (9.1)–(9.3). To achieve the IO switch detection, we need to segment each of these three measurement sequences into N shorter sequences using a sliding window. Thus, we can obtain three measurement sequence sets $S_{GNSS} = \left\{S_{GNSS}^1, \ldots, S_{GNSS}^N\right\}$ from Φ_{GNSS}, $S_{Mag} = \left\{S_{Mag}^1, \ldots, S_{Mag}^N\right\}$ from Φ_{Mag}, and $S_{Pressure} = \left\{S_{Pressure}^1, \ldots, S_{Pressure}^N\right\}$ from $\Phi_{Pressure}$.

To achieve robust IO detection, we extract 9 different features from satellite measurement sequences, including the number of visible GNSS satellites, the mean, variance, standard deviation, maximum, minimum, median, range, and interquartile of visible satellite carrier-to-noise ratio (CNR). Similarly, 8 different features are obtained from magnetometer sequences, including mean, variance, standard deviation, maximum, minimum, median, range, and interquartile. Based on these extracted features, we use the popular ResNet [22] neural network to detect IO switch due to its excellent performance. However, the original ResNet was proposed for dealing with images, which is not directly suitable for our case. Therefore, we modify the popular ResNet network from 2D network into 1D network to adapt the WiFi-based floor identification.

The ground floor identification algorithm using the GNSS and magnetometer measurements is described in Algorithm 1. This algorithm takes as input the sensor reading sequences Φ_{GNSS}, Φ_{Mag}, and $\Phi_{Pressure}$, and outputs the reference pressure pg and time Tg on the ground floor. Firstly, the GNSS measurement sequence, magnetometer sequence, and pressure sequence are segmented into sequences using a sliding window. Based on the resulting sequences of GNSS and magnetometer measurements, we extract time-series features and statistical features. After extracting features, the ResNet1D method is used to detect IO scenarios. If the user is detected moving from indoor to outdoor or from outdoor to indoors, the median timestamp in the several GNSS measurement sequences, which corresponds to being indoor before or behind the switch happens, is considered as the reference time T_g on the ground floor. Note that n_1 and n_2 between line 11–15 are constant parameters to avoid detection error, which are both set to 3 in this work. Finally, based on the reference time, the sequence of pressure whose timestamps is closest to the reference time is selected, and the average value of this pressure sequence is considered as the reference pressure P_g on the ground floor.

Algorithm 1: Ground Floor Detection

Input : Trace of the GNSS readings Φ_{GNSS}, trace of the magnetometer readings Φ_{Mag}, and trace of barometric pressure $\Phi_{Pressure}$
Output: Reference barometric pressure P_g and time T_g on the ground floor
Initialize $T_g \leftarrow 0$, $P_g \leftarrow 0$, $R \leftarrow \emptyset$;
Segment traces into sequences and obtain S_{GNSS}, S_{Mag}, and $S_{Pressure}$;
//IO detection by using the GNSS and Magnetometer measurements
for $(i \leftarrow 0; i++; i < |S_{GNSS}|)$ **do**
    Extract features f_{GNSS} from the GNSS measurement sequence S^i_{GNSS};
    Extract features f_{Mag} from the magnetometer measurement sequence S^i_{Mag};
    $R(i) \leftarrow ResNet1D(f_{GNSS}, f_{Mag})$;
end
//Determine the moment on the ground floor
for $(i \leftarrow 0; i++; i < |R|)$ **do**
    if $R(i-n_1 : i) == Indoor \&\& R(i : i+n_2) == Outdoor$ **then**
        Record the median time t_{GNSS} in the sequences corresponding to $R(i-n_1 : i)$;
    end
    if $R(i-n_1 : i) == Outdoor \&\& R(i : i+n_2) == Indoor$ **then**
        Record the median time t_{GNSS} in the sequences corresponding to $R(i : i+n_2)$;
    end
end
//Obtain the reference pressure and time on the ground floor
$j \leftarrow findSequence\,(S_{Pressure}, t_{GNSS})$;
$P_g \leftarrow mean\left(S^j_{Pressure}\right)$.

9.3.2 *Automatic Fingerprint-Floor Level Association Via Crowdsourcing*

Fingerprint-floor level association consists of two stages: initial labeling and fingerprint expansion. During the initial labeling stage, RSS measurements from sequences that traverse the ground floor and multiple floors are labeled using the reference pressure and time data collected on the ground floor. In the fingerprint expansion stage, RSS measurements from all sequences, even those not including ground floor data, are labeled through semi-supervised learning.

(1) **Initial labeling**

We first obtain the reference pressure and time on the ground floor, and then associate the RSS measurements in the sequence with corresponding floor levels according to the change of barometric pressure. Figure 9.4 shows the change of barometric pressure across multiple floors. It is observed that different phones witness different barometric pressure values though they are placed on the same floor and the barometric pressure varies obviously as the floor changes. This implies that one cannot directly use the absolute values of barometric pressure to identify floors. However, the pressure difference between two floors is relatively stable and independent of

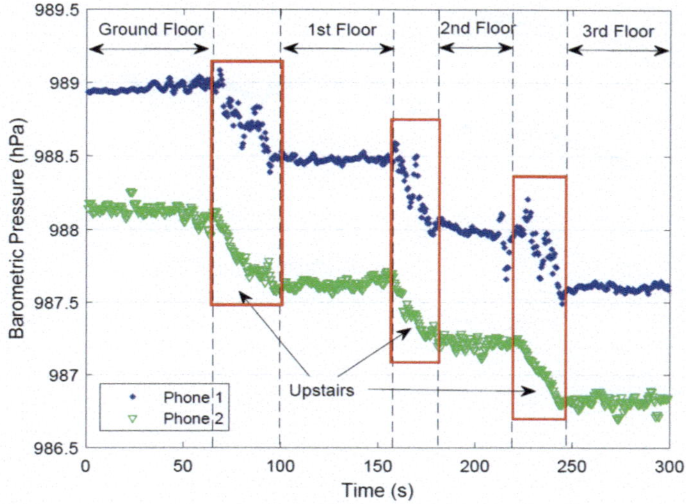

Fig. 9.4 The change of barometric pressure across multiple floors (the user carrying two phones of different models at the same time and walking across multiple floors). It shows that the relative pressure difference is stable between different floors regardless of phone models

phone models. This means that we could recognize different floors by using the relatively stable pressure difference. It should be also noted that the barometer readings are affected by the environment temperature, humidity, altitude (or height), and the used devices [11]. Figure 9.5 shows that the barometric pressure varies over time and different devices report different barometric values even when they are put on the same desk to measure the barometer readings during the same period of time.

Fortunately, the barometric pressure in indoor environments is relatively stable during a short period of time (e.g., 10 min). The pressure variation (around 0.1 hPa) on the same floor during the short period of time is much smaller than the difference (around 0.4 hPa given the floor height is about 3 m) between different floors. Thus, it is feasible to use the pressure difference during a short period of time to recognize the floor change.

(2) **Fingerprint Expansion**

After obtaining a small set of labeled fingerprints from initial labeling stage, we use semi-supervised learning to obtain more labeled fingerprints in order to achieve a higher floor identification accuracy (fingerprint expansion). In the fingerprint expansion stage, only WiFi RSS measurements and barometer readings collected from the traces crossing multiple floors are required.

Here we first give the pseudo-code of the fingerprint expansion algorithm in Algorithm 2, and then present our analysis. This algorithm takes as input a dataset consisting of RSS measurements (represented by Φ_{RSS}) and barometer readings (represented by $\Phi_{Pressure}$), along with labeled fingerprints acquired during the initial

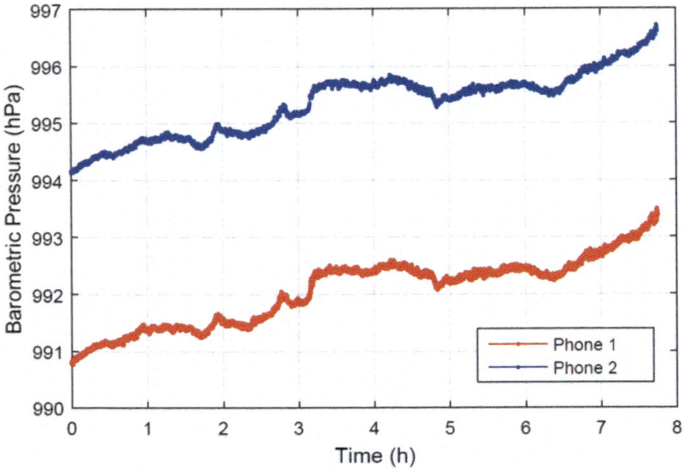

Fig. 9.5 The change of barometric pressure over time at the same location point. It implies that the barometric pressure is only relatively stable at the same point during a short period of time (e.g., 10 min)

labeling stage (denoted by L). Initially, it identifies the entry and exit points of stairs and elevators using pressure measurements, with the respective timestamps noted as T_b. Subsequently, data captured while traversing stairs and elevators are excluded to ensure the fidelity and accuracy of the fingerprint repository. The remaining RSS data are then divided into sequences $\{S_{RSS}\}_{i=1}^{N}$, where N signifies the sequence count. The relative floor labels F_r are derived from pressure readings and are sequentially numbered either upwards or downwards (increasingly for descending and decreasingly for ascending floors). The estimated floor labels for each RSS sequence are recorded as F_i. Here, F_r is a vector of length N and F_i is a vector of length n_i, where n_i is the number of samples in the ith RSS sequence.

It should be noted that the acquired floor labels F_i may be prone to inaccuracies due to classification errors, necessitating refinement through neighboring constraints from F_r. For instance, fingerprints collected on a single floor might be erroneously categorized into different floors. To mitigate this, the algorithm simultaneously evaluates the proximity relationships inferred from pressure measurements and those derived from the classified results of fingerprint sequences. During the refinement process, each floor label within the initial sequence's floor set is sequentially considered as the starting floor. The corresponding probability of accurate classification (denoted by p_j) is calculated by dividing the number of fingerprints with labels matching the presumed label l_1 by the total number of fingerprints in the sequence. Subsequently, this probability p_j is iteratively adjusted through a similar process applied to subsequent sequences. Upon updating all probabilities, the initial floor with the highest probability is identified. Consequently, the absolute labels corresponding to the unlabeled fingerprints are obtained and integrated into the fingerprint database L alongside the existing fingerprints.

Algorithm 2: Fingerprint Expansion via Semi-supervised Learning

Input : a set of unlabeled traces $\mathcal{D}_{\mathbf{u}} = \{< \Phi^u_{RSS}, \Phi^u_{Pressure} >\}^U_{u=1}$, where U is the number of traces, consisting of unlabeled RSS and barometric pressure measurements, and a set of initial labeled fingerprint database $\mathcal{L}$

Output: An updated fingerprint set $\mathcal{L}$ with more fingerprints

for $u \leftarrow 1$ *to* U **do**
- $\mathbf{T}_b$ = detectFloorBoundary($\Phi^u_{Pressure}$);
- Remove sensor readings between two floors (e.g., during stairs) by using $\mathbf{T}_b$;
- Segment the RSS data Φ^u_{RSS} into sequences $\{S_{RSS}\}^N_{i=1}$ by using $\mathbf{T}_b$ and obtain corresponding relative floor labels F_r numbering from 0;
- **for** $i \leftarrow 1$ *to* N **do**
  - $F_i \leftarrow$ KNN($\mathcal{L}$, $S_{RSS}(i)$); //Obtain the initial labels
- **end**
- //Refine the estimated labels using F_r
- Calculate the unique values of F_1 and denote as $\mathbf{U}_1$;
- **for** $j \leftarrow 1$ *to* $|\ \mathbf{U}_1\ |$ **do**
  - ℓ_1 = $\mathbf{U}_1$(j); // Assume one of the initial labels is correct
  - $p_j \leftarrow \sum_{i=1}^{|F_1|}(\ell_1 == F_1(i))/|F_1|$; //Corresponding probability of being correctly classified
  - **for** $k \leftarrow 2$ *to* N **do**
    - ℓ = ℓ_1 - F_r(k);
    - $p_j \leftarrow p_j \cdot \sum_{i=1}^{|F_k|}(\ell == F_k(i))/|F_k|$;
  - **end**
- **end**
- $\hat{\ell}_1$ = $\arg\max_{\ell_1}(p_j)$; //Refined floor label for the first sequence
- Refine all the labels for other sequences according to $\hat{\ell}_1$ and F_r;
- Add new labeled data into $\mathcal{L}$;

end

Return $\mathcal{L}$;

9.3.3 *Floor Identification*

After constructing the required fingerprint database, we can identify the floor level for incoming RSS measurements. Let x represent a fingerprint, which is a vector of WiFi RSS measurements, namely

$$x = (rss_1, rss_2, \ldots rss_M) \tag{9.4}$$

With M denoting the total number of visible WiFi APs in the environment, Since raw WiFi RSS values are negative, we adopt the positive RSS data description method in [23] and transform raw RSS values into positive values, namely

$$prss_i = \begin{cases} (rss_i - rss_{\min}), & \text{if AP}_\text{i} \in x \text{ and } rss_i > \tau \\ 0, & otherwise \end{cases} \tag{9.5}$$

where τ is a RSS threshold, which is set to -100 dBm in this study, indicating whether a WiFi AP is detected in a fingerprint, and $rss_{\min}$ is the minimum RSS from WiFi APs. These APs with RSS lower than τ are considered as not-detected. Thus, x can be re-written as:

$$x = (prss_1, prss_2, \ldots prss_M) \tag{9.6}$$

To achieve high-accuracy floor identification, we develop a deep learning-based method. It includes four 1-dimensional convolutional layer (Conv1D), two fully-connected (FC) layers, one dropout layer and one voting layer. Note that each convolutional layer is followed by a batch normalization layer to make the network faster and more stable. Following the normalization layer, the ReLU function is used for capturing useful features. The network also includes two skip connections to better capture useful features. Since the amount of available dataset for floor identification is usually not very large, the depth of the proposed network is enough to achieve high-accuracy floor identification. We also include one dropout layer to make the network more robust to noisy data or data from different devices.

To train the network, we adopt the Adam optimizer [24] to minimize the cross-entropy loss. The initial learning rate of network is set to 0.001 until the iteration reaches up to 100, after which the learning rate is reduced by a factor of 0.5 every 10 iterations. The total number of iterations is set to 200, and the batch size is set to 4. To conduct a robust evaluation, we run the model for multiple rounds with different random seeds, and report the mean value of accuracy.

9.4 Experiments and Results

The proposed method was evaluated by experiments conducted in an office building and a shopping mall. The office building has eight floors, including elevators, staircases, corridors, common rooms, and office rooms; while the shopping mall has seven floors, and includes elevators, staircases, and stores. The experiments conducted in both the office building and the shopping mall cover seven floors and the number of visible APs is about 1505 and 1466, respectively. Two smartphones of different brands were used to collect sensor data. The training data include WiFi RSS measurements, GNSS measurements, and barometric readings, while the test data include WiFi RSS measurements only. During the experiments, three participants were asked to walk along different paths crossing multiple floors. Note that the participants might skip some floors, meaning that the floors visited during a trajectory may not be consecutive. The time spent on collecting one trajectory of data lasts about 10 min to half one hour, and the variation of pressure for the same floor is below 0.15 hPa. The data collection for each scene lasts two or three days. In total, six sets of data were collected in the office building, where four sets were used for training (namely initial

Table 9.1 Dataset profile

Scenarios	Training trajectory no.	Training sample no.	Test sample no.
Office building	1	699	4878
	2	2414	
	3	3537	
	4	5293	
Shopping mall	1	2294	8429
	2	3577	
	3	5096	
	4	5842	

Note that both training datasets and test datasets have no floor label information. Training datasets include WiFi RSS measurements, barometric readings, magnetic measurements and GNSS measurements, which are required for initial labeling and fingerprint expansion. Test datasets include only WiFi RSS measurements

labeling and fingerprint expansion) and two sets for testing. In the shopping mall, nine sets of data were collected, where five were used for training, and four for testing. The dataset profiles are given in Table 9.1.

9.4.1 Accuracy of IO Switch Detection

We first evaluate the proposed IO switch detection method with other popular methods including Support Vector Machines (SVM), Random Forests (RF), and Naive Bayes (NB) based on the measurements of GNSS and magnetometer. These measurements are first segmented into sequences using a window of 3 s, and the resulting sequences are then split into training data (80%) and test data (20%). The proposed IO detection is based on the popular ResNet network (specifically ResNet-18). However, ResNet is used to deal with images and cannot be directly used for dealing with GNSS and magnetic data. Therefore, we modify the popular ResNet network from 2D network to 1D network to adapt our case, and re-train the whole network. We use accuracy as the performance metric for IO switch detection and floor identification. The accuracy is defined as the probability of detection, which is defined as the ratio of correctly predicted samples over the total number of samples.

Table 9.2 shows that the proposed method performs the best, achieving an accuracy of 97.7%, which is higher than other methods. The proposed method can also make use of multi-modal data, which is not true for RF, NB, and SVM methods. It can be also seen that GNSS-based methods usually have better performance than magnetometer-based methods. This might be attributed to that GNSS measurements in outdoor environments are more distinguishable from indoor environments due to the obstacles such as buildings and trees. The high-accuracy IO detection ensures that

Table 9.2 Accuracy comparison of IO detection

Sensor	Method	Accuracy (%)
GNSS	SVM	93.5
	NB	91.2
	RF	96.6
	ResNet1D	96.8
Magnetometer	SVM	70.5
	NB	60.8
	RF	71.2
	ResNet1D	71.4
GNSS + Magnetometer	SVM	93.3
	NB	90.6
	RF	96.3
	Ours	97.7

we can accurately detect the ground floor by voting the results from several sequences, and further enables us to automatically and accurately label WiFi fingerprints.

9.4.2 Floor Identification Performance

We have first conducted experiments to evaluate the floor association accuracy, and experimental results show that the proposed floor association method can correctly associate all the fingerprints with corresponding floors with an accuracy of 100%. This is due to the use of the stable characteristics of the barometer readings that the pressure difference between two floors is stable during a short period of time.

We then show the floor identification performance of the proposed method using different amount of training data and compare its accuracy with that of state-of-the-art methods. Note that this training data is not labeled manually, but is automatically labeled by detecting the ground floor and using the relatively stable air pressure difference between two floors. The baseline methods are K-Nearest Neighbors (KNN) [25] (the number of k is set to 3), SVM [26], NB [27], RF [28], and Autoencoder (AE) [29]. For the AE, we use two layers for pretraining with 512 and 256 neurons, respectively, and add a softmax layer on the top to identify floors.

Figure 9.6 shows the performance of the proposed method and the five baseline methods on the data collected in the office building. We can see that the floor identification accuracy of different methods generally increases when the amount of used training data rise. When using the data collected from one trajectory for training, KNN performs the best (84.4%), which is followed by AE (84.1%) and our method (83.8%). When using more training data, the performance of the proposed method is

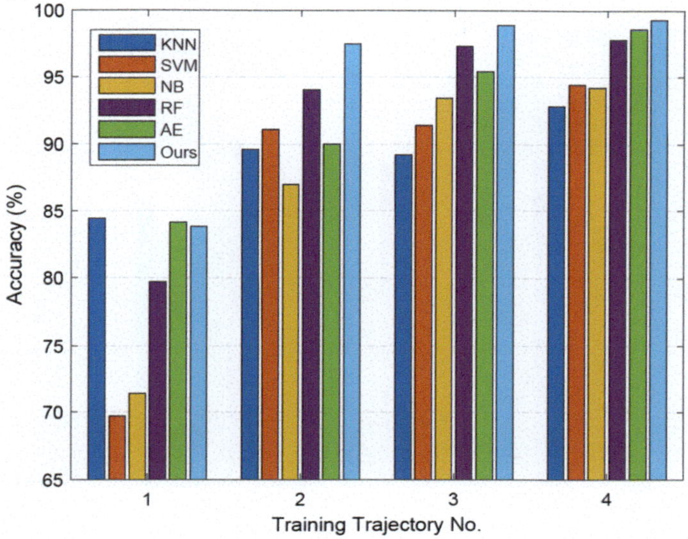

Fig. 9.6 The floor identification accuracy of different methods in the office building

significantly improved, and reaches up to 99.3% with training data from four trajectories, which is higher than the baseline methods. This is because our method uses more layers and needs more data to be well trained.

Figure 9.7 demonstrates the accuracy of these methods on the data collected in the shopping mall. One can find that the proposed method outperforms the baseline methods except for the case of using the data from three trajectories for training, where the AE method (91.9%) performs slightly better than our method (91.3%). When using the data from four trajectories for training, our method can achieve an accuracy of 98.6%, which is significantly higher than AE (86.1%), RF (85.8%), SVM (83.8%), KNN (80.7%), and NB (62.4%). However, using more training data does not necessarily improve the floor identification accuracy of some methods since more training data means that there are more APs visible, resulting in a higher input dimension. This can be justified by the decrease in the floor identification accuracy of KNN and RF when the training data increases from three trajectories to four trajectories.

9.4.3 Effect of Different Smartphones

We analyze the effect of different smartphones on the floor identification algorithms. In the first case, only one set of training data is used. The training data were collected by phone 2, and the test data were collected by the same phone 2 and another phone (phone 1). In the second case, multiple sets of training data are used, and both phones

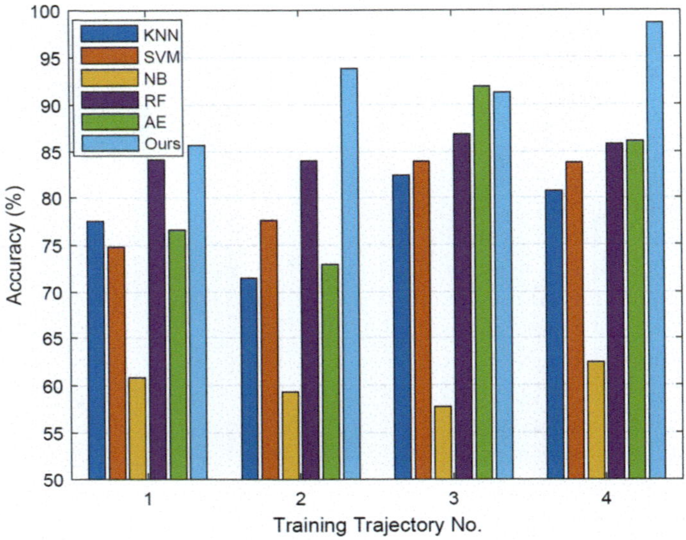

Fig. 9.7 The floor identification accuracy of different methods in the shopping mall

were used for collecting training data and test data. During the data collection, both phones were held in hand together by the same participant in the two cases. Figures 9.8 and 9.9 show the floor identification accuracy of different methods using one set of training data and multiple sets of training data, respectively. We can see that all the methods perform better when the test data are collected from the same phone used to collect training data. Moreover, the proposed method outperforms significantly the baseline methods, and is much more robust to hardware diversity/heterogeneity of different phones. When using multiple sets of training data, the effect of hardware diversity is significantly reduced, which is justified by the accuracy improvement for both methods shown in Fig. 9.9 compared to that shown in Fig. 9.8. It is also observed that the achieved accuracy with phone 2 is higher than that with phone 1. This might be because the WiFi sampling rate of phone 2 is about 1.5 times higher than that of phone 1 and the WiFi signal strength of phone 2 is more stable.

9.4.4 Computational Cost

We finally compare the computational cost of the proposed method with the baseline methods. These algorithms were run on a PC with an Intel i9-10900K CPU and a NVIDIA GeForce RTX 2080 GPU. From Fig. 9.10, we can see that AE witnesses the lowest computational cost due to its shallow structure, which is followed by NB and SVM. The consumed time of the proposed method is about 5.8 s, which is about 1.5 times higher than the KNN method that has highest computational cost among

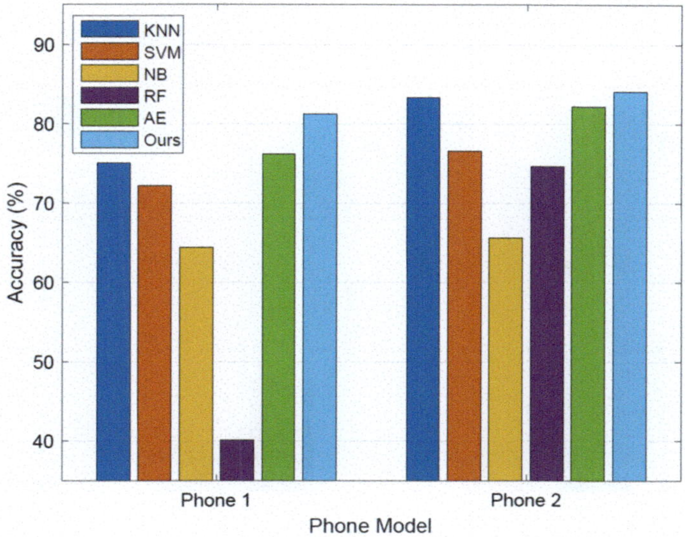

Fig. 9.8 The floor identification accuracy of different methods using one set of training data

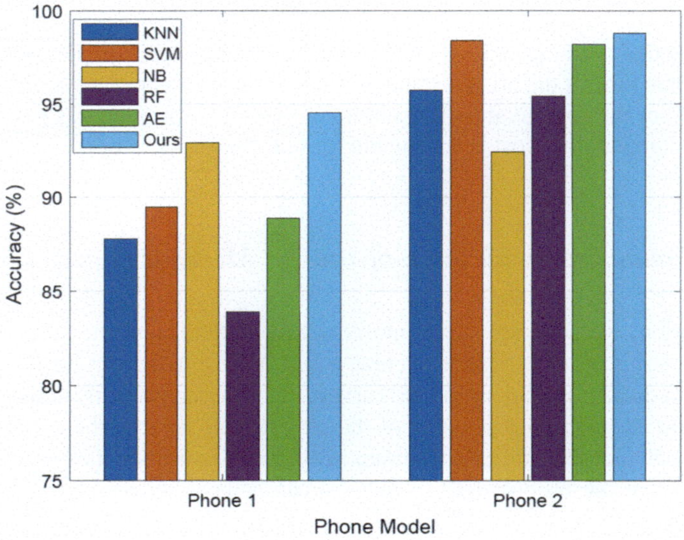

Fig. 9.9 The floor identification accuracy of different methods using multiple sets of training data

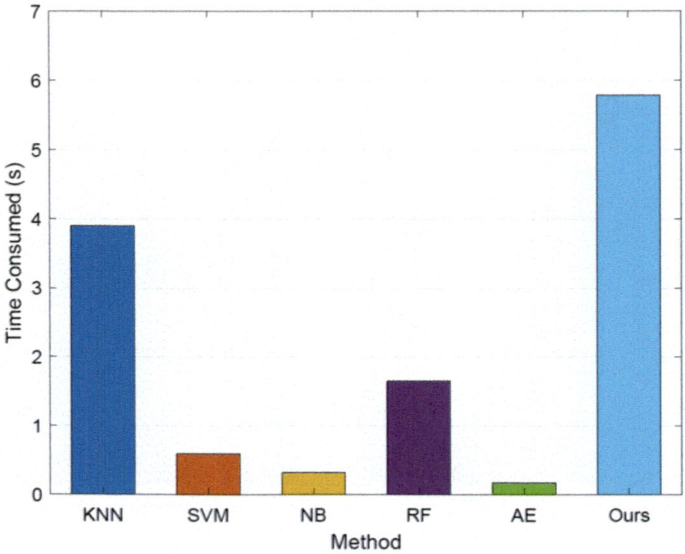

Fig. 9.10 Computational cost comparison of different methods

the five baseline methods. Given that the number of test data samples is 4878, the computational time of the proposed method is about 1.2 ms per sample, which is still very low. Since the computational capability of modern smartphones is powerful, we think the computational cost is acceptable for real-time localization.

9.4.5 Comparison with State-of-the-Art Methods

In this section, we compare the proposed UnFI method with the state-of-the-art methods in terms of identification accuracy, sensors required for collecting training data and test data, requirement for site survey, and other constraints or assumptions. Note that the accuracy for the state-of-the-art methods is from the corresponding papers and obtained from different datasets in different environments. From Table 9.3, we can see that the proposed UnFI method can achieve a very competitive floor identification accuracy (about 99%) and has no requirement for site survey. Existing methods have either the requirement for site survey, which is time-consuming and labor-intensive, or other constraints such as initial floor knowledge and user encounters. For example, B-Loc achieves a similar accuracy as our method, but it assumes that users meet each other at the elevators, which is not a realistic assumption. Also, it needs to give the initial floor information during the construction of barometric map.

Table 9.3 Comparison with the state-of-the-art methods

Method	Accuracy (%)	Sensors		Site survey requirement	Other constraints
		Training	Test		
SkyLoc [5]	73	Cellular	Cellular	Yes	None
Method in [7]	99	WiFi, Baro, Acc	WiFi, Baro, Acc	Yes	None
Method in [14]	97	WiFi, Baro	WiFi, Baro	Yes	Initial floor knowledge
FloorLoc-SL [8]	87	WiFi, Acc, Baro	WiFi	No	Initial floor knowledge for training
F-Track [18]	90	Acc, Bluetooth	Acc, Bluetooth	Yes	User encounters, initial floor knowledge
F-Loc [19]	95	WiFi, Acc	WiFi	No	Initial floor knowledge for training
B-Loc [11]	98	Baro	Baro	No	User encounters, initial floor knowledge
BarFi [12]	96	WiFi, Baro	WiFi	No	Initial floor knowledge for training
StoryTeller [17]	98	WiFi	WiFI	Yes	None
DAE + LSTM [6]	93	Cellular	Cellular	Yes	None
Method in [9]	92	WiFi	WiFi	Yes	None
UnFI (Ours)	99	WiFi, Baro, GNSS, Mag	WiFi	No	None

9.5 Conclusion

In this chapter, we present a novel method that can achieve high-accuracy floor identification without any effort from the user. Different from existing methods, which suffer from varying limitations, our method does not require site survey, user encounters, initial floor knowledge, and other assumptions. Experimental results show that the proposed UnFI can achieve an accuracy of about 99% in floor identification outperforming a number of state-of-the-art methods.

Funding This paper is supported by the National Natural Science Foundation of China (No. 42174050, 41874031, 42111530064), and Venture & Innovation Support Program for Chongqing Overseas Returnees (No. cx2021047).

References

1. Gu F, Hu X, Ramezani M, Acharya D, Khoshelham K, Valaee S, Shang J (2019) Indoor localization improved by spatial context-a survey. ACM Comput Surv 52(3):64–16435
2. El-Sheimy N, Li Y (2021) Indoor navigation: state of the art and future trends. Satell Navig 2(1):1–23
3. Luo J, Zhang Z, Wang C, Liu C, Xiao D (2019) Indoor multifloor localization method based on wifi fingerprints and lda. IEEE Trans Indus Inf 15(9):5225–5234
4. El-Sheimy N, Youssef A (2020) Inertial sensors technologies for navigation applications: state of the art and future trends. Satell Navig 1(1):1–21
5. Varshavsky A, LaMarca A, Hightower J, De Lara E (2007) The skyloc floor localization system. In: IEEE fifth annual ieee international conference on pervasive computing and communications (PerCom'07), pp 125–134
6. Zhang Y, Ma L, Wang B, Qin D (2020) Building floor identification method based on dae-lstm in cellular network. In: 2020 IEEE 91st vehicular technology conference (VTC2020-Spring), pp 1–5
7. Ai H, Liu M, Shi Y, Zhao J (2016) Floor identification with commercial smartphones in wifi-based indoor localization system. Int Arch Photogram Remote Sens Spatial Inf Sci 41:573
8. Khaoampai K, Na Nakorn K, Rojviboonchai K (2015) Floorloc-sl: floor localization system with fingerprint self-learning mechanism. Int J Distrib Sens Netw 11(11), article id: 523403
9. Qi H, Wang Y, Bi J, Cao H, Si M (2019) Fast floor identification method based on confidence interval of wi-fi signals. Acta Geod Geoph 54(3):425–443
10. Ibrahim M, Torki M, ElNainay M (2018) Cnn based indoor localization using rss time-series. In: 2018 IEEE symposium on computers and communications (ISCC), pp 01044–01049
11. Ye H, Gu T, Tao X, Lu J (2016) Scalable floor localization using barometer on smartphone. Wirel Commun Mob Comput 16(16):2557–2571
12. Shen X, Chen Y, Zhang J, Wang L, Dai G, He T (2015) Barfi: Barometer-aided wi-fi floor localization using crowdsourcing. In: IEEE 12th international conference on mobile ad hoc and sensor systems (MASS), pp 416–424
13. Cao X, Zhuang Y, Yang X, Sun X, Wang X (2021) A universal wi-fi fingerprint localization method based on machine learning and sample differences. Satell Navig 2(1):1–15
14. Li Y, Gao Z, He Z, Zhang P, Chen R, El-Sheimy N (2018) Multi-sensor multi-floor 3d localization with robust floor detection. IEEE Access 6:76689–76699
15. Zhao F, Luo H, Zhao X, Pang Z, Park H (2015) Hyfi: Hybrid floor identification based on wireless fingerprinting and barometric pressure. IEEE Trans Industr Inf 13(1):330–341
16. Gu F, Blankenbach J, Khoshelham K, Grottke J, Valaee S (2019). Zeefi: zero-effort floor identification with deep learning for indoor localization. In: IEEE global communications conference (GlobeCom)
17. Elbakly R, Youssef M (2020) The storyteller: Scalable building-and ap-independent deep learning-based floor prediction. Proc ACM Interact Mobile Wearable Ubiquitous Technol 4(1):1–20
18. Ye H, Gu T, Zhu X, Xu J, Tao X, Lu J, Jin N (2012) Ftrack: infrastructure-free floor localization via mobile phone sensing. In: IEEE international conference on pervasive computing and communications (PerCom), pp 2–10
19. Ye H, Gu T, Tao X, Lu J (2014) F-loc: floor localization via crowdsourcing. In: 20th IEEE international conference on parallel and distributed systems (ICPADS), pp 47–54
20. Haque F, Dehghanian V, Fapojuwo AO, Nielsen J (2018) Wi-fi rss and mems barometer sensor fusion framework for floor localization. IEEE Sens J 19(2):623–631
21. Zhou P, Zheng Y, Li Z, Li M, Shen G (2012). Iodetector: a generic service for indoor outdoor detection. In: Proceedings of the 10th Acm conference on embedded network sensor systems, pp 113–126
22. He K, Zhang X, Ren S, Sun J (2016) Deep residual learning for image recognition. In: Proceedings of the IEEE conference on computer vision and pattern recognition, pp 770–778

23. Torres-Sospedra J, Montoliu R, Trilles S, Belmonte OB, Huerta J (2015) Comprehensive analysis of distance and similarity measures for wi-fi fingerprinting indoor positioning systems. Expert Syst Appl 42(23):9263–9278
24. Kingma DP, Ba J (2014) Adam: a method for stochastic optimization. arXiv:1412.6980
25. Hu X, Shang J, Gu F, Han Q (2015) Improving wi-fi indoor positioning via ap sets similarity and semi-supervised affinity propagation clustering. Int J Distrib Sens Netw 11(1):109642
26. Zhang S, Guo J, Wang W, Hu J (2018) Floor recognition based on svm for wifi indoor positioning. In: China satellite navigation conference, pp 725–735
27. Ashraf I, Hur S, Shafiq M, Park Y (2019) Floor identification using magnetic field data with smartphone sensors. Sensors 19(11):2538
28. Zhang X, Sun W, Zheng J, Xue M, Tang C, Zimmermann R (2022) Towards floor identification and pinpointing position: a multistory localization model with wifi fingerprint. Int J Control Autom Syst 20:1484–1499
29. Gu F, Khoshelham K, Valaee S, Shang J, Zhang R (2018) Locomotion activity recognition using stacked denoising autoencoders. IEEE Internet Things J 5(3):2085–2093

Chapter 10
Indoor Floor Detection and Localization Based on Deep Learning and Particle Filter

Chenxiang Lin and Yoan Shin

Abstract In this chapter, we present an infrastructure-independent multi-floor indoor localization scheme. The proposed scheme has several notable features. First, we utilize the strong feature extraction capability of the sequence-to-sequence (Seq2Seq) deep learning model for sequential data to implement real-time step action prediction. We also develop a floor decision algorithm to extract vertical movement information from the step action sequence under a variety of user activities. Second, we configure calibration nodes on the map based on prior knowledge from the environmental information to extend the localization to three-dimensional applications and achieve calibration of the estimation. Third, we introduce a clustering method to improve localization performance in uncertain measurements. The experimental results show that the Seq2Seq model has good robustness to noisy data. Under the long path multi-floor scenario, our scheme achieved a localization accuracy of over 96% within a 2 m error boundary.

10.1 Introduction

The growing demand for location-based services (LBS) has catalyzed active research in the field of localization. Although satellite-based schemes offer robust LBS outdoors, they are inefficient indoors due to obstructions caused by buildings. Because most people spend the majority time inside buildings, the research related to indoor localization become an important research topic. As the most commonly used electronic devices, smartphones possess a combination of micro-electromechanical systems (MEMS) sensors and progressively improving computational capabilities. This

The present chapter is an updated adaptation of the journal article [1].

C. Lin · Y. Shin (✉)
Soongsil University, Seoul, South Korea
e-mail: yashin@ssu.ac.kr

C. Lin
e-mail: CXLin@soongsil.ac.kr

K. Yu (ed.), *Positioning and Navigation Using Machine Learning Methods*, Navigation: Science and Technology 14, https://doi.org/10.1007/978-981-97-6199-9_10

makes them promising platforms for the development of indoor localization solutions. Among these, the most straightforward method to obtain a smartphone user's location is to use received wireless signal information, including ultra-wideband (UWB), Bluetooth low energy (BLE), and Wi-Fi, to name a few [2–7]. The fundamental idea is to utilize the received signal strength (RSS) readings taken online from anchor nodes, and the smartphone then matches these readings to the location of the sample point bearing the closest RSS fingerprint [4]. However, these technologies are not always available in specific situations due to dependence on infrastructure (e.g., disasters, power outages), and manual setup and maintenance are time-consuming and labor-intensive.

As an alternative viable in indoor localization systems, pedestrian dead reckoning (PDR) methods refer to techniques that can perform localization tasks without reliance on external infrastructure. PDR utilizes the inertial measurement unit (IMU) components (e.g., accelerometer, gyroscope, magnetometer) on a smartphone as input devices to iteratively update the user's location. However, since PDR's current estimation is derived from previous states, the estimation errors accumulate as the location changes. Thus, conventional PDR is used in conjunction with a dedicated IMU to ensure accuracy in practical applications. To improve the performance of cheap IMU-based PDR, fusion algorithms have been proposed, such as the gradient descent algorithm (GDA) [8], complementary filter (CF) [9, 10] and Kalman filter (KF) [11–13]. These fusion algorithms efficiently combine data from multiple sensors, mitigating individual sensor errors and reducing noise. They can improve the accuracy of estimated parameters, such as location and orientation, and hence result in a more precise and reliable localization outcome.

Moreover, environmental information can be used as prior knowledge for boundary constraints and location calibration. Since the map information fits conveniently into particle propagation, an efficient way that combines the particle filter (PF) and spatial information was successfully implemented in [14–16]. However, the PF approach suffers from two notorious problems: multimodality and sample impoverishment [17–19]. The multimodality issue in positioning arises from building structures that allow multiple propagation possibilities, especially when the initial location is not provided [20]. On the other hand, particles tend to collapse at one or a few locations when the PF relies too heavily on measurement data, resulting in a loss of diversity, known as sample impoverishment. This can cause failure when particles cross walls due to inevitable noise and get eliminated. Resampling further concentrates particle distribution [21]. These problems are more likely to occur when measurements are insufficient.

Furthermore, the initial state is vital for relative methods like PDR. Although the initial location and height can be acquired when the user enters a building, in most cases the localization process begins indoors where the user's exact location is typically unknown. PF can compute user location without an initial state, by distributing particles on a floor plan, updating them via PDR estimation, and assigning weights until convergence [15]. This requires no infrastructure but high computation because it needs enough particles to cover relevant state space areas. Additionally, since

IMU data lacks absolute location information, many iterations are needed for particle convergence. These problems arise due to insufficient information from PDR, necessitating the inclusion of more user movement data.

Another drawback of PDR is that it only calculates the two-dimensional (2D) location, while most buildings are multi-floor structures. The user's altitude in a multi-floor building can be represented as a floor number, and a three-dimensional (3D) location can be derived by integrating the 2D location with the floor number. Nowadays, most modern smartphones come equipped with barometers, enabling the widespread use of barometric-based methods for floor detection. Numerous floor localization methods based on barometric pressure have been developed [22–25], which can broadly be classified as reference station-based or pressure difference measurement-based floor localization methods [26]. Since station-based methods require the deployment of infrastructure within the environment, they are not considered. On the other hand, a relationship between atmospheric pressure and height h was constructed in [27] as

$$h(p, p_0) = 44330 \cdot \left[1 - \left(\frac{p}{p_0}\right)^{\frac{1}{5.255}}\right], \tag{10.1}$$

where p and p_0 are the current barometer reading and standard pressure at sea level in mbar. However, due to atmospheric pressure error caused by weather factors, the altitude calculated by (10.1) is inaccurate. Therefore, it is generally accepted to estimate the change in altitude through the pressure difference, instead of directly calculating the altitude. Another challenge of the pressure difference measurement-based method is that barometric measurement is also affected by smartphone usage and environmental factors. Although the pressure distribution of each floor, even containing various noises, has a range in an overall view, waiting for enough data to characterize the pressure distribution in a specific region will result in a significant delay, which is not conducive to cooperating with 2D locations. Moreover, height information can not only be combined with 2D location to provide 3D location, but can also be used to optimize 2D localization since both are derived from the user's motion measurements. However, only a few solutions took advantage of this feature to optimize the performance [17, 28]. The reasons could be as follows.

- Delays that result from slow floor transition detection lead to a loss of correlation between height information and 2D location.
- Most floor detection solutions merely compute altitude without the capacity to extract features associated with floor transitions, while altitude alone does not suffice to correct the user's 2D location.

Therefore, the use of height information to improve 2D location requires the ability to detect fast floor transitions and extract vertical motion features. Moreover, despite the wide usage of the PF in indoor localization, handling the state of particles when the user changes floors is a tricky issue, and its optimal strategy for multi-floor scenarios remains an open problem.

To address the aforementioned issues, [29] propose an indoor multi-floor localization scheme that is infrastructure-independent and solely relies on the onboard sensors of a smartphone. Because of the scarcity of wireless signal information, the main challenge for our scheme is determining how to extract as much motion information as possible from the limited sensor data and combine them comprehensively to provide stable, fast, and mobile-friendly indoor multi-floor localization. This chapter is an extended version of our previous work [29]. We introduced a deep learning (DL)-based floor detection that exploits the sequence-to-sequence (Seq2Seq) model to predict the user's step action from time-series barometric data. A floor decision algorithm is developed to not only identify floor transitions and estimate floor numbers from the step action sequence but also extract vertical movement features of a step. Based on the Seq2Seq model's stable performance for noisy data, our floor detection can work regardless of how the smartphone was worn. Next, we design and implement a PF with clustering to fuse sensor data, map information, and floor detection prediction for estimating 2D locations. We introduce mean shift and calibration nodes (CN) matching-based location correction to improve the PF's performance. The mean shift is applied as a clustering algorithm to detect PF divergence, improve location estimation, and reduce the computational complexity without sacrificing performance. The CN matching-based location correction is used to combine prior knowledge from the map and the vertical user movement obtained from the proposed floor detection to accelerate particle convergence, correct the particle state, and provide an effective way to extend the 2D PF to 3D scenarios.

The remainder of this study is structured as follows. Section 10.2 provides a review of related works in the field. Sections 10.3 and 10.4 elaborate on the methodology of our indoor multi-floor localization scheme. Section 10.5 is devoted to the evaluation and analytical examination of our scheme, followed by Sect. 10.6 which presents a detailed discussion, conclusion, and potential avenues for future work.

10.2 Related Works

10.2.1 Localization Based on PDR

PDR has gained widespread adoption on smartphone platforms due to its lightweight implementation and usability in areas lacking wireless signal coverage. A typical PDR system comprises three components: step detection, step length calculation, and heading direction estimation. It provides polar coordinates, such as {step length, heading direction}, to represent steps. The location of a step can be calculated by summing these vectors. Step detection is usually achieved by identifying peaks in vertical acceleration [30]. Other methods such as threshold detection and adaptive detection are also in use [31, 32]. Step length estimation can be derived based on features of the acceleration data, such as peaks/valleys, variance, walking speed, etc. [33–35]. Heading estimation through IMU sensors is generally achieved in two ways: (1) by integrating gyroscope readings to compute the change in angle over a short period of time, and (2) by determining the absolute orientation relative to the north using readings from accelerometers and magnetometers [36]. However, PDR is a recursive method, which inevitably leads to errors accumulating over time. These errors come from missing steps, inappropriate step length compensation coefficients, and distorted directions. Particularly, consumer-grade sensors are subject to cost and size limitations, which results in smartphone sensor-based PDR only maintaining good accuracy over short paths. To overcome these limitations, data fusion techniques are commonly used to combine various sensor inputs from smartphones and additional accessible information about the user.

10.2.2 Floor Detection

When pedestrians move across different areas, smartphone sensor readings exhibit corresponding distinct patterns, which can be utilized for floor detection. There are a few PF approaches that assist in determining the user's vertical movement by propagating particles to predefined vertical transition areas [15, 37]. However, particles tend to move to broader areas, while floor transition zones are generally narrow. When a user moves to another floor, only a small number of correct particles enter the transition zone. Reference [38] demonstrated that magnetic field signals from the smartphone sensor exhibit temporal stability and spatial resolution, proving significantly beneficial for magnetic field cartography. Considering the ubiquitous presence, reliability, and low cost of magnetic field signals, floor detection based on magnetic field measurements have been presented in [39–41]. Altitude calculation based on features extracted from different IMU modes during the user's vertical movement was successfully implemented in [42–44]. Reference [42] proposed two acceleration integration methods to determine height difference, and a mapping table was formed from distinct movement patterns for floor change estimation using travel time and step count [43]. An inherent issue with IMU-based floor detection is that unpredictable actions from the user severely impact IMU measurements. Consequently, these systems typically maintain optimal performance under constant user behaviors. The barometer sensor avoids this problem because its measurement is dominated by atmospheric pressure instead of user motion. Reference [22] used crowdsourcing to develop a barometric fingerprint map for floor localization. This map clustered barometric readings from each floor, using shared timestamps to gather real-time fingerprints. Reference [44] tracked changes in floor location by identifying user's ascending or descending activities based on changes in atmospheric pressure and altitude. Because a floor number only provides a rough range of information, it needs to be combined with other data (e.g., 2D locations) when higher location precision is required. Therefore, besides height estimation, the challenge of the floor detection approach also lies in its effective incorporation with other solutions when implemented within varied systems.

10.2.3 DL in Localization Systems

In recent years, DL has been widely used in the analysis and processing of sensor data, creating significant advances in data feature engineering and providing many solutions in LBS [2, 45–50]. Reference [2] combined UWB localization with a long short-term memory (LSTM) network to predict user locations based on distance information derived from a time of arrival (ToA) distance model, offering enhanced accuracy in UWB localization systems. Reference [48] utilized a bidirectional LSTM architecture to map IMU signals to varying rates of motion, offering robust and accurate velocity estimates even under dynamically changing IMU orientations. Experimental results show that it achieved an error rate less than 0.10 m/s for instantaneous velocity and less than 29 m/km for travelled distance. Reference [49] presented a hierarchical Seq2Seq model, termed DeepHeading, which utilizes spatial transformer networks (STNs) and LSTM technologies. DeepHeading's encoder operates by taking in sensor data over time intervals of one step, while its decoder predicts heading based on state vectors received from the encoder. Reference [50] presented a StepNet, a suite of deep learning-based approaches for predicting step-length or change in distance, which surpassed traditional methods in the trajectories examined. An important aspect of the DL schemes in LBS is that, since the platform for LBS is generally a mobile device, the power consumption and computational complexity must be considered.

10.3 Proposed DL-Based Floor Detection

Figure 10.1 illustrates an overview of the proposed scheme, which consists of two modules: DL-based floor detection and PDR-PF with clustering. As illustrated in Fig. 10.1, the scheme reads barometer and IMU sensor data, while the user is walking in a building. The DL-based floor detection receives barometric data as the input to perform the floor tracking. Meanwhile, the PF incorporates the PDR estimation based on the IMU data, the prediction from floor detection, and the data from the smartphone database to calculate the 2D location. Finally, the results of floor detection and PF are combined to achieve indoor multi-floor localization. The DL-based floor detection is introduced in this section, which is responsible for floor transition detection and floor number calculation.

10.3.1 User Activity Analysis and Floor Detection Scenario

We divide multi-floor localization into two stages: the first stage where the user enters the building and moves around freely, and the second stage where the localization begins. To provide the height information required for initializing 2D localization,

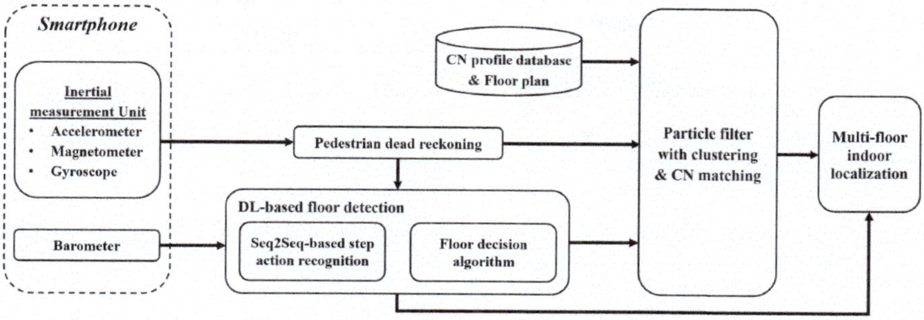

Fig. 10.1 Architecture of our scheme

it is necessary during the first stage to obtain the floor number from the entrance or other technologies (e.g., GPS) [51, 52] and track the floor, while the user moves around with the smartphone. The floor number will be used to provide the correct floor plan in the second stage.

The barometer measurement, primarily based on altitude, is also affected by short-term noise from user activity and the ambient environment, as well as long-term drift caused by weather in practical applications. We presume that for most cases, the height of the device relative to the user's body is within a specific range, as shown in Fig. 10.2. There are several representative modes of smartphone usage listed: (a) calling, (b) typing, (c) swinging, and (d) pocket [53]. Here, (a) represents the highest case of a smartphone, (c) represents the lowest case, and (d) represents the different surroundings. The data in Fig. 10.3 show an example of barometric data collected including the above cases. From Fig. 10.3, means and variances of barometric data collected on the same floor differ due to height and environmental changes, with outliers appearing during usage changes. Our previous work [29]

Fig. 10.2 Height range of the smartphone relative to the body

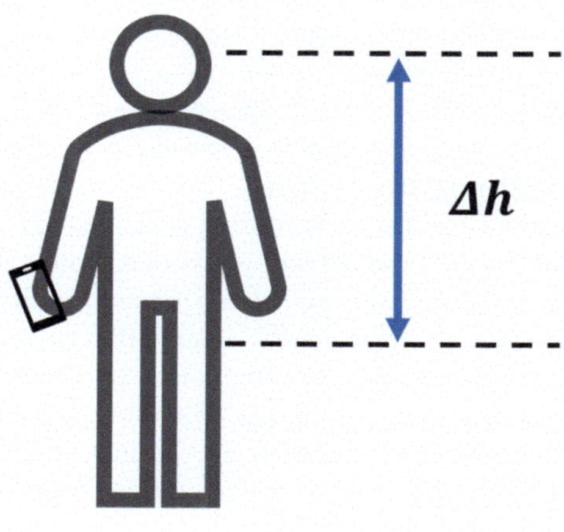

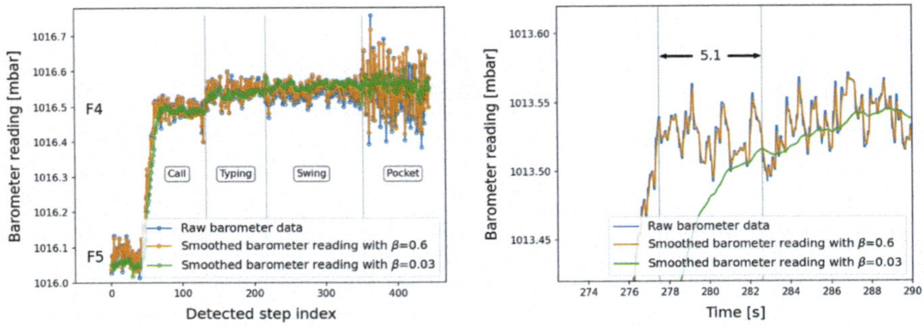

(a) Raw and smoothed barometric data

(b) Smoothing and delay effect

Fig. 10.3 Examples of raw and smoothed barometric data and associated time lag effect

leveraged time-series pressure data for step action recognition, proposing a multi-layered perceptron (MLP)-based method to detect floor transitions from noisy data. However, this assumed the user consistently held the smartphone in front, making the user behavior is easily recognized as the stair step under the free activity scenarios. The typical approach to smoothing these transient pressure fluctuations is to utilize a lowpass filter, such as simple moving average (SMA) or weight smoothing, as follows [54].

$$x_t^d = \frac{\sum_{i=t-m+1}^{t} x_i}{m}, \tag{10.2}$$

$$x_t^d = (1 - \beta) \cdot x_{t-1}^d + \beta \cdot x_t. \tag{10.3}$$

Here, x_t and x_t^d indicate the tth sampling data and smoothed sampling data, respectively, m is the size of the average window, β is the smooth factor, and they are used to control the smoothing effect. The trade-off between delay and smoothing effect is a known problem with the smoothing algorithm. Figure 10.3a and b demonstrate the smoothing and delay effect of (10.3) with different values of β, and they were smoothed under a sampling frequency of 20 Hz. In Fig. 10.3b, the barometer reading with $\beta = 0.6$ only smoothed out a few severe outliers, and caused a delay of less than 0.05 s; while the smoothed barometer reading with $\beta = 0.03$ exhibited a clear height correlation, but the trade-off is causing a delay of 5.1 s, which means that the user may take 6–8 steps before the pressure measurement shows the characteristics of the flat floor. These missing steps constitute a significant error in our system because the floor transition signal generated from floor detection is exploited in the PF component to correct the PF's estimation (to be described in Sect. 10.4.2.6). The time difference between the step action and the 2D position should be as small as possible to ensure their correlation. Therefore, we design a Seq2Seq model that can predict the correct step action from barometric data containing noise and outliers instead of heavily relying on the smoothing filter. In addition, because a delay of 0.05 s is acceptable, we utilize (10.3) with $\beta = 0.6$ to preliminarily smooth the data.

10.3.2 DL Model Selection, Data Processing, and Training Results

We found the potential of the Seq2Seq DL model for handling noisy time-series pressure data. The Seq2Seq is an encoder-decoder framework model using recurrent neural network (RNN) [55] and consists of three components: encoder, decoder, and state vector that connects them. The encoder is responsible for compressing the input sequence into a state vector as the initial hidden state of the decoder, and then the decoder predicts the probability of each class from the state vector. The Seq2Seq model can deal with various tasks such as many-to-many, many-to-one, and one-to-many, as shown in Fig. 10.4. We apply the Seq2Seq model to the many-to-one task. Unlike MLP that is the most widely employed DL model, the output of Seq2Seq is determined by both current and previous inputs, thus it is well suited for handling sequences such as time-series data. In this chapter, the Seq2Seq model receives time-series pressure data that includes the previous and current barometer readings to confirm whether the current barometric fluctuation is caused by noise or height change.

The training data was collected from Hyeongnam Engineering Building which is a typical large multi-floor building at Soongsil University, with a height of 17 cm for each stair. Regarding the data collection, a barometer reading is recorded in the smartphone database once a step is detected. The step's label is determined based on the region where the user is located. For example, if a user enters the upstairs at the 30th step and exits the staircase at the 60th step, then the labels for the 30th to the 60th steps would be assigned as "Going up." Moreover, unlike IMU sensors, barometric measurements are primarily driven by changes in altitude. Thus, the impact of individual user characteristics (e.g., weight, gender, height) on barometric measurements is negligible compared to the noise induced by user activity. The primary impact of different users on barometric measurements comes from their walking styles on the staircase. Therefore, we ensure the inclusion of

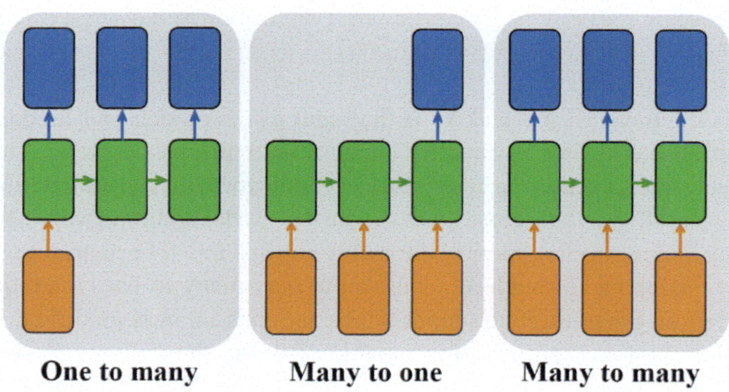

Fig. 10.4 Various tasks of the RNN DL model

features from various movement patterns in the training data by randomly taking one or two stairs while climbing stairs during data collection. There were 3,726 barometric data collected for model training, which included 14 events of ascending stairs and 14 events of descending stairs. Next, we applied a data augmentation method to the collected data, as follows.

$$\delta_i = x^d_{i+1} - x^d_i, \tag{10.4}$$

$$x^a_1 = x^d_{last} - \delta_1, \tag{10.5}$$

$$x^a_i = x^a_{i-1} - \delta_i, \tag{10.6}$$

where δ is the pressure difference of adjacent steps, x^d_k is the kth smoothed pressure data, x^d_{last} is the last pressure data of the collected dataset, and x^a is the barometer data generated by data augmentation. Through data augmentation, the ascending and descending pressure data can be mutually transformed. A thorough analysis on the reasoning behind, feasibility, and optimization strategies for ensuring high dataset quality in this data augmentation method can be found in [29]. We concatenated x^d and x^a as training data x^w. By performing data augmentation, the size of the training dataset was increased to 7,098, and the number of events for ascending and descending stairs was expanded to 28.

Subsequently, a sliding window method was used to convert the data into learnable forms, as follows.

$$X^w_k = \{x^w_{k-s+1}, \ldots, x^w_k\}, \tag{10.7}$$

where s stands for the window size. X^w_k is a subset of the dataset, which contains the pressure change from the previous s steps. Its label is determined by the label of the kth step. It is recommended that the value of s be between 10 and 20 to ensure that the barometer sequence of size s is sufficient to represent the pressure change information over a short period, and $s = 15$ in this study. Next, mean centering is used to shift the feature's center to 0.

$$X_k = X^w_k - \mu_k = \{x_{k-s+1}, \ldots, x_k\}, \tag{10.8}$$

where μ_k is the mean of X^w_k and X_k is the input of the model. Our model predicts the step action based on the pressure changes over the past s steps. The reason for employing mean centering instead of normalization or standardization lies in our desire to shift the data close to 0 to aid the model training, while refraining from scaling operations that modify the data's original units. Furthermore, the main advantage of using fixed-length X_k as the input (i.e., many-to-one) to predict a step action instead of generating the output whenever each input is read (i.e., many-to-many), is that the Seq2Seq model is capable of fitting the trajectories of different lengths well and eliminates the effect of outliers that accumulate over time and the weather factors. The performance of the many-to-many approach becomes unstable

Table 10.1 Hyperparameters used in model training

Model parameters	Values
Number of trainable parameters	2,355
Activation function of hidden layers	Tanh
Activation function of output layer	Softmax
Initializer	Xavier uniform
Loss function	Cross entropy
Optimizer	Adam
Learning rate	0.001
Number of epochs	30

in long path scenarios, which results from the accumulation of outliers in previous inputs. Additionally, early barometer data have little correlation with the current step action. In contrast, the fixed-length input means that the model's prediction only depends on past s measurements and has approximate performance for a sequence with arbitrary lengths. Furthermore, since pressure fluctuations typically require tens of minutes to hours to produce a significant altitude drift [56], a pressure sequence with a size of 15 which corresponds to a pressure change over a 10 s-period, enables the avoidance of long-term errors arising from weather factors.

Table 10.1 lists the hyperparameters adopted in the Seq2Seq model. The hyperbolic tangent (tanh) was used instead of sigmoid for faster and better training. The Xavier uniform initialization was utilized as a weight initializer to make the variance of the output of each layer roughly equal to the variance of its input, to prevent the gradients from becoming too large or too small during training [57]. They are both hyperparameters commonly used in RNN training, and their definitions are given in (10.9) and (10.10), where the fan_{in} and fan_{out} indicate the number of input units and output units in the weight tensor, respectively.

$$\tanh(x) = \frac{e^{2x} - 1}{e^{2x} + 1}, \tag{10.9}$$

$$W_{i,j} \sim \mathcal{U}\left(-\sqrt{\frac{6}{fan_{in} + fan_{out}}}, \sqrt{\frac{6}{fan_{in} + fan_{out}}}\right). \tag{10.10}$$

Figure 10.5 provides an overview of the proposed Seq2Seq DL model. The decoder and encoder of the model are both composed of an LSTM layer with 16 hidden units [58]. The model is initialized using the Xavier uniform initialization, then sequentially receives past k barometric data. As the model processes the time-series data, it utilizes the hidden state transitions of the Seq2Seq model to extract the temporal dependencies in the data, thereby effectively extracting the inherent features

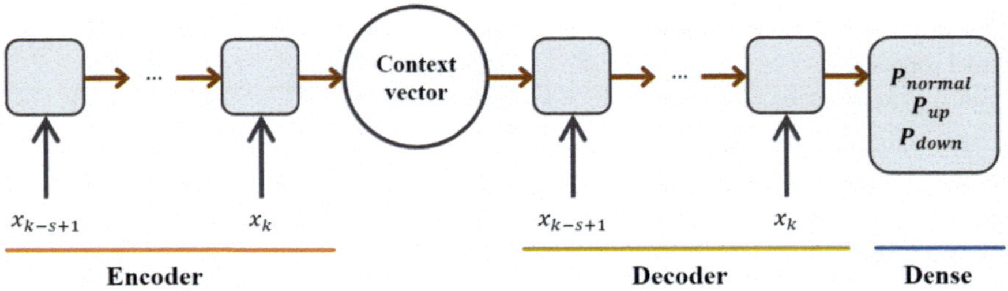

Fig. 10.5 Seq2Seq DL model overview

associated with ascending or descending stairs. Subsequently, the dense (or fully-connected) layer outputs the probabilities of three distinct classes: "Normal," "Going up," and "Going down." The model then updates its weights according to the ground truth. Each of these actions represents a potential pattern of walking behavior.

Figure 10.6 shows simple step action recognition examples. The test data was collected within a building with a floor height of 3.5 m, where the tester ascended the stairs and then returned via the elevator. From Fig. 10.6, the barometric data exhibits notable noises even when the tester is walking on a flat floor, making it difficult to differentiate between pressure differences caused by changes in altitude and those caused by noise. For such data, the Seq2Seq model (left in Fig. 10.6) correctly identifies the majority of step actions in real-time. Additionally, the proposed Seq2Seq model exhibits sensitivity to sudden changes in barometric data, which allows to immediately detect the first step after the user takes the elevator, as indicated by

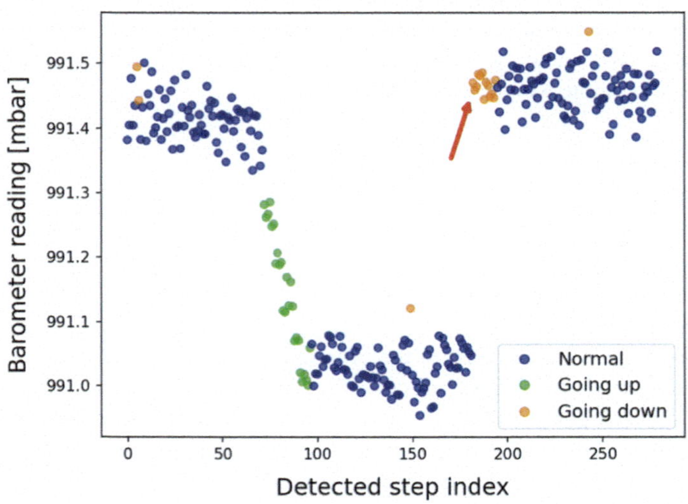

Fig. 10.6 Step action recognition examples

Algorithm 1 Floor decision algorithm

Input: predicted action act of current step
Output: floor detection of a step

```
Initialization: n_wait, p_elevator ← 3, 0.35
if act is same as previous & queue is empty then
    Calculate pressure value p with a lowpass filter
    return floor detection result same as previous step
else
    Enqueue current IMU and barometer data into queue
    Count number of consecutive occurrences n_con
    // Decide whether to change walking state
    if n_con > n_wait then
        Generate a floor transition signal
        Calculate average pressure value p_a of the step data in queue
        Obtain pressure difference by p_delta ← p − p_a
        if act ≠ Normal then
            // Decide transition type
            if p_delta < p_elevator then
                transition type ← stairs
            else
                transition type ← elevator
            end if
            Update and dequeue the steps in queue
            return transition type and direction
        else
            // Update floor number when back to floor
            floor number ← UpdateFloor()
            Update floor number of current step
            Update and dequeue the steps in queue
            return floor number
        end if
    else
        Wait for prediction of next step
    end if
end if
```

Algorithm 2 UpdateFloor()

Input: relative pressure map RM, pivot floor, p_{delta}
Output: floor number

```
// Obtain pressure value of pivot floor from RM
p_pivot ←RM[pivot floor]
// Calculate new pressure value after floor transition
p_new ← p_pivot + p_delta
floor number ← the floor with the closest pressure value to p_new in RM.
return floor number
```

the red arrow in Fig. 10.6. This characteristic is crucial for our approach because it enables the immediate detection of an elevator event from the first step through the floor decision algorithm described in the following subsection.

10.3.3 Floor Decision Algorithm with Relative Pressure Map

In our method, the user's vertical movement is represented by step actions instead of directly calculating the altitude. The advantage of this approach is that, due to factors such as pressure drift and user behavior, the barometer reading can vary even if the altitude is the same (i.e., the user is on the same floor). These short-term and long-term noises in barometer measurements cause the height calculation to be inaccurate. The step action sequence is a form of data without atmospheric pressure value, thus avoiding the above problems. The only requirement is to ensure that DL predictions are accurate and robust enough. Based on the step action sequence, we know the exact step of the floor transition that occurred. Therefore, this study estimates the height change based on barometric pressure difference only when the region changes are detected.

Algorithm 1 explains the proposed floor decision algorithm. We obtain the step action sequence according to the prediction of the Seq2Seq model, which implies the user's vertical movement. For example, a sequence of "Going up" steps means that the user is climbing the stairs. We can update the floor number according to the number of such steps. However, this method has two drawbacks: (1) the number of steps required to walk up a floor varies depending on the user's climbing method, and (2) false floor transitions on a flat floor are also recognized as floor transitions. Therefore, it is necessary to calculate the height difference through the barometer reading. Although the atmospheric pressure drifts due to weather conditions, the pressure difference in a short time interval is credible [59].

Before applying the step action sequence, incorrect DL model predictions need to be eliminated. Outlier data is typically isolated and unordered, whereas height-induced pressure changes are persistent and ordered. Therefore, confirming a floor transition through multiple step actions can eliminate most of the incorrect predictions. When detecting a different step action with previous steps, the proposed method does not immediately confirm the region changed, but instead enqueues the step data in memory and waits for the prediction of new step actions until the queue length exceeds n_{wait}. At the point the current region is changed, a floor transition signal is generated.

The floor decision algorithm first calculates the pressure difference p_{delta} when a region change is confirmed. In particular, the pressure value p of the previous floor is calculated through a lowpass filter (line 2 in Algorithm 1), while the current pressure value p_a is obtained from the average of the data in the queue (line 10 in Algorithm 1). This ensures that the time interval in calculating the pressure difference is minimized to avoid the impact of long-term drift errors and smoothens the pressure values of the previous floor and current floor against short-term noises. At line 12 in Algorithm 1, a step action not equal to "Normal" indicates that the user moves from the flat floor to an elevator or stairs, and the decision between elevator or staircase is made based on the pressure difference p_{delta}. Otherwise, if the step action is "Normal," it means the user is returning to the flat floor, and the floor number must be updated accordingly.

The new floor number is obtained from a function UpdateFloor(), as shown in Algorithm 2. UpdateFloor() reads the pressure value of the pivot floor and calculates p_{new} by adding the pressure difference. The floor with the closest pressure value to p_{new} is mapped as the new floor number. Next, the floor number of the current step and steps in the queue is updated. The proposed floor detection can immediately estimate floor number when a step is detected, with only a delay of n_{wait} steps when the region changes.

10.4 PDR-PF with Clustering

The 2D location calculation is introduced in this section, which is activated when a user starts localization. As shown in Fig. 10.7, the PF fuses data from floor detection, smartphone database, and PDR to calculate the 2D location of the user's kth step. The PF first updates the state of particles based on the transition signal from the DL-based floor detection or PDR estimation, and then utilizes mean shift to cluster them. Afterward, the location estimation and resampling are performed based on

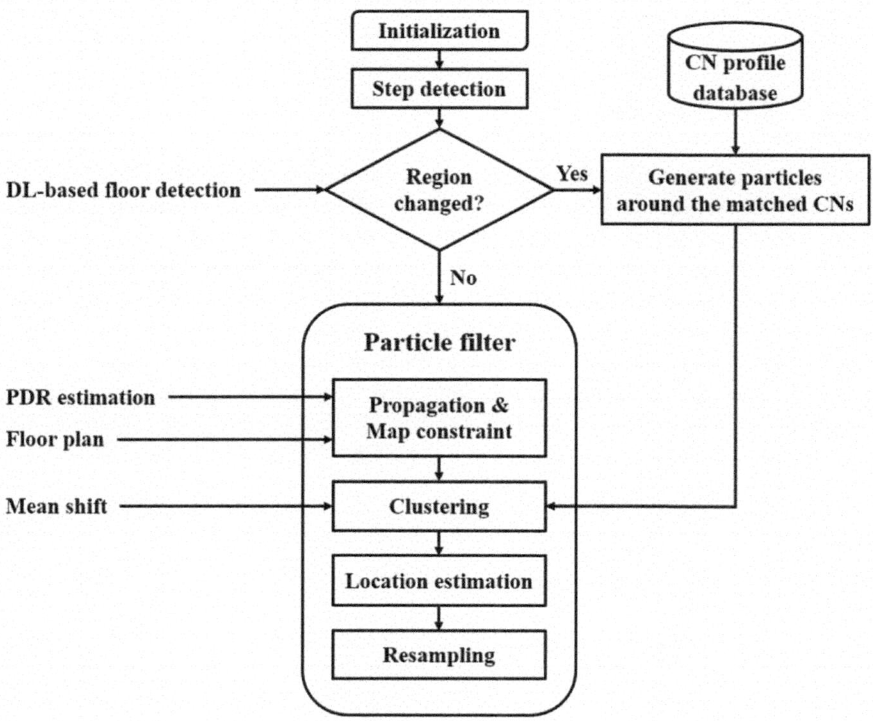

Fig. 10.7 Flowchart of the PF used in our scheme

the clustering results. PDR is described first because it drives the entire scheme. Then, PF will be introduced in detail, including map constraint, clustering, and CN matching-based location correction.

10.4.1 Smartphone Sensor-Based PDR

The Android and iOS operating systems respectively provide the SensorEvent and CoreMotion classes to report motion information from the onboard sensors of devices [60, 61], which enables us to estimate the location through PDR. The PDR approach suggests that a step can be expressed as a distance and an angle referring to the previous state, i.e., the current location is determined by the current displacement and previous location. Thus, the location of the kth step $P_k(x_k, y_k)$ can be expressed as

$$P_k(x_k, y_k) = \begin{bmatrix} x_k \\ y_k \end{bmatrix} = \begin{bmatrix} x_{k-1} \\ y_{k-1} \end{bmatrix} + \lambda_k \begin{bmatrix} \sin(\alpha_k) \\ \cos(\alpha_k) \end{bmatrix}, \tag{10.11}$$

where λ_k and α_k are the stride length and heading direction of the kth step. Next, we introduce the step detection, stride length calculation, and heading direction estimation of PDR.

10.4.1.1 Step Detection

When a pedestrian walks, the vertical acceleration presents periodic sine waves, with each step represented by a local peak or valley in the acceleration. This pattern enables step detection by recognizing these peaks and valleys in vertical acceleration. To counteract the impact of device tilts on sensor measurements, rotation transformation needs to be performed to convert the accelerometer readings from the local coordinate system (LCS) to the global coordinate system (GCS). The rotation matrix R can be calculated through several methods, including quaternions and sensor fusion [62, 63]. We utilized getRotationMatrix() function in SensorManager class to compute the rotation matrix by cross-product of accelerometer and magnetometer measurements. The acceleration vector in the GCS A_t^G can then be determined as

$$A_t^G = R_t \cdot A_t^L, \tag{10.12}$$

where A_t^L are the acceleration vectors in LCS. Subsequently, a valid step is defined as

$$\{a_t > a^{upper},\ a_{t+\Delta t} < a^{lower},\ 0.15\text{s} < \Delta t < 0.6\text{s}\}, \tag{10.13}$$

where a_t is the vertical acceleration, and Δt is the time interval between the peak and valley. We established the amplitude thresholds $a^{upper} = 1.0\,\text{m/s}^2$ and $a^{lower} = -0.8\,\text{m/s}^2$ in this study.

10.4.1.2 Stride Length Calculation

We use a method in [36] to estimate the stride of a step, which establishes a relationship between vertical acceleration and step length, as follows.

$$\lambda_k = \tau_k \cdot \sqrt[4]{a_{max,k} - a_{min,k}}, \tag{10.14}$$

where $a_{max,k}$ and $a_{min,k}$ denote the maximum and minimum vertical acceleration value during the kth step, and τ_k is the coefficient that can be specified for different subjects. Due to the influence of gravity, the variance of acceleration for steps taken in stairs is typically greater than that for steps taken on flat ground, and causes errors in the stride estimation within stairs. To improve the step length calculation, we adjust the value of τ_k based on the region where the user is located using (10.15). The region information is obtained through the floor detection method.

$$\tau_k = \begin{cases} \rho \cdot \tau, & \text{if stairs} \\ \tau, & \text{otherwise} \end{cases}. \tag{10.15}$$

Here, ρ denotes a scale factor used to compensate for step length calculation in stairs. It is recommended that the value of ρ ranges from 0.5 to 0.8, and we empirically set $\tau = 0.43$ and $\rho = 0.6$ in this study. In fact, the exact value of ρ is not strictly required in our scheme, since a correction will be made both when detecting entering and exiting stairs (to be described in Sect. 10.4.2.6).

10.4.1.3 Heading Direction Estimation

The accuracy of heading direction in PDR is crucial because the main source of error comes from distortions of direction. The accelerometer/magnetometer orientation α^m is obtained from getOrientation() in SensorManager class. For the iOS platform, α^m can be retrieved from the CLHeading class in the CoreLocation framework [64, 65]. The orientation calculated by integrating the gyroscope reading tends to slowly drift away from the actual orientation, while the orientation derived from the accelerometer/magnetometer can be easily distorted by surrounding electronic devices. Thus, the common practice is to fuse these two measurements based on certain criteria rather than relying on a single angle source. A typical orientation fusion can be expressed as follows [8].

$$\begin{aligned} \alpha_k &= \gamma \cdot (\alpha_{k-1} + \omega_k \cdot \Delta t) + (1 - \gamma) \cdot \alpha_k^m \\ &= [\alpha_k^x \ \alpha_k^y \ \alpha_k^z]^T \end{aligned}, \tag{10.16}$$

where ω is the gyroscope reading, and γ is a coefficient that determines the fusion proportion, with its value ranging between 0 and 1. A larger value of γ indicates a stronger influence of the angle calculated by the gyroscope in the direction update.

10.4.2 PF with Clustering and Correction

The PF is a sequential importance sampling (SIS) method used to estimate a system's state from noisy and incomplete measurements. A PF represents the distribution of possible states of the system using a set of random particles. These particles are propagated through the state space based on measurement data, and the weights of the particles are updated each iteration [66]. The weights reflect the likelihood that each particle represents the true state of the system, given the measurement data. For more description of the PF in general, see [21, 67]. Next, we introduce the components of the PF as depicted in Fig. 10.7.

10.4.2.1 Particle Initialization

The PF is activated when the user starts localization (e.g., press the localization button). Since the user's location is not given, the PF first acquires the current floor plan based on our floor detection, and then N_0 particles are uniformly dispersed on the entire map. The attribute of the ith particle at the kth step, including 2D coordinates, heading, weight, and cluster number, is as follows.

$$\mathbf{A}_k^{(i)} = [P_k^{(i)}, \theta_k^{(i)}, w_k^{(i)}, c_k^{(i)}], \tag{10.17}$$

where $P_k^{(i)} = (x_k^{(i)}, y_k^{(i)})$ denotes the 2D location and $\theta_k^{(i)}$ denotes the heading direction. We assume that the orientation measured by accelerometer/magnetometer sensors is Gaussian-distributed around the true orientation and thus generates $\theta_0^{(i)} \sim \mathcal{N}(\alpha^m, \sigma^{ori})$ as the initial heading of the ith particle. $w_k^{(i)}$ and $c_k^{(i)}$ stand for the particle weight and cluster label, respectively, and they are initialized as $1/N_0$ and -1.

10.4.2.2 Propagation, Update, and Map Constraint

In this subsection, the location $P^{(i)}$, heading $\theta^{(i)}$, and weight $w^{(i)}$ of particles are updated. Whenever a step is detected, the step length and direction are calculated by the PDR and fed into the PF. The particles propagate based on the current state and PDR estimation. In addition, Gaussian errors with zero mean and standard deviation σ^l and σ^o respectively are added to step length and heading direction update to simulate the effect of the uncertainty and noises of measurements, as well as avoid the loss of diversity among the particles.

The weight of the particles is determined by the system evaluation function, with the commonly used evaluation parameters including direction or distance [68, 69]. We apply Gaussian distribution to calculate the weight of the particles, as follows [70].

$$w_k^{(i)} = w_{k-1}^{(i)} \cdot \frac{1}{\sigma\sqrt{2\pi}} \cdot e^{\left\{\frac{\left(\Delta x_k - \Delta x_k^{(i)}\right)^2 + \left(\Delta y_k - \Delta y_k^{(i)}\right)^2}{-2\sigma^2}\right\}}, \tag{10.18}$$

where Δ represents the displacement. Furthermore, map information is used to detect collisions with walls and eliminate invalid particles to guide the propagation of particles. The digital map is derived from the floor plan and comprises blocks and lines. The blocks represent unreachable areas and are used to eliminate impossible particles during particle generation (e.g., initialization and resampling). The lines represent the walls and are used to kill particles that cross them. The weight of invalid particles is set as zero, while the valid particles are retained. After the map constraint, the weights are normalized to ensure their sum is one.

10.4.2.3 Clustering Using Mean Shift

We use clustering to group the surviving particles and find the centroid of each cluster. With the clustering results, we (1) confirm the convergence of particles, (2) optimize the location estimation from multiple modes of particle distribution (to be described in Sect. 10.4.2.4), and (3) adjust the number of particles dynamically subject to reduce computational burden without sacrificing performance (to be described in Sect. 10.4.2.5). Before explaining these, we introduce the clustering algorithm.

We utilized mean shift as the clustering algorithm [71]. Mean shift is a non-parametric and centroid-based technique that defines a region around each data point and moves the center (or mean) of that region toward the densest part of the region until it converges to the local maximum. The motivation for using mean shift is that it is simple, fast, and can delineate arbitrarily shaped clusters and count the number of clusters automatically, which is well suited for the dynamic and irregularly shaped particle cloud. In the proposed scheme, mean shift evaluates the similarity between particles based on their location in the orthogonal coordinate system. In particular, normalization is not performed as we want to retain the unit of features to express the real distance. The main parameter of mean shift is the bandwidth B, which is set as 3 m in our scheme. This is a relatively large value, which implies that the spatial separation between particles may need to approximate the distance across a room for them to be classified into distinct clusters. The purpose of utilizing mean shift is to describe particle distribution and discover dispersed particle clusters rather than dividing a converged particle cloud into several clusters. Therefore, a slightly larger value of B is recommended.

In the initialization phase, particles are evenly distributed across the state space area. As the new step is detected, the particle cloud converges to where the user might be. The early particles are meaningless until the filter gathers over several iterations to represent the user's possible location. The clustering results can be exploited to explain the current particle distribution. The more dispersed the particles are, the more clusters and the more distant the centroid are from each other. Therefore, we assume that the PF has converged enough to provide valid location information when only one cluster exists or the largest cluster's weight exceeds 80% of the total weight.

10.4.2.4 Localization Estimation

In PF, the estimate of user location is obtained by taking the weighted average of the surviving particle's location. Mathematically, they can be expressed as follows.

$$\begin{bmatrix} x_k \\ y_k \end{bmatrix} = \begin{bmatrix} \sum_i^{n^s} w_k^{(i)} x_k^{(i)} \\ \sum_i^{n^s} w_k^{(i)} y_k^{(i)} \end{bmatrix}, \tag{10.19}$$

where n^s indicates the number of surviving particles. As we mentioned before, due to the building structure and uncertainty of the measurement, the particles may become spread out over multiple modes during propagation. Furthermore, particles can be regenerated in one or a few locations based on the user's motion information and CN profile in our scheme (to be discussed in Sect. 10.4.2.6). As a result, it is difficult to accurately estimate the user's location using the entire particle set. Therefore, we calculate the location from the selected particles through the results of clustering. First, we ignore tiny clusters whose weight is less than 5–10% of the total weight. Second, when the weight of the largest cluster exceeds 70% of the total weight, the location is calculated only using that cluster. Otherwise, all the particles are used to estimate location.

10.4.2.5 Resampling

One drawback of the SIS method is the degeneracy of weight, where the importance weights concentrate on a few particles while the majority of particles have weights close to 0 after multiple iterations. Resampling is a common solution to handle this issue which ignores the particles with low weights and multiplies the particles with high weights. However, resampling causes the particles to lose diversity, resulting in sample impoverishment [14, 21, 72]. A typical approach to handle this issue is to implement resampling only at certain iterations [28]. Hence, instead of resampling every iteration, we only perform it when (1) five iterations have passed since the last resampling, or (2) the number of surviving particles is less than $N_p/5$, where N_p is the current maximum number of particles, and its value is dynamically adjusted based on the clustering result.

Theoretically, a large number of particles can reduce the variance of the estimated posterior distribution, leading to more accurate state estimation. However, the PF is a computationally intensive method, and increasing the number of particles increases the computational cost of the algorithm. This problem is magnified in real-time tasks and on the mobile platform. Therefore, balancing the number of particles with computational efficiency is important for optimizing the performance of a PF. We assume that fewer particles can still achieve good localization results when the particles are concentrated in one area, such as a closed corridor, while if the particle distribution is dispersed, more particles are needed to explore the feasible paths. Thus, the proposed scheme dynamically adjusts the number of particles N_p based on the number of clusters to achieve good performance with lower overhead, as follows.

$$N_p = \min(15, n_{cluster}) \times N_c, \tag{10.20}$$

where $n_{cluster}$ is the number of clusters, and N_c stands for the number of particles assigned for one cluster. In the initialization phase, there can be many clusters due to the dispersion of particles. To prevent generating too many particles, the maximum value of N_p is set to $15 \times N_c$. A large number of particles N_0 are only used at the first iteration since PF has to cover the interesting state space areas. Then, N_p is determined by (10.20).

10.4.2.6 CN Matching-Based Location Correction

The performance of PF is improved according to the map constraint. However, there are still three practical problems that have to be considered:

- In the initialization phase, the convergence of PF is slow due to inadequate measurement, which can take over a hundred steps in large buildings.
- When a user changes the floor, the particle's state may not be available on the new map because they were iteratively updated on the previous floor plan. Figure 10.8 illustrates an example. It is an important issue when we want to efficiently extend PF to 3D scenarios.
- Due to the absence of absolute location information (e.g., RSS), the PF cannot correct itself when particles converge to the wrong location.

Fig. 10.8 After the user took the elevator, some particles appeared on the other side of the wall. The gray particles were eliminated due to falling within the block, the purple particles remained valid, and the navy particles were not eliminated but were incorrect

To solve these problems, we present CN matching-based location correction, which combines the floor transition signal from the floor decision algorithm and prior knowledge from the map. Whenever a step is detected, the floor decision algorithm is first checked for the presence of a floor transition signal. If a transition signal is detected, particles are corrected based on the matched CNs.

When the user changes floors, the floor decision algorithm outputs a transition signal that includes the vertical motion feature. This vertical motion feature is valuable for the probabilistic approach such as PF, and it is used to match the CNs on the floor plan to narrow down the possible area where the user is located to one or few places. CNs are established around vertical transition facilities (e.g., elevator, stairwell, etc.) based on their usage and location. The CN's profile is designed as follows.

$$CN_{set} = \{f_j, P_j^{cn}, Type_j, D_j, \theta_j^{cn}, j = 1, \ldots, n^{cn}\}. \tag{10.21}$$

Here, f_j and P_j^{cn} respectively indicate the floor level and location of the CN, $Type_j$ is the floor transition type (i.e., stairs or elevator), D_j is the transition direction, θ_j^{cn} stands for the possible direction range, and n^{cn} is the total number of CNs. Because the structure of the facilities restricts the direction of user movement (e.g., an elevator having only one exit direction), each CN is assigned a θ^{cn} that is used for initializing the heading of the regenerated particles.

There is an example of CN matching-based location correction. A user takes an elevator from the first floor to the third floor, and the floor decision algorithm returns the first floor as the previous region and the third floor as the current region, with the mode of vertical transportation being recognized as the elevator. Therefore, the CNs with $f = 3$, $Type =$ elevator, and $D =$ ascending are matched for location correction. Location correction is performed according to the following criteria.

- If the PF has not converged, then N_c particles are generated around each matched CN. Note that we established CNs near each vertical transportation, thus there is at least one CN that exists for matching.
- If the PF has converged, the closest CN is treated as the main CN based on the estimated user's location, and then $n_{cluster} \times N_c$ particles are generated near this CN. The remaining CNs are considered as sub-CNs, and $(n_{cluster} - 1) \times N_{sub}$ particles are generated near each sub-CN. We set $N_c = 150$ and $N_{sub} = 15$; thus sub-CNs will be identified as tiny clusters and do not affect the location estimation.

In this way, we not only extend PF to 3D scenarios but also correct particle states using information from matched CNs. Additionally, this method accelerates particle convergence, because the collision detection algorithm can easily eliminate incorrect clusters due to the narrow transition zone. Furthermore, small clusters generated based on the sub-CN provide an opportunity for rectification when the PF converges to the incorrect place: the correct cluster can grow after the other cluster disappears by colliding with the wall.

Note that while there are variants of the PF to improve the performance, the optimal strategy remains an open question. Our focus is not on providing the most accurate localization solution, but rather on offering sustainable and reliable long path tracking services with minimal human efforts and limited measurement data. Finally, the 2D location and the floor number obtained from the floor decision algorithm are combined to represent the user's location in a multi-floor scenario.

10.5 Experiment Results

In this section, we presented the effectiveness of our suggested method using various experiments. The data was collected via an Android app on a Samsung Note 10+ with a barometer and IMU sensors, all at a 20 Hz sampling rate. We approach indoor multi-floor localization in two stages: (1) entry and movement inside the building, and (2) initiating localization. The first determines the initial floor upon entry using entrance data or other technologies, while our DL-based method tracks the user's floor without limiting their actions. During the second stage, the smartphone is assumed to be directed forward relative to the user's body. To reflect the performance under real-world usages, the tester exhibited complex mobility patterns during the experiments, including different walking speeds and unconventional stair navigation. To demonstrate the performance of our Seq2Seq scheme, we used the MLP model proposed in [29] for comparison. In this context, F# indicates floors above ground and B# denotes those below; for instance, F5 is the fifth floor.

10.5.1 DL-Based Floor Detection

When a step is detected, the Seq2Seq model first predicts the step action, and then the floor decision algorithm calculates the floor number and user's vertical movement information based on the step action, barometer reading, and relative pressure map. The floor decision algorithm describes a step using one of the following classes: a specific floor, "Stairs up," "Stairs down," "Elevator up," or "Elevator down." A flat step is referred to a step on a flat floor in this section.

The accuracy rate (AR) is adopted to evaluate the accuracy of floor number calculation as follows.

$$AR^{FN} = \frac{\#\{\hat{f}_i \mid \hat{f}_i = f_i\}}{n_{floor}} \ (\times 100\%), \tag{10.22}$$

where $\#\{\hat{f}_i \mid \hat{f}_i = f_i\}$ represents the count of steps where the predicted floor number $\hat{f}_i$ is identical to the actual floor number f_i, and n_{floor} indicates the total number of steps whose actual label is "Normal" (i.e., a floor number). AR^{FN} quantifies the proportion of correctly identified floor numbers out of all the steps on a flat floor. Since steps in staircases and elevators do not represent an exact floor number, they are excluded in the computation of AR^{FN}.

Table 10.2 Paths and activities during experiments

#	Building	Paths and activities
1	Sung-deok Hall	F1 (P-T-C) $\xrightarrow{E}$ F2 (C-S) $\xrightarrow{S}$ F3 (S-T) $\xrightarrow{E}$ F6 (T-S) $\xrightarrow{E}$ F2 (S-P) $\xrightarrow{S}$ F1 (P)
2	Sung-deok Hall	B2 (T) $\xrightarrow{E}$ F2 (T-P) $\xrightarrow{S}$ F6 (P-C) $\xrightarrow{E}$ F3 (C-T) $\xrightarrow{S}$ F1 (T)
3	Jilli Hall	F1 (P) $\xrightarrow{E}$ F5 (P-C) $\xrightarrow{S}$ F2 (C-T) $\xrightarrow{E}$ F1 (T-P) $\xrightarrow{S}$ F3 (P)
4	Jilli Hall	F3 (T) $\xrightarrow{S}$ F1 (T-C) $\xrightarrow{S}$ F4 (C-P) $\xrightarrow{E}$ F5 (P-T) $\xrightarrow{E}$ F1 (T)
5	Cho Man-sik Memorial Hall	F1 (T) $\xrightarrow{E}$ F7 (T-C) $\xrightarrow{S}$ F4 (C-P) $\xrightarrow{E}$ F2 (P-T) $\xrightarrow{S}$ F1 (T)
6	Cho Man-sik Memorial Hall	F4 (P) $\xrightarrow{S}$ F7 (P) $\xrightarrow{E}$ F1 (P-C) $\xrightarrow{S}$ F3 (C-T) $\xrightarrow{E}$ F7 (T)

There were 9 floor detection experiments conducted in Sung-deok Hall, Jilli Hall, and Cho Man-sik Memorial Hall at Soongsil University. Each floor of the buildings had a height of about 3.0–3.5 m and was equipped with both elevators and stairs. Table 10.2 provides a detailed description of the paths and activities performed during the experiments, where C, T, S, and P stand for calling, typing, swinging, and pocket cases, respectively, while $\xrightarrow{E}$ and $\xrightarrow{S}$ represent elevators and stairs. For example, F1 (C) $\xrightarrow{E}$ F2 (C-P) means the user goes upstairs from F1 to F2 by elevator, during which he finishes a phone call and puts his smartphone in his pocket. Tables 10.3, 10.4 and 10.5 present the evaluation confusion matrices and AR^{FN} scores for floor detection using MLP and Seq2Seq models. From Table 10.3, the Seq2Seq model accurately recognized all elevator steps and misclassified some stairs and floor steps. On the other hand, from Table 10.4, the MLP model outperforms Seq2Seq in correctly-identifying floor steps. However, the comparison of Tables 10.3 and 10.4 shows the relative purity of results for the Seq2Seq model. For example, Seq2Seq's false negatives for "Floor" were exclusively misidentified as "Stairs," with no instances of

Table 10.3 Confusion matrix for floor transition detection using Seq2Seq

	Ground truth action			
Recognized action		Stairs (%)	Elevator (%)	Floor (%)
	Stairs	**97.08**	0	6.58
	Elevator	0	**100**	0
	Floor	2.92	0	**93.42**

Table 10.4 Confusion matrix for floor transition detection using MLP

	Ground truth action			
Recognized action		Stairs (%)	Elevator (%)	Floor (%)
	Stairs	**94.36**	1.71	3.93
	Elevator	0	**99.64**	0.36
	Floor	4.94	0.12	**94.94**

Table 10.5 The AR scores of Seq2Seq and MLP in different buildings

Model	Sung-deok hall (%)	Jilli hall (%)	Cho Man-sik memorial hall (%)	Total (%)
Seq2Seq	93.81	92.56	93.77	93.42
MLP	94.56	43.69	85.40	75.86

confusion with "Elevator." This contrasts with the MLP model which presents a more complex confusion matrix, with instances of misclassification between all classes. Table 10.5 shows AR^{FN} scores of the floor calculation in the three experimental buildings. The Seq2Seq model yields a total accuracy of over 90%, indicating that the majority of estimated floor numbers are consistent with the actual floor numbers under conditions of complex user activity. Although some errors exist, they primarily occur when the user enters and exits the transition zone or during changes in activity. These false positive and negative errors result in a delay of n_{wait} steps but do not cause inaccuracies in the computation of the floor level. In a scenario of unrestricted user activity, our goal is not to guarantee perfect accuracy in step action detection, but rather to prevent these potential errors from leading to incorrect floor number calculations. On the other hand, the AR scores of the MLP model are noticeably lower than those of the Seq2Seq model, indicating that the unstable step action recognition reflected in Table 10.4 has an considerable impact on floor calculation. This implies that the Seq2Seq model demonstrates significantly better stability when dealing with noisy data compared to the MLP model.

10.5.2 Multi-floor Indoor Localization

We conducted long-path tracking experiments on the third floor to sixth floor of Cho Man-sik Memorial Hall, with each floor divided into two sections measuring 75×18 m and 18×42 m including study spaces, corridors, rooms, staircases, stairwell, and elevators. We employ the 2D location's AR to compute the proportion of localization errors that fall below a particular threshold ϵ, as follows [73].

$$d_k = \sqrt{(P_k - \hat{P}_k)^2}, \tag{10.23}$$

$$AR^{Loc} = \frac{\#\{d_k \mid d_k \leq \epsilon\}}{n} \times (100\%), \tag{10.24}$$

where $P_k = (x_k, y_k)$ and $\hat{P}_k = (\hat{x}_k, \hat{y}_k)$ indicate the estimated location and actual location of kth step, respectively, and d_k indicates the Euclidean distance between them. $\#\{d_k \mid d_k \leq \epsilon\}$ represents the count of steps where d_k is less than or equal to ϵ, and n is the total number of steps. Furthermore, root-mean-square error (RMSE) was employed to compute the localization loss as

$$\text{RMSE} = \sqrt{\frac{\sum_{k=1}^{n}(P_k - \hat{P}_k)^2}{n}}. \tag{10.25}$$

We collected 1,000 step data for the experiment. To demonstrate the performance of our scheme, we performed six approaches to calculate the location, which are: (a) the proposed scheme, (b) the proposed scheme with MLP, (c) PF (1k particles) with CN matching, (d) PF (1k particles), (e) PDR with CF, (f) calibrated PDR with CF, and (g) PDR with Acc & Mag. The setup of each approach is described as follows.

(a) The PF with CN matching-based location correction and dynamic adjustment of particle numbers via (10.20).
(b) The PF with CN matching-based location correction and dynamic adjustment of particle numbers via (10.20). Notably, the DL model utilized in floor detection is MLP.
(c) The PF with CN matching-based location correction uses a fixed number of 1,000 (1k) particles generated in the resampling phase, instead of using (10.20).
(d) Conventional PF generates a fixed number of 1k particles in the resampling phase. When a floor transition occurs, the particle information from the previous region is used directly.
(e) Step locations are calculated with (10.11). The step length is calculated by (10.14) and (10.15), and the heading direction is calculated using (10.16), where $\gamma = 0.99$.
(f) Step locations are calculated in the same way as (e). In addition, whenever a floor transition is detected, the location of matched CN is set as the current location to correct the location [29].
(g) Step locations are calculated using (10.11). The step length is calculated by (10.14) and (10.15), and the heading direction is calculated from the accelerometer and magnetometer sensors, i.e., α^m.

The vertical movement and altitude information were obtained from the floor detection. Because traditional PDR can not function without an initial state, the start location of (e), (f), and (g) was manually annotated. Furthermore, the probabilistic nature of the PF, slight variations can occur in its computational results each time it

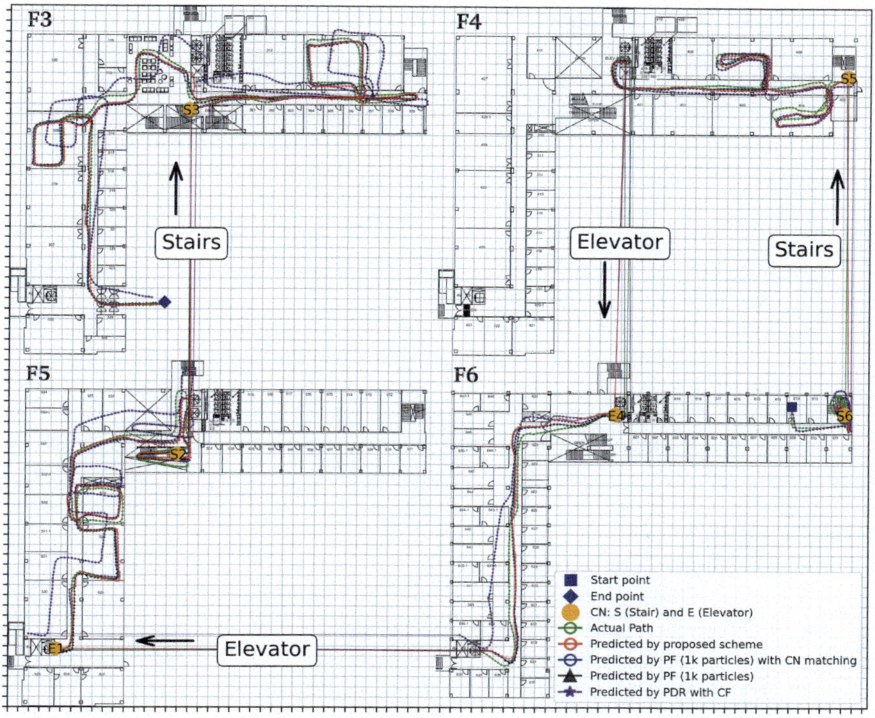

Fig. 10.9 Results of long-path tracking experiment

is run. Hence, the scores of (a), (b), (c), and (d) are obtained from the average of ten separate computations.

Figure 10.9 presents the results of the long path tracking. Before analyze the experimental results, there are some notes for the visualization of the tracking trajectory. To illustrate the details between different floors and areas, we plot the actual location of the user on each floor's floor plan. The start and end points marked by a navy square and diamond, respectively. The orange circles represent CNs, labeled "S" for stairs and "E" for elevators. Only CNs identified as main are depicted. Each grid on the map equates to 2 m of space, and text blocks and arrows indicate floor transition types and directions. For clarity, only results from (a), (c), (d), and (e) were illustrated in Fig. 10.9. Since the estimated locations before particle convergence are meaningless and could potentially hinder visual comprehension, the results for PF methods are plotted after convergence is confirmed.

In Fig. 10.9, we consider a realistic trajectory consisting of a sequence of movements through various complex sections within the building, as follows.

- F6 activities: Start from room 613 → go downstairs through the right stairwells. Since convergence had not yet been reached, N_p particles are generated around each matched CN.
- F4 activities: Enter rooms 404 and 407 consecutively → move along the corridor → return to F6 via elevator. In this segment, S5 was matched as the main CN.
- F6 activities: Move along the corridor → go downstairs via elevator. In this segment, E4 was matched as the main CN.
- F5 activities: Move along the corridor → enter rooms 525 and 524 consecutively → go downstairs through the staircase near the open corridor. In this segment, E1 and S2 were matched as the main CNs.
- F3 activities: Move along the right side corridor → enter room 311 and 329 consecutively → reach the exit → detected as leaving the building. In this segment, S3 was matched as the main CN.

Note that the user's walking trajectory may vary between rooms due to the presence of obstacles (e.g., tables). The first floor transition occurred at the 32nd step, and convergence was confirmed at the 42nd step for the PF with CN matching and at the 150th step for the PF without CN matching.

Figure 10.9 shows PDR with CF tracking well initially, but deviating over time. PF (1k particles) offered an enhanced result by eliminating cumulative errors in long path tracking through boundary constraints, but faced issues: (1) Prone to failure: It failed in 6 out of 10 trials due to incorrect particle convergence, (2) Slow convergence: In successful trials, correct location was achieved after roughly 150 steps, and (3) Inaccurate localization: It incorrectly entered an adjacent room (310) when entering room 311 on F3, rendering it inaccurate and impractical in real applications. Fortunately, they are overcome by our solutions. All ten instances of PF (1k particles) with CN matching successfully located the user. Particularly, there were two instances where convergence to incorrect locations was observed, and both were rectified using our CN matching-based location correction. Additionally, it detected all rooms correctly. Moreover, the proposed scheme demonstrated performance comparable to that of PF (1k particles) with CN matching, i.e., achieving successful tracking throughout, convergence at approximately the 42nd step, and correct detection of all rooms.

Table 10.6 also shows the AR and RMSE values for the approaches. Since we conducted multiple computations for PF approaches, values in Table 10.6 were obtained from the average of separate computations. Furthermore, because the convergence speed is environment-dependent and unrelated to estimated location accuracy, the AR score and RMSE loss of PF approaches were calculated when the convergence is confirmed, i.e., at the 42nd step. This implies that slower convergence in the case of PF without CN matching would yield a lower AR score. Furthermore, because the estimated locations before convergence are random and would distort the informative value of the RMSE loss, the RMSE loss for PF without CN matching is calculated from the 150th step. In Table 10.6, PDR with Acc & Mag achieved an AR value of 8.1% within a 2.0 m error boundary, illustrating the inaccuracies in the conventional

Table 10.6 Evaluation results for each approach

Approach	AR $\epsilon = 0.5$ m (%)	AR $\epsilon = 1.0$ m (%)	AR $\epsilon = 1.5$ m (%)	AR $\epsilon = 2.0$ m (%)	RMSE (m)
Proposed scheme	17.6	58.9	86.6	96.7	1.1
Proposed scheme with MLP	17.0	58.3	85.2	95.7	1.2
PF (1k particles) with CN matching	16.7	61.5	87.1	97.8	1.1
PF (1k particles)	16.6	46.1	68.8	86.9	1.2
PDR with CF	3.2	8.5	19.7	27.3	3.8
Calibrated PDR with CF	11.3	24.1	31.6	41.2	3.8
PDR with ACC & Mag	0.7	2.0	3.9	8.1	11.2

PDR methods using consumer-grade processors. PDR with CF and its calibrated version improved results by fusing orientations, but performance was still inadequate due to inherent path deviation when IMU data is used over long trajectories. On the other hand, the PF methods show a better performance. Among them, PF (1k particles) without CN matching obtained low AR scores due to slow convergence. By comparing the RMSE values, it can be observed that even excluding the factor of convergence speed, the PF without CN matching still underperforms compared to the PF with CN matching. Moreover, our proposed scheme's performance is better than the proposed scheme with MLP and is comparable to PF (1k particles) with CN matching, which obtained an AR score of 96.7% within the error boundary $\epsilon = 2.0$m. Furthermore, the average particle count per iteration $N_{average}$ can be computed as

$$N_{average} = \frac{\sum_{i=42}^{1000} N_{p,i}}{n_{conv}}, \tag{10.26}$$

where $N_{p,i}$ is the particle number of the ith step and n_{conv} is the number of steps after convergence. By computing (10.26), we obtain an $N_{average}$ of 198.7, signifying our scheme's comparable performance to the PF using 1k particles, but with less than 1/5 of particles. Overall, the results show our floor detection method's benefits for 2D localization in multi-floor scenarios, and confirm the proposed scheme's superior performance and computational efficiency.

10.6 Conclusion and Discussion

In this chapter, we contend that the merit of a localization system should not be solely evaluated based on location accuracy, and a robust system should exhibit long-term stability and the capability to efficiently process and operate on a limited amount of measurement data. We propose an indoor multi-floor localization scheme that leverages only a smartphone's IMU and barometer sensors. Our scheme consists of two components: DL-based floor detection and PF with clustering. Our scheme is designed to facilitate indoor localization without relying on the infrastructure and calculate the user's location without a given initial state. We conducted multiple extensive experiments in typical university buildings to evaluate the proposed floor detection and multi-floor indoor localization. The experimental results show the promising performance of our scheme. The DL-based floor detection accurately tracked the floor number and efficiently extracted vertical movement information under a variety of user activities. The indoor multi-floor long-path tracking scheme achieved an average localization accuracy of over 96% within a 2 m error boundary with a limited number of particles in the PF.

Our DL-based floor detection not only tracks the floor level but also extracts the vertical movement information of a step. The floor level can be used to extend a 2D localization to the 3D application, and the vertical movement features are particularly useful for probabilistic methods such as the PF. Furthermore, the proposed CN matching-based location correction also holds value within some infrastructure-dependent systems. For instance, CN is well-suited to serve as a substitute for anchor nodes in areas such as stairwells that lack adequate signal coverage.

While our scheme performs well in typical medium-sized buildings, its efficiency may be challenged in large, open spaces (e.g., airports) due to the lack of map constraints. Magnetic field information, providing universal absolute location data, could be a solution, as suggested by recent research [47, 74, 75]. In future work, we will optimize DL models and floor decision algorithms for better vertical movement data, and accommodate various ways of carrying smartphones to make our scheme applicable to more scenarios.

Acknowledgements This work was supported by the National Research Foundation of Korea (NRF) grant funded by the Korea government (MSIT) (No. RS-2023-00251595).

References

1. Lin C, Shin Y (2023) Multi-floor indoor localization scheme using a Seq2Seq-based floor detection and particle filter with clustering. IEEE Access 11:66094–66112
2. Poulose A, Han DS (2020) UWB indoor localization using deep learning LSTM networks. Appl Sci 10(18):6290–6312
3. Zhao H, Cheng W, Yang N, Qiu S, Wang Z, Wang J (2019) Smartphone-based 3D indoor pedestrian positioning through multi-modal data fusion. Sensors 19(20):4554–4573

4. Chen Z, Zhu Q, Soh Y (2016) Smartphone inertial sensor-based indoor localization and tracking with iBeacon corrections. IEEE Trans Industr Inf 12(4):1540–1549
5. Nguyen TLN, Vy TD, Kim KS, Lin C, Shin Y (2021) Smartphone-based indoor tracking in multiple-floor scenarios. IEEE Access 9:141048–141063
6. Brovko T, Chugunov A, Malyshev A, Korogodin I, Petukhov N, Glukhov O (2021) Complex Kalman filter algorithm for smartphone-based indoor UWB/INS navigation systems. In: Proceedings of Ural symposium on biomedical engineering, radioelectronics and information technology (USBEREIT), pp 0280–0284
7. Tan P, Tsinakwadi TH, Xu Z, Xu H (2022) Sing-Ant: RFID indoor positioning system using single antenna with multiple beams based on LANDMARC algorithm. Appl Sci 12(13):6751–6765
8. Madgwick SOH, Harrison AJL, Vaidyanathan R (2011) Estimation of IMU and MARG orientation using a gradient descent algorithm. In: Proceedings of IEEE international conference on rehabilitation robotics, pp 1–7
9. Mahony R, Hamel T, Pflimlin JM (2008) Nonlinear complementary filters on the special orthogonal group. IEEE Trans Autom Control 53(5):1203–1218
10. Xie L, Tian J, Ding G, Zhao Q (2017) Holding-manner-free heading change estimation for smartphone-based indoor positioning. In: Proceedings of IEEE 86th vehicular technology conference (VTC-Fall), pp 1–5
11. Jiménez AR, Seco F, Prieto JC, Guevara J (2010) Indoor pedestrian navigation using an INS/EKF framework for yaw drift reduction and a foot-mounted IMU. In: Proceedings of workshop on positioning, navigation and communication, pp 135–143
12. Wang C, Liang H, Geng X, Zhu M (2014) Multi-sensor fusion method using Kalman filter to improve localization accuracy based on android smart phone. In: Proceedings of IEEE international conference on vehicular electronics and safety, pp 180–184
13. Jiawei C, Wenchao Z, Dongyan W, Xiaofeng S (2022) Research on indoor constraint location method of mobile phone aided by magnetic features. In: Proceedings of IEEE international conference on indoor positioning and indoor navigation (IPIN), pp 1–7
14. Racko J, Brida P, Perttula A, Parviainen J, Collin J (2016) Pedestrian dead reckoning with particle filter for handheld smartphone. In: Proceedings of IEEE international conference on indoor positioning and indoor navigation (IPIN), pp 1–7
15. Fetzer T, Ebner F, Bullmann M, Deinzer F, Grzegorzek M (2018) Smartphone-based indoor localization within a 13th century historic building. Sensors 18(12):4095–4126
16. Pipelidis G, Tsiamitros N, Gentner C, Ahmed D, Prehofer C (2019) A novel lightweight particle filter for indoor localization. In: Proceedings of IEEE international conference on indoor positioning and indoor navigation (IPIN), pp 1–8
17. De Cock C, Joseph W, Martens L, Trogh J, Plets D (2021) Multi-floor indoor pedestrian dead reckoning with a backtracking particle filter and Viterbi-based floor number detection. Sensors 21(13):4565–4593
18. Wang X, Li T, Sun S, Corchado J (2017) A survey of recent advances in particle filters and remaining challenges for multitarget tracking. Sensors 17(12):2707–2727
19. Qian J, Pei L, Ma J, Ying R, Liu P (2015) Vector graph assisted pedestrian dead reckoning using an unconstrained smartphone. Sensors 15(3):5032–5057
20. Wu Y, Zhu HB, Du QX, Tang SM (2019) A survey of the research status of pedestrian dead reckoning systems based on inertial sensors. Int J Autom Comput 16(1):65–83
21. Ristic B, Arulampalam S, Gordon N (2003) Beyond the Kalman filter: particle filters for tracking applications. Artech house
22. Ye H, Gu T, Tao X, Lu J (2014) B-Loc: scalable floor localization using barometer on smartphone. In: Proceedings of IEEE international conference on mobile Ad Hoc and sensor systems, pp 127–135
23. Yi C, Choi W, Jeon Y, Liu L (2019) Pressure-pair-based floor localization system using barometric sensors on smartphones. Sensors 19(16):3622–3640
24. Ichikari R, Ruiz L, Kourogi M, Kurata T, Kitagawa T, Yoshii S (2015) Indoor floor-level detection by collectively decomposing factors of atmospheric pressure. In: Proceedings of IEEE international conference on indoor positioning and indoor navigation (IPIN), pp 1–11

25. Ye HB, Gu T, Tao XP, Lv J (2015) Infrastructure-free floor localization through crowdsourcing. J Comput Sci Technol 30(6):1249–1273
26. Wang Q, Fu M, Wang J, Luo H, Sun L, Ma Z, Li W, Zhang C, Huang R, Li X, Jiang Z, Huang Y, Xia M (2023) Recent advances in floor positioning based on smartphone. Measurement 214:112813–112836
27. Willemsen T, Keller F, Sternberg H (2014) Concept for building a MEMS based indoor localization system. In: Proceedings of IEEE international conference on indoor positioning and indoor navigation (IPIN), pp 1–10
28. Nurminen H, Ristimäki A, Ali-Löytty S, Piché R (2013) Particle filter and smoother for indoor localization. In: Proceedings of IEEE international conference on indoor positioning and indoor navigation (IPIN), pp 1–10
29. Lin C, Shin Y (2022) Deep learning-based multifloor indoor tracking scheme using smartphone sensors. IEEE Access 10:63049–63062
30. Nilsson JO, Gupta AK, Händel P (2014) Foot-mounted inertial navigation made easy. In: Proceedings of IEEE international conference on indoor positioning and indoor navigation (IPIN), pp 24–29
31. Ryu U, Ahn K, Kim E, Kim M, Kim B, Woo S, Chang Y (2013) Adaptive step detection algorithm for wireless smart step counter. In: Proceedings of international conference on information science and applications (ICISA), pp 1–4
32. Zhang Y, Zhu Z, Wang S (2018) Multi-condition constraint adaptive step detection method based on the characteristics of gait. In: Proceedings of ubiquitous positioning, indoor navigation and location-based services (UPINLBS), pp 1–5
33. Lee JH, Shin B, Kim SLJH., Kim C, Lee T, Park J (2014) Motion based adaptive step length estimation using smartphone. In: Proceedings of IEEE international symposium on consumer electronics (ISCE), pp 1–2
34. Shin S, Park C, Kim J, Hong H, Lee J (2007) Adaptive step length estimation algorithm using low-cost MEMS inertial sensors. In: Proceedings of IEEE sensors applications symposium, pp 1–5
35. Abadleh A, Al-Hawari E, Alkafaween E, Al-Sawalqah H (2017) Step detection algorithm for accurate distance estimation using dynamic step length. In: Proceedings of IEEE international conference on mobile data management (MDM), pp 324–327
36. Weinberg H (2002) Using the ADXL202 in pedometer and personal navigation applications. Analog Devices AN-602 application note 2(2):1–6
37. Jaworski W, Wilk P, Zborowski P, Chmielowiec W, Lee A, Kumar A (2017) Real-time 3D indoor localization. In: Proceedings of IEEE International conference on indoor positioning and indoor navigation (IPIN), pp 1–8
38. Ouyang G, Abed-Meraim K (2021) Analysis of magnetic field measurements for mobile localisation. In: Proceedings of IEEE international conference on indoor positioning and indoor navigation (IPIN), pp 1–8
39. Haque F, Dehghanian V, Fapojuwo AO (2017) Sensor fusion for floor detection. In: Proceedings of IEEE annual information technology, electronics and mobile communication conference (IEMCON), pp 134–140
40. Zhao M, Qin D, Guo R, Wang X (2020) Indoor floor localization based on multi-intelligent sensors. ISPRS Int J Geo Inf 10(1):6–22
41. Li Y, Gao Z, He Z, Zhang P, Chen R, El-Sheimy N (2018) Multi-sensor multi-floor 3D localization with robust floor detection. IEEE Access 6:76689–76699
42. Boim S, Even-Tzur G, Klein I (2021) Height difference determination using smartphones based accelerometers. IEEE Sens J 22(6):4908–4915
43. Ye H, Gu T, Zhu X, Xu J, Tao X, Lu J, Jin N (2012) FTrack: infrastructure-free floor localization via mobile phone sensing. In: Proceedings of IEEE international conference on pervasive computing and communications, pp 2–10
44. Itzik K, Yaakov L (2019) Step-length estimation during movement on stairs. In: Proceedings of mediterranean conference on control and automation (MED), pp 518–523

45. Gao B, Yang F, Cui N, Xiong K, Lu Y, Wang Y (2022) A federated learning framework for fingerprinting-based indoor localization in multibuilding and multifloor environments. IEEE Internet Things J 10(3):2615–2629
46. Rihan N, Abdelaziz M, Soliman S (2022) A hybrid deep-learning/fingerprinting for indoor positioning based on IEEE P802.11az. In: Proceedings of international conference on communications, signal processing, and their applications (ICCSPA), pp 1–6
47. Abid M, Compagnon P, Lefebvre G (2021) Improved CNN-based magnetic indoor positioning system using attention mechanism. In: Proceedings of IEEE international conference on indoor positioning and indoor navigation (IPIN), pp 1–8
48. Feigl T, Kram S, Woller P, Siddiqui R, Philippsen M, Mutschler C (2019) A bidirectional LSTM for estimating dynamic human velocities from a single IMU. In: Proceedings of IEEE international conference on indoor positioning and indoor navigation (IPIN), pp 1–8
49. Wang Q, Luo H, Ye L, Men A, Zhao F, Huang Y, Ou C (2019) Pedestrian heading estimation based on spatial transformer networks and hierarchical LSTM. IEEE Access 7:162309–162322
50. Klein I, Asraf O (2020) StepNet-deep learning approaches for step length estimation. IEEE Access 8:85706–85713
51. Kim Y, Lee S, Lee S, Cha H (2012) A GPS sensing strategy for accurate and energy-efficient outdoor-to-indoor handover in seamless localization systems. Mob Inf Syst 8(4):315–332
52. Yu M, Xue F, Ruan C, Guo H (2019) Floor positioning method indoors with smartphone's barometer. Geo-Spat Inf Sci 22(2):138–148
53. Wang L, Dong Z, Pei L, Qian J, Liu C, Liu D, Liu P (2015) A robust context-based heading estimation algorithm for pedestrian using a smartphone. In: Proceedings of international technical meeting of the satellite division of the institute of navigation (ION GNSS+), pp 2493–2500
54. Milette G, Stroud A (2012) Professional android sensor programming. Wiley
55. Sutskever I, Vinyals O, Le Q (2014) Sequence to sequence learning with neural networks. Advances in neural information processing systems 27
56. Tanigawa M, Luinge H, Schipper L, Slycke P (2008) Drift-free dynamic height sensor using MEMS IMU aided by MEMS pressure sensor. In: Proceedings of workshop on positioning, navigation and communication, pp 191–196
57. Glorot X, Bengio Y (2010) Understanding the difficulty of training deep feedforward neural networks. In: Proceedings of international conference on artificial intelligence and statistics, pp 249–256
58. Hochreiter S, Schmidhuber J (1997) Long short-term memory. Neural Comput 9(8):1735–1780
59. Muralidharan K, Khan A, Misra A, Balan R, Agarwal S (2014) Barometric phone sensors: more hype than hope!. In: Proceedings of workshop on mobile computing systems and applications, pp 1–6
60. Android Developer. SensorEvent (2023). https://developer.android.com/reference/android/hardware/SensorEvent
61. Apple Developer. CoreMotion (2023). https://developer.apple.com/documentation/coremotion
62. Kang W, Han Y (2014) SmartPDR: smartphone-based pedestrian dead reckoning for indoor localization. IEEE Sens J 15(5):2906–2916
63. Valenti R, Dryanovski I, Xiao J (2015) Keeping a good attitude: a quaternion-based orientation filter for IMUs and MARGs. Sensors 15(8):19302–19330
64. Apple Developer. CoreLocation (2023). https://developer.apple.com/documentation/corelocation
65. Poulose A, Senouci B, Han D (2019) Performance analysis of sensor fusion techniques for heading estimation using smartphone sensors. IEEE Sens J 19(24):12369–12380
66. Kitagawa G (1993) A Monte Carlo filtering and smoothing method for non-Gaussian nonlinear state space models. In: Proceedings of US-Japan joint seminar on statistical time series analysis, pp 110–131
67. Doucet A, Johansen AM (2009) A tutorial on particle filtering and smoothing: fifteen years later. In: Handbook of nonlinear filtering, vol 12, issue 3, pp 656–704

68. Medina D, Schwaab M, Plaia D, Romanovas M, Traechtler M, Manoli Y (2015) A foot-mounted pedestrian localization system with map motion constraints. In: Proceedings of workshop on positioning, navigation and communication, pp 1–6
69. Perttula A, Leppäkoski H, Kirkko-Jaakkola M, Davidson P, Collin J, Takala J (2014) Distributed indoor positioning system with inertial measurements and map matching. IEEE Trans Instrum Meas 63(11):2682–2695
70. Wang H, Lenz H, Szabo A, Bamberger J, Hanebeck U (2007) WLAN-based pedestrian tracking using particle filters and low-cost MEMS sensors. In: Proceedings of workshop on positioning, navigation and communication, pp 1–7
71. Comaniciu D, Meer P (2002) Mean shift: a robust approach toward feature space analysis. IEEE Trans Pattern Anal Mach Intell 24(5):603–619
72. Elfring J, Torta E, Molengraft R (2021) Particle filters: a hands-on tutorial. Sensors 21(2):438–465
73. Vy TD, Nguyen TLN, Shin Y (2021) A precise tracking algorithm using PDR and Wi-Fi/iBeacon corrections for smartphones. IEEE Access 9:49522–49536
74. Frassl M, Angermann M, Lichtenstern M, Robertson P, Julian B, Doniec M (2013) Magnetic maps of indoor environments for precise localization of legged and non-legged locomotion. In: Proceedings of IEEE/RSJ international conference on intelligent robots and systems, pp 913–920
75. Antsfeld L, Chidlovskii B (2021) Magnetic field sensing for pedestrian and robot indoor positioning. In: Proceedings of IEEE international conference on indoor positioning and indoor navigation (IPIN), pp 1–8

Chapter 11
An Indoor Wi-Fi Localization Algorithm Using BP Neural Network

Yiruo Lin and Kegen Yu

Abstract This chapter presents a Wi-Fi localization algorithm based on BP neural network. This localization algorithm first transforms the received signal strength indicator (RSSI) data by translation and scaling. A BP neural network is utilized to develop a ranging model based on the transformed RSSI data, which estimates the distance between the target point and each reference point. To improve the accuracy of the ranging model, the initial weights and biases of the BP neural network are optimized using a genetic algorithm (GA). Subsequently, localization is achieved using the ranging model alongside Sequential Quadratic Programming (SQP), an iterative nonlinear optimization technique. For brevity, the ranging model is referred to as GTBPD, and the localization method is referred to as GTBPD-LSQP. The performance of the ranging and localization algorithms are evaluated through conducting experiments in three areas of two academic office buildings.

11.1 Introduction

Wi-Fi positioning technology utilizes existing indoor access points (APs), and it has advantages of simple system deployment, no requirement of additional hardware equipment, and large coverage area. However, Wi-Fi signal has significant volatility, due to signal scattering, reflection and diffraction, which directly affects the accuracy of Wi-Fi positioning. Existing research has shown that machine learning methods can suppress the effect of Wi-Fi signal fluctuation on localization [1], such as support vector machines (SVM), artificial neural network (ANN), back propagation (BP) neural network, and convolutional neural network (CNN) methods.

To achieve a more accurate and adaptable RSSI-based ranging model, the RSSI data is transformed by translation and scaling to reduce its fluctuation (see Sect. 11.3).

Y. Lin (✉) · K. Yu
School of Environmental Science and Spatial Informatics, China University of Mining and Technology, Xuzhou 221116, China
e-mail: yiruo.lin@foxmail.com

K. Yu
e-mail: kegen.yu@cumt.edu.cn

K. Yu (ed.), *Positioning and Navigation Using Machine Learning Methods*, Navigation: Science and Technology 14, https://doi.org/10.1007/978-981-97-6199-9_11

Following this, the transformed RSSI data and a BP neural network are employed to construct the ranging model, termed GTBPD for simplicity (see Sect. 11.4). The GTBPD ranging model has three main advantages: (1) Distance estimation can be performed using the Wi-Fi RSSI signal received by a smartphone, without the need for specialized equipment. (2) Unlike most existing ranging models that estimate the distance between the receiver and the transmitter and rely on a path loss model (PLM), which requires pre-deployment of the transmitter, our proposed GTBPD ranging model estimates the distance between any two receivers. Once the GTBPD model is trained with RSSI data from the smartphone, it can estimate the distance between two indoor locations using just the smartphone collected RSSI data at those locations. (3) The BP neural network, grounded in deep learning, constructs the distance measurement model, which excels in nonlinear fitting and effectively adapts to RSSI fluctuations in distance estimation. Additionally, a new localization algorithm based on GTBPD model is introduced, which is named GTBPD-LSQP, and is highly adaptable to different indoor environments (see Sect. 11.5). Before proceeding further on the details of the proposed ranging model and the localization algorithm, we first briefly study the fundamentals of Wi-Fi localization in the following section.

11.2 Wi-Fi Localization

Wi-Fi localization can be broadly categorized into three methods: fingerprint-based, range-based and angle-based. The basic theory of each method are briefly described in the remainder of this section.

11.2.1 Fingerprint-Based Wi-Fi Localization

Fingerprinting technique comprises an offline phase and an online phase, as shown in Fig. 11.1.

(1) Offline phase: firstly, the reference points are deployed in the location area of indoor environment, and the positions of the points are precisely measured. A device (e.g. a smartphone) is used to collect the RSSI of a number of APs at each reference point, so a fingerprint is formed as a vector of RSSI, labeled by the reference point position. Then, using the fingerprints of all the reference points, an RSSI offline database is built and stored in a server or processing center.
(2) Online phase: when the pedestrian moves in the location area, the smartphone carried by the pedestrian collects the RSSI of each of the same APs in the offline phase and sends the online RSSI vector to the server. The server then compares the online RSSI vector with the RSSI vectors of the database and uses a positioning algorithm to estimate the current location of the user and returns the estimated coordinates to the user.

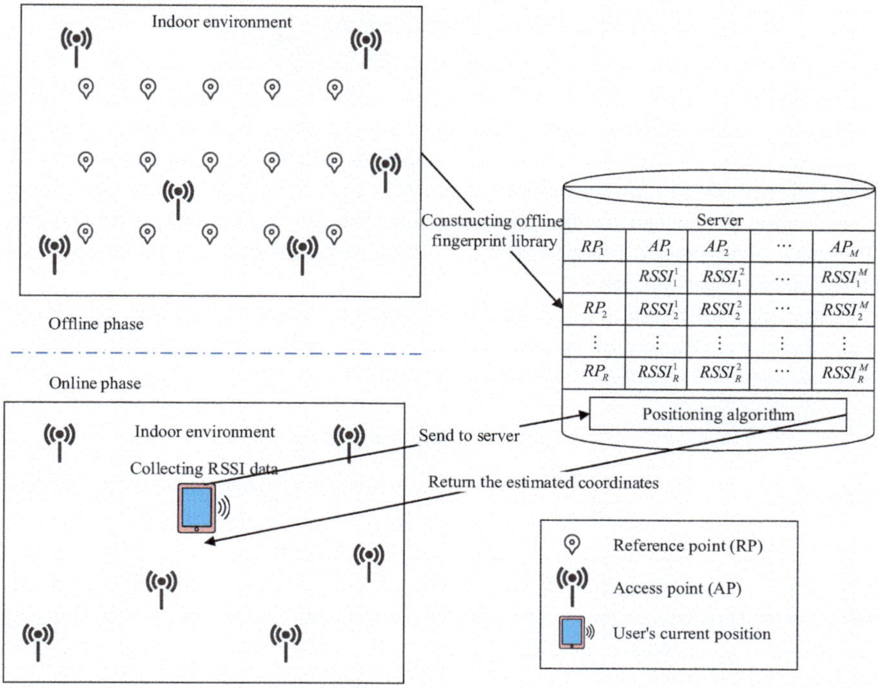

Fig. 11.1 Principle of fingerprint-based Wi-Fi positioning

Weighted K-nearest neighbors (WKNN) algorithm is a widely used Wi-Fi positioning algorithm [2], which selects k ($k > 1$) reference points with minimum Euclidean distances between the offline and online RSSI vectors, and generates the estimate of the user's location by:

$$\begin{bmatrix} \hat{x}_u \\ \hat{y}_u \end{bmatrix} = \sum_{i=1}^{k} w_i \begin{bmatrix} x_i \\ y_i \end{bmatrix} \tag{11.1}$$

Here $[\hat{x}_u, \hat{y}_u]^T$ is the estimated user position, $[x_i, y_i]^T$ is position of the ith reference point out of the k reference points selected, and w_i is the ith reference point's weight, which is determined based on RSSI Euclidean distance [3]. In addition, in order to overcome the shortcomings of the WKNN algorithm, Shin et al. introduced the Enhanced Weighted K-Nearest Neighbor (EWKNN) algorithm [4], and a Bayesian localization algorithm was introduced in [5], aiming to improve accuracy.

11.2.2 Ranging-Based Wi-Fi Localization

The range-based approach is similar to pseudo-satellite positioning and contains a ranging phase and a localization phase. The ranging phase is the estimation of the distance between the transmitter and receiver such as by RSSI based PLM [6] or signal propagation time based on round-trip time (RTT) [7]. The localization phase uses the estimated distances from the ranging phase to determine the unknown position. The classical algorithm for estimating position based on distance is the trilateral measurement algorithm [8].

In the trilateral measurement algorithm, which is based on the known positions of a number of APs and the estimated distances, the user position estimate $(\hat{x}_u, \hat{y}_u)$ can be determined by solving the nonlinear equations:

$$\begin{cases} (\hat{x}_u - x_1) + (\hat{y}_u - y_1) = \hat{d}_1^2 \\ (\hat{x}_u - x_2) + (\hat{y}_u - y_2) = \hat{d}_3^2 \\ (\hat{x}_u - x_3) + (\hat{y}_u - y_3) = \hat{d}_3^2 \end{cases} \tag{11.2}$$

Here, $\hat{d}_i(i = 1, 2, 3)$ is the distance estimate from the ith AP to the receiver. Figure 11.2. illustrates the basic principle of the algorithm in the absence of distance measurement error, where the three distance circles intersect at one point which is the true location of the user. However, in reality, there are usually errors in distance measurements. As a consequence, as shown in Fig. 11.3, such an intersection at a point will not happen, but two different situations would occur, and solving Eq. (11.2) will produce an estimate of the user's position. To reduce the effect of ranging errors, redundant distance observations related to more APs can be used. Linear least-squares algorithm and optimization algorithms can be utilized to determine the user's position [1].

11.2.3 Angle-Based Wi-Fi Localization

Angle-based positioning method utilizes the geometric relationship between the APs and the receiver to estimate position, as shown in Fig. 11.4. Unlike ranging-based method, the measurement is not distances, but angles. θ_i is the angle between the positive x-axis and the direction from the ith AP to the user (smartphone). And θ_i can be measured by the Angle of Arrival (AoA) method [9] or the Angle of Departure (AoD) method [10]. The relationship between θ_i and the user position $(\hat{x}_u, \hat{y}_u)$ is given as:

$$\tan\theta_i = \frac{\hat{y}_u - y_i}{\hat{x}_u - x_i} \tag{11.3}$$

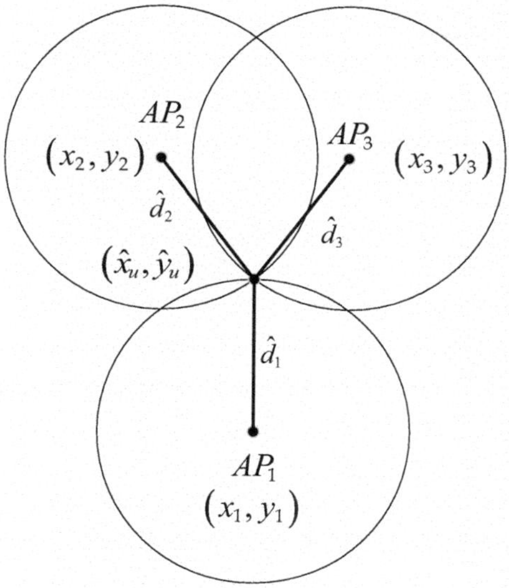

Fig. 11.2 Principle of trilateral measurement algorithm with no measurement errors

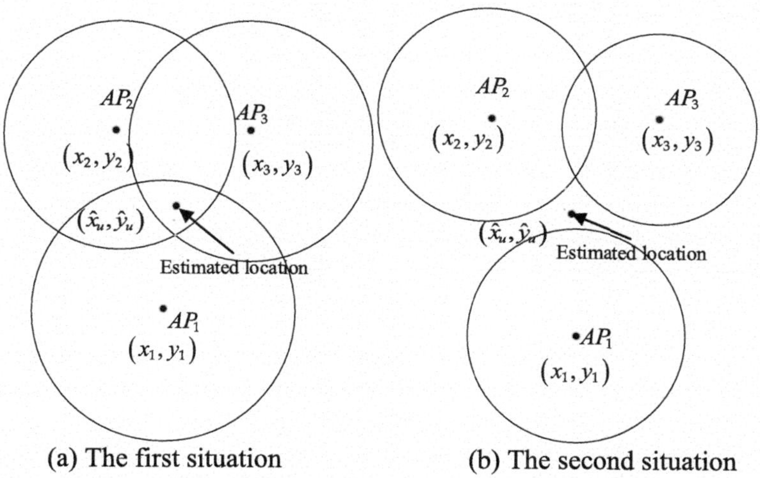

Fig. 11.3 Principle of trilateral measurement algorithm with measurement errors

where (x_i, y_i) is the position of the ith AP. If the measurements are made at M APs, according to (11.3), the following equation is generated:

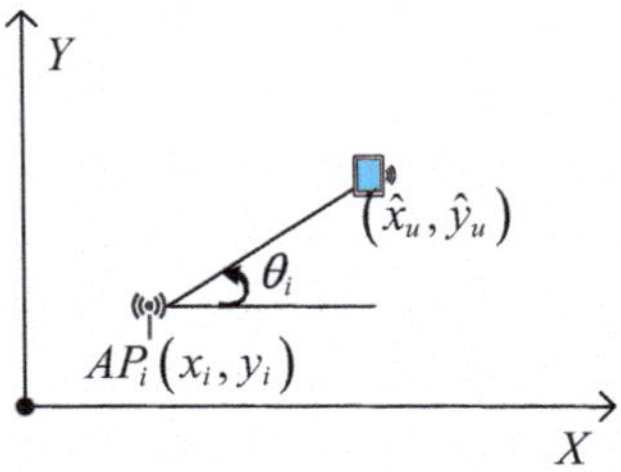

Fig. 11.4 The principle of angle-based positioning

$$\begin{bmatrix} y_1 - x_1 \tan\theta_1 \\ y_2 - x_2 \tan\theta \\ \vdots \\ y_M - x_M \tan\theta M \end{bmatrix} = \begin{bmatrix} -\tan\theta_1 & 1 \\ -\tan\theta_2 & 1 \\ \vdots & \vdots \\ -\tan\theta_M & 1 \end{bmatrix} \begin{bmatrix} \hat{x}_u \\ \hat{y}_u \end{bmatrix} \tag{11.4}$$

Then the linear least-squares algorithm can be used to determine the unknown position $(\hat{x}_u, \hat{y}_u)$.

11.2.4 Machine Learning-Based Wi-Fi Localization

In recent years, machine learning-based algorithms for indoor positioning have garnered significant attention. A new localization model is introduced by [11], utilizing a combination of Convolutional Neural Networks (CNN) and stacked auto-encoders to create a CNN-based indoor localization system. Reference [12] presents a server-based positioning model optimized by a genetic algorithm (GA) and Artificial Neural Networks (ANN) is exploited. Additionally, reference [13] details a mathematical model of a GA-ANN indoor positioning algorithm optimized through regularity coding. The ANN model is optimized by GA in [12, 13]. However, these localization models establish a direct relationship between RSSI values and position coordinates, with RSSI data as the network's input and position coordinates as the output. This approach leads to a significant decline in localization performance if signals from certain APs are not received. The experimental results in [14] show that, compared to DNN, CNN and Recurrent Neural Network (RNN [15]), the BP neural network has better robustness in Wi-Fi localization. BP neural network builds a nonlinear mapping between the input and output data through repeated training, which enables them to automatically extract solution rules and demonstrate their learning capabilities.

Furthermore, there are several reviews on machine learning-based Wi-Fi localization. Roy et al. [16] reviewed the applicability of machine learning techniques in the field of indoor localization, including Wi-Fi indoor localization techniques. Shang et al. [17] studied Wi-Fi fingerprint identification and machine learning methods for indoor localization and analyzed the advantages and disadvantages of these methods for indoor localization applications. Feng et al. [18] presented a timely and comprehensive review of the most interesting deep learning methods for Wi-Fi fingerprint recognition, with the goal of identifying the most effective neural networks under various localization evaluation metrics.

11.3 Translation and Scaling of RSSI Vector

When utilizing RSSI for positioning, averaging multiple RSSI samples is a common practice to mitigate the impact of signal fluctuations. However, as demonstrated in Fig. 11.5, repeated RSSI measurements from the same AP at the same location reveal that Wi-Fi signals are highly vulnerable to external environmental factors. Signal fluctuations can achieve 10 dBm, significantly hindering the accuracy of RSSI-based ranging model. To reduce the effect of RSSI fluctuations, we perform transform the RSSI vector by translation and scaling, as follows.

The RSSI vector of the AP signal received at the ith location point is converted using z-score standardization by (11.5), to mitigate the effects of signal variations.

$$\boldsymbol{TR}_i = \frac{\boldsymbol{RSSI}_i - v_i}{\sigma_i}, i = 1, 2, \ldots \tag{11.5}$$

Here $\boldsymbol{RSSI}_i$ represents the $RSSI$ vector received at the ith location point, which can be described as:

$$\boldsymbol{RSSI}_i = [r_{i,1}, r_{i,2}, \ldots, r_{i,M}] \tag{11.6}$$

where, M is the total number of APs in the indoor location area, and $r_{i,q}$ (q = 1,2, ..., M) represent the RSSI of the qth AP signal received at the ith location

Fig. 11.5 Example of signal fluctuation

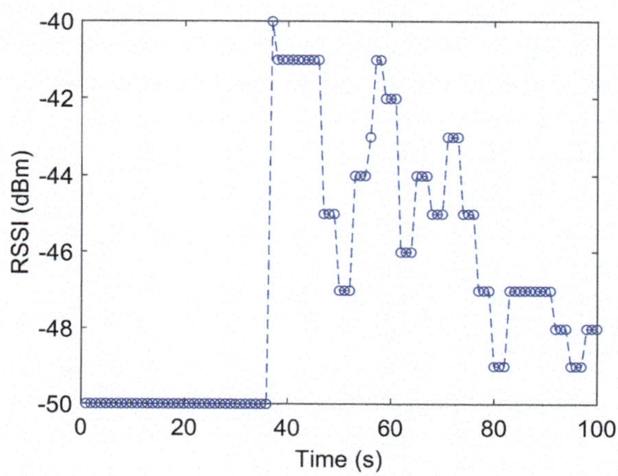

point. Then, υ_i and σ_i are calculated by:

$$\begin{cases} \upsilon_i = \frac{1}{M}\sum_{q=1}^{M} r_{i,q} \\ \sigma_i = \sqrt{\frac{1}{M}\sum_{q=1}^{M}(r_{i,q} - \upsilon_i)^2} \end{cases} \tag{11.7}$$

Z-score standardization employs the mean and standard deviation of all components within the vector, making the data comparable across different dimensions and allowing for more reliable extraction of data characteristics [19]. After standardization, each element in the RSSI vector adheres to a standard normal distribution with a mean of 0 and a standard deviation of 1. This process diminishes discrepancies among corresponding elements caused by noise and multipath interference, while preserving the essential characteristics of the RSSI vector. Then a ranging model based on this transformed RSSI vector is proposed.

11.4 Constructing the Ranging Model

We develop a new ranging model using a BP neural network and the transformed RSSI vector. The number of hidden layers of BP network can be regarded as either multiple or single. While increasing the number of hidden layers might slightly enhance estimation accuracy, it also raises the risk of overfitting, where the network performs well on training data but poorly on online test data. Furthermore, a multi-layer hidden network structure is more complex, leading to longer training time. To create a training time-efficient ranging model, we employ a BP neural network with one hidden layer, as illustrated in Fig. 11.6. Satisfactory estimation accuracy can be achieved with a sufficient number of nodes in a single hidden layer.

The differential transformed RSSI vector between the ith and jth location points can be calculated by ($i = j$ means at the same location):

$$\begin{aligned} \boldsymbol{TR}_{ij} &= \boldsymbol{TR}_i - \boldsymbol{TR}_j \\ &= \left[\frac{r_{i,1} - v_i}{\sigma_i} - \frac{r_{j,1} - v_j}{\sigma_j}, \ldots, \frac{r_{i,M} - v_i}{\sigma_i} - \frac{r_{j,M} - v_j}{\sigma_j}\right], i, j = 1, 2, \ldots \end{aligned} \tag{11.8}$$

Here $\boldsymbol{TR}_\mathbf{i}$ and $\boldsymbol{TR}_j$ are calculated by (11.5). Then, each element in the $\boldsymbol{TR}_{ij}$ vector is squared and define as $\boldsymbol{TR}^2_{ij}$:

$$\boldsymbol{TR}^2_{ij} = \left\{Input_1^{(ij)}, \ldots, Input_M^{(ij)}\right\}$$

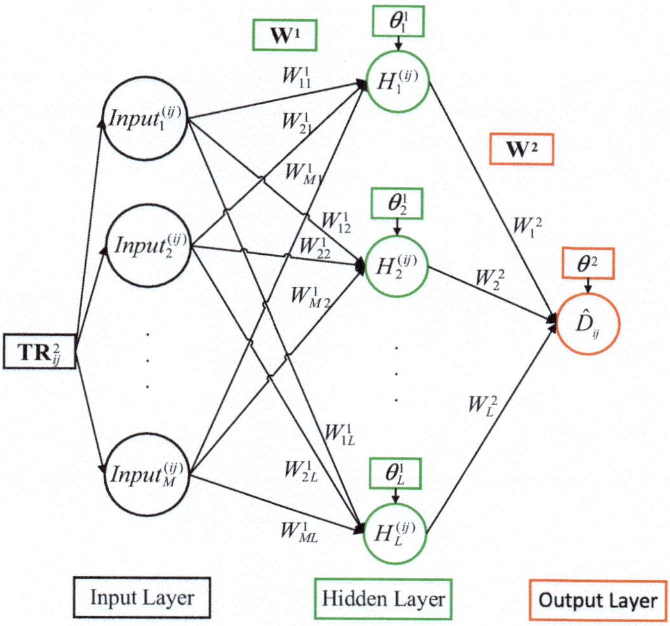

Fig. 11.6 Design of BP neural network-based ranging model

$$= \left\{ \left[\frac{r_{i,1} - v_i}{\sigma_i}, \frac{r_{j,1} - v_i}{\sigma_j} \right]^2, \ldots, \left[\frac{r_{i,M} - v_i}{\sigma_i} - \frac{r_{j,M} - v_j}{\sigma_j} \right]^2 \right\} \tag{11.9}$$

which is used as an input vector of the BP network.

Figure 11.6 represents the network structure, and the input vector is $\mathbf{TR}_{ij}^2$, the corresponding output is $\hat{D}_{ij}$, which is the estimated distance from the ith location point to the jth location point. In Fig. 11.6 $\mathbf{W}^1 = [W_{11}^1, \ldots, W_{1L}^1, W_{21}^1, \ldots, W_{2L}^1, \ldots, W_{M1}^1, \ldots, W_{ML}^1]$ and $\mathbf{W}^2 = [W_1^2, W_2^2, \ldots, W_L^2]$ denote the weight vector from the input layer to the hidden layer and the weight vector from the hidden layer to the output layer, respectively. Here L denotes the number of nodes in the hidden layer, which is It is an empirical value [20]. In this paper, L is determined by an empirical formula [21]:

$$L = \left\lfloor \sqrt{M + O} \right\rfloor + a \tag{11.10}$$

Here $\lfloor\rfloor$ is the rounding down operation, M represents the number of nodes in the input layer, equivalent to the total number of APs involved, and O represents the number of nodes in the output layer. Given that the model's output is solely the distance, O is set to 1. The parameter a is a positive integer constant typically ranging from 1 to 10

[22]. In this study, the value of a is determined using data from previous extensive experiments.

In Fig. 11.6 $H_k^{(ij)}$ ($k = 1,2, \ldots, L$) is the value of the kth node in the hidden layer. θ_k^1 represents the bias of the kth node in the hidden layer, and θ^2 represents the bias of the node in the output layer. In fact, the values of the weights and the biases are given randomly at first. Then $H_k^{(ij)}$ and $\hat{D}_{ij}$ are calculated as follows:

$$\begin{cases} H_k^{(ij)} = f\left(\sum_{q=1}^{M}\left(W_{qk}^1 \times Input_q^{(ij)}\right) - \theta_k^1\right), \quad k = 1, 2, \ldots, L \\ \hat{D}_{ij} = f\left(\sum_{k=1}^{L}\left(W_k^2 \times H_k^{(ij)}\right) - \theta^2\right) \end{cases} \tag{11.11}$$

Here, $f(x)$ is the activation function, which is set to be $f(x) = (\arctan(x) + 1)^{-1}$ from the input layer to the hidden layer, and set to be $f(x) = x$ from the hidden layer to the output layer. Then, $H_k^{(ij)}$ and $\hat{D}_{ij}$ become:

$$\begin{aligned} H_k^{(ij)} &= \left\{\arctan\left(\sum_{q=1}^{M}\left(W_{qk}^1 \times Input_q^{(ij)}\right) - \theta_k^1\right) + 1\right\}^{-1} \\ \hat{D}_{ij} &= \sum_{k=1}^{L}\left(W_k^2 \times H_k^{(ij)}\right) - \theta^2 \end{aligned} \tag{11.12}$$

BP network needs input samples and output samples to train the values of weights and biases. The RSSI data collected from reference points in the location area of interest is used to train the ranging model. Assume that η reference points are laid out in the location area, indexed by $i, j = 1, 2, \ldots, \eta$ in (11.12) and the total number of training samples, denoted by S, equals $\eta \times \eta$. To ensure that the weights and biases in the network can adapt to different input $\mathbf{TR}_{ij}^2$, the BP neural network undergoes iterative training using the BP algorithm [23]. This process updates the values of the weights and biases so that the loss function E, defined by:

$$E = \frac{1}{S}\sum_{i=1}^{\eta}\sum_{j=1}^{\eta}(\hat{D}_{ij} - D_{ij})^2 \tag{11.13}$$

is finally minimized. Here, D_{ij} is the true distance from the ith location point to the jth location point:

$$D_{ij} = \sqrt{(X_i - X_j)^2 + (Y_i - Y_j)^2} \tag{11.14}$$

where (X_k, Y_k) is the true location of the kth reference point. The final values of the network's weights and biases are established when the loss function E falls below a threshold or the maximum number of iterations is reached, thereby completing the construction of the ranging mode. Notably, we perform 100 iterations of network training for each given value of a using pre-measured data. The loss function values after 100 iterations are presented in Table 11.1. The training results indicate that the loss function is minimized when a equals 10. Consequently, in this chapter, the BP neural network constructs the ranging model with a set to 10. However, the initial weights and biases in the network are assigned randomly, which can impact the accuracy of the ranging model. Suboptimal initial values may cause the loss function E to settle at a local minimum rather than the global minimum [24]. Therefore, optimizing the initial weights and biases is essential for enhancing the accuracy of the BP network-based ranging model.

To optimize these initial values, accelerate network convergence, and achieve the global optimal solution, this chapter employs the GA for optimizing BP neural networks. While other algorithms, such as particle swarm optimization and simulated annealing, can also optimize the initial weights and biases, GA offers distinct advantages. Firstly, GA operates using a coding scheme and can optimize multiple parameters simultaneously, making it highly operable. Secondly, GA iteratively optimizes parameters based on probabilistic transfer rules, which provides superior global optimization capabilities. Thus, we utilize GA to optimize the initial weights and biases in the BP neural network, thereby constructing a high-precision ranging model. GA, a metaheuristic algorithm inspired by natural selection, belongs to the broader category of evolutionary algorithms [25]. The process of multi-objective optimization using GA involves six steps, as outlined below.

(1) Initialization

Firstly, the chromosome representing the individual is needed to define, which is set as a string of binary characters with length C:

$$C = (M \times L + L + L \times O + O) \times 4 \tag{11.15}$$

Here, O stands for the number of output layer nodes, L stands for the number of hidden layer nodes, and M stands for the number of input layer nodes. We represent every 4 characters with a value between −7 and 7, with the first bit indicating whether the number is positive or negative, to simplify calculations and accelerate optimization. The chromosome is divided into 4 parts, as illustrated in Fig. 11.7. Specifically, Part 1 contains $4ML$ characters and represents the values in the weight vector $\mathbf{W}^1$. Part 2 consists of $4L$ characters, representing the bias values θ_k^1 of the hidden layer ($k = 1, 2, \ldots, L$). Part 3 includes $4LO$ characters, representing the values in the

Table 11.1 The value of loss function E when a takes different values

a	1	2	3	4	5	6
E	0.0171	0.0141	0.0163	0.0140	0.0146	0.0143
7	8	9	10	11	12	13
0.011	0.0137	0.0134	0.0102	0.0118	0.0127	0.0133

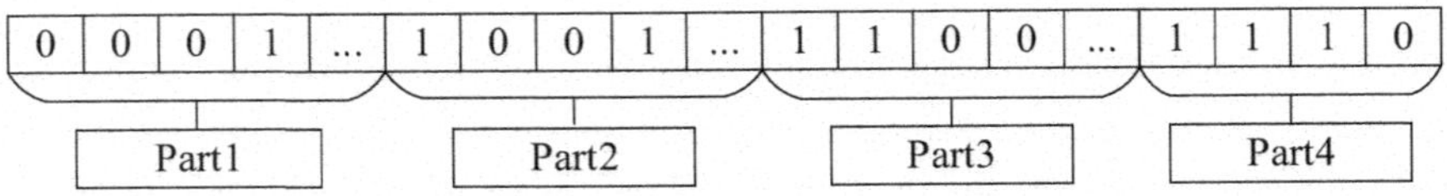

Fig. 11.7 Chromosome Definition and structure

weight vector $\mathbf{W}^2$. Part 4 has $4O$ characters, representing the bias value θ^2 of the output layer.

(2) Fitness

At the start, each character value of the chromosome is randomly generated, creating an individual that represents all the weights and biases of the BP neural network. Let the population size be B, meaning there are B chromosomes in total. For the weights and biases in the neural network represented by the bth chromosome ($b = 1, 2, \ldots, B$), and using all input samples, the estimated distance of all input samples under the bth chromosome are obtained using Eq. (11.12). Use the data collected at the reference point as sample data, then define the estimated distance vector $\hat{\mathbf{R}}_b$ for all input samples under the bth chromosome as:

$$\widehat{\boldsymbol{R}}_b = \left[\widehat{G}_{11}^b, \widehat{G}_{12}^b, \ldots, \widehat{G}_{21}^b, \widehat{G}_{22}^b, \ldots, \widehat{G}_{\eta\eta}^b\right], \quad b = 1, 2, \ldots, B \tag{11.16}$$

Here $\hat{G}_{ij}^b(i, j = 1, 2, \ldots, \eta)$ is the estimated distance of the input sample TR_{ij}^2 under the bth chromosome. Then fitness of the bth individual is caulated as follows:

$$F_b = \sqrt{\sum_{i=1}^{\eta}\sum_{j=1}^{\eta}(\hat{G}_{ij}^b - D_{ij})^2} \tag{11.17}$$

Here D_{ij} is defined in (11.14). The smaller the individual's fitness, the more accurate the prediction, indicating better initial values for the network's weights and biases [11].

Then the average fitness of the population is calculated by:

$$\overline{F} = \frac{\sum_{b=1}^{B} F_b}{B} \tag{11.18}$$

(3) Selection

The next generation of new individuals is generated by mutation and crossover, and the roulette wheel selection [26] is employed to decide whether to retain the resulting new individuals. And, the probability of selecting the individual to remain is given by:

$$P_b = \frac{F_b^{-1}}{\sum_{b=1}^{B} F_b^{-1}} \tag{11.19}$$

Then, the cumulative probability Q_b is calculated by:

$$Q_b = \sum_{j=1}^{b} P_j \tag{11.20}$$

Figure 11.8 illustrates the relationship between individual selection probabilities and cumulative probabilities. Subsequently, a random number ℓ $(0 < \ell < 1)$ is generated. If ℓ is less than Q_1, the first individual is selected. Otherwise, if ℓ falls between Q_{b-1} and Q_b, the bth individual is chosen. This process is repeated B times until all individuals are selected, thus forming a new population.

(4) Mutation

Mutation entails a sudden change in a binary character's value, flipping it from 0 to 1 or from 1 to 0. As illustrated in Fig. 11.9, when the bth individual undergoes mutation, a new individual is generated. As a result, the weights and biases corresponding to the bth individual are also altered.

(5) Crossover

Crossover between two chromosomes occurs in segments of 4 binary characters, as depicted in Fig. 11.10. During crossover, the last 4 characters of the bth individual and the $(b + 1)$th individual in the population are interchanged. This exchange also affects the respective output layer bias values they represent, resulting in the creation

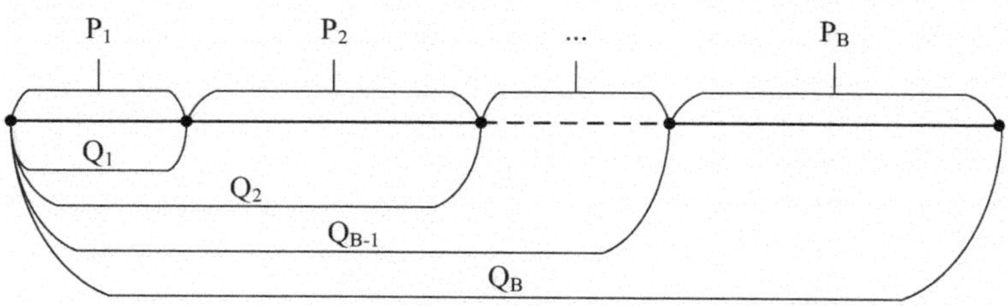

Fig. 11.8 Individual selection probability and cumulative probability relationship

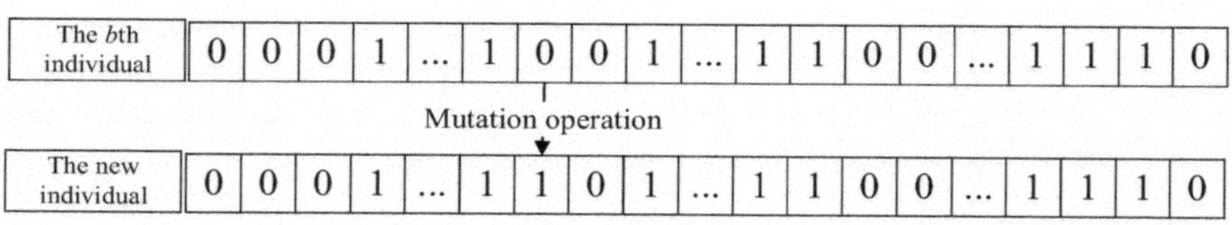

Fig. 11.9 Mutation in the bth individual

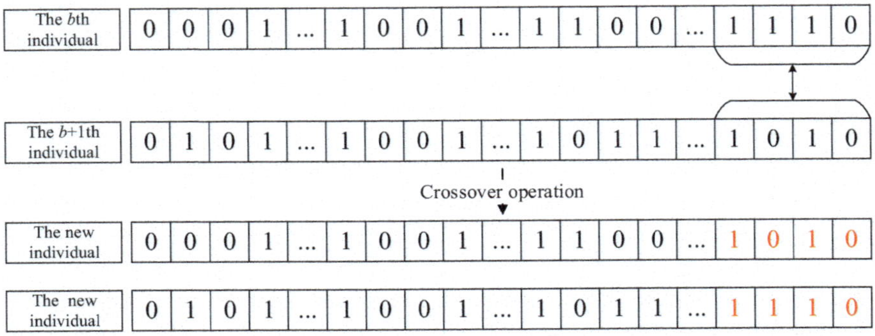

Fig. 11.10 The bth individual is crossed with the $(b+1)$th individual

of two new individuals. Each individual engages in crossover with only one other individual. Subsequently, after B/2 crossover operations, a new population is formed.

(6) Iteration

Continue repeating steps (2) through (5) until either $\overline{F}$ is below a threshold or the predefined number of iterations is reached. Following this, select the values of weights and biases represented by the chromosome with the minimum fitness value in the population as the initial values for the BP neural network's weights and biases, used to construct the ranging model.

Using measured data from the indoor environment, the initial values for the ranging model's weights and biases are determined through GA. Subsequently, the ranging model undergoes iterative training via the BP algorithm until the loss function E falls below a threshold or the predefined number of iterations is reached. The flowchart illustrating the construction of our proposed GTBPD model is depicted in Fig. 11.11. The training data's input samples $\mathbf{TR}_{ij}^2$ are calculated using Eq. (11.9), and the output samples of the training data $\{D_{ij}\}$ are calculated using Eq. (11.14). Upon completion of the offline phase, the ranging model is established and utilized for position determination during the online phase.

11.5 Position Determination

During the online phase, the RSSI vector collected at the target point undergoes translation and scaling according to (11.5). Subsequently, input vectors of GDBPD model from the target point to all reference points are computed using (11.9). The established GDBPD model is then employed to derive the estimated distance vector $\widehat{\boldsymbol{D}} = [\widehat{d}_1, \widehat{d}_2, \ldots, \widehat{d}_\eta]$ from the target point to each reference point. If the estimated distance is excessively large, the model may yield distance estimates with significant errors. Similarly, if the reference points are situated far from the target point,

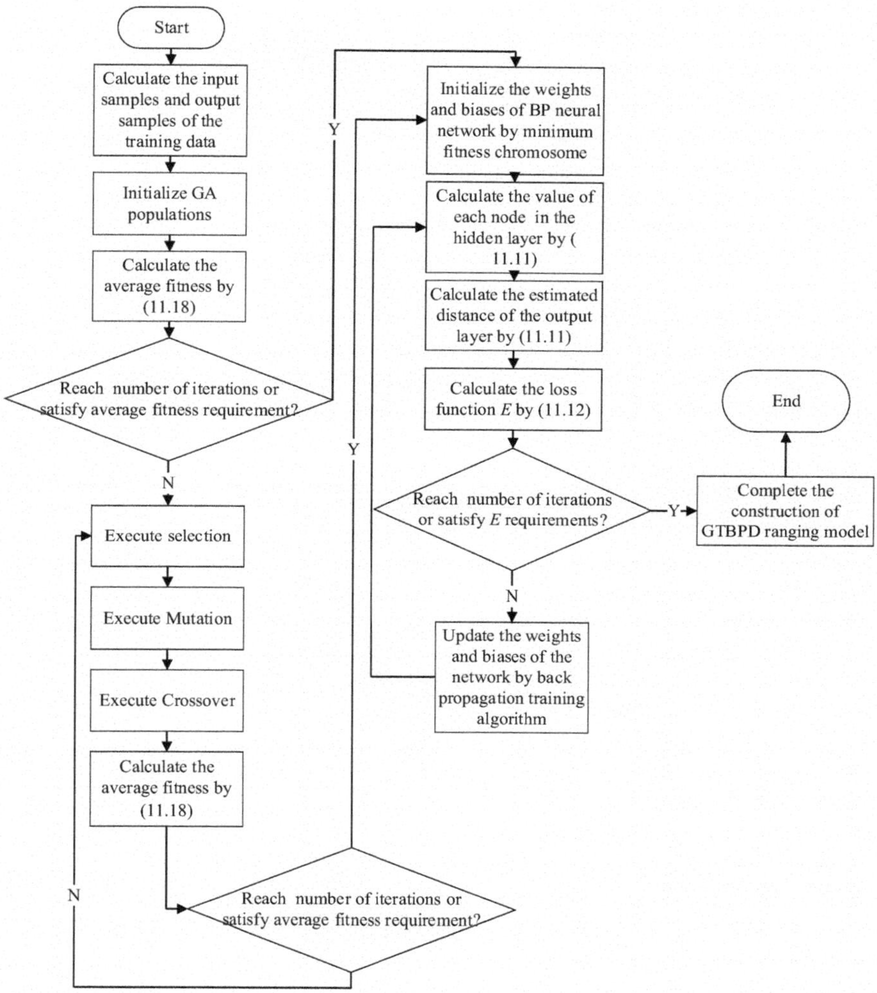

Fig. 11.11 Flowchart of GTBPD ranging model construction

the distance estimates are typically unreliable. To mitigate these issues, a distance threshold θ_d is employed to determine the acceptability of the distance estimate. The threshold θ_d is primarily determined by the dimensions of the indoor environment. If θ_d is too large, both aforementioned situations may arise; conversely, setting it too small may result in insufficient estimated distances for accurate position determination. The impact of the θ_d will be assessed using experimental data. Excluding estimated distances greater than the θ_d yields the new estimated distance vector $\widehat{\boldsymbol{\xi}} = [\widehat{\xi}_1, \widehat{\xi}_2, \ldots, \widehat{\xi}_\varepsilon]$, while the true distance vector is denoted by $\widetilde{\boldsymbol{d}} = [\tilde{d}_1, \tilde{d}_2, \ldots, \tilde{d}_\varepsilon]$, where $\varepsilon \leq \eta$. Utilizing the estimated distance vector $\widehat{\boldsymbol{\xi}} = [\widehat{\xi}_1, \widehat{\xi}_2, \ldots, \widehat{\xi}_\varepsilon]$, a least squares estimator is employed to calculate the initial position estimate of the target. Specifically, squaring both sides of the distance equations yields:

$$\begin{cases} (\hat{x}_0 - \tilde{x}_1)^2 + (\hat{y}_0 - \tilde{y}_1)^2 = \hat{\xi}_1^2 \\ (\hat{x}_0 - \tilde{x}_2)^2 + (\hat{y}_0 - \tilde{y}_2)^2 = \hat{\xi}_2^2 \\ \vdots \\ (\hat{x}_0 - \tilde{x}_\varepsilon)^2 + (\hat{y}_0 - \tilde{y}_\varepsilon)^2 = \hat{\xi}_\varepsilon^2 \end{cases} \tag{11.21}$$

Here, $(\hat{x}_0, \hat{y}_0)$ represents the initial estimated coordinates of the target point, $(\tilde{x}_i, \tilde{y}_i)$ is the true position of the ith reference point, and $\hat{\xi}_i$ $(i = 1,2, \ldots, \varepsilon)$ is the estimated distance from the target point to the ith reference point. By subtracting the last equation from each of the other equations in (11.21) and rearranging the resulting equations, we obtain:

$$\mathbf{A}\mathbf{p}_0 = \mathbf{b} \tag{11.22}$$

Here $\mathbf{b}$ is constant vector, $\mathbf{p}_0$ is the initial estimated position vector, and $\mathbf{A}$ is constant matrix, which are defined as:

$$\boldsymbol{A} = \begin{bmatrix} 2(\tilde{x}_1 - \tilde{x}_\varepsilon) & 2(\tilde{y}_1 - \tilde{y}_\varepsilon) \\ 2(\tilde{x}_2 - \tilde{x}_\varepsilon) & 2(\tilde{y}_2 - \tilde{y}_\varepsilon) \\ \vdots & \vdots \\ 2(\tilde{x}_{\varepsilon-1} - \tilde{x}_\varepsilon) & 2(\tilde{y}_{\varepsilon-1} - \tilde{y}_\varepsilon) \end{bmatrix} \tag{11.23}$$

$$\boldsymbol{b} = \begin{bmatrix} \tilde{x}_1^2 - \tilde{x}_\varepsilon^2 + \tilde{y}_1^2 - \tilde{y}_\varepsilon^2 - \xi_1^2 + \xi_\varepsilon^2 \\ \tilde{x}_2^2 - \tilde{x}_\varepsilon^2 + \tilde{y}_2^2 - \tilde{y}_\varepsilon^2 - \xi_2^2 + \xi_\varepsilon^2 \\ \vdots \\ \tilde{x}_{\varepsilon-1}^2 - \tilde{x}_\varepsilon^2 + \tilde{y}_{\varepsilon-1}^2 - \tilde{y}_\varepsilon^2 - \xi_{\varepsilon-1}^2 + \xi_\varepsilon^2 \end{bmatrix} \tag{11.24}$$

$$\mathbf{p}_0 = \begin{bmatrix} \hat{x}_0 \\ \hat{y}_0 \end{bmatrix} \tag{11.25}$$

Therefore, the initial position estimate using the least squares estimator is:

$$\mathbf{p}_0 = (\mathbf{A}^T\mathbf{A})^{-1}\mathbf{A}^T\mathbf{b} \tag{11.26}$$

Linearization by squaring both sides of the distance equation leads to performance degradation; especially when the distance error is large, the position estimation error can be large. In order to improve the position estimation of the target point, refined position estimates can be obtained using sequential quadratic programming (SQP)

[27], which is a very effective algorithm for nonlinear constrained optimization problems. The position estimates given in (11.26) are used as initial position guesses for the SQP algorithm.

The nonlinear distance error equations can be described as:

$$\begin{cases} v_1 = \sqrt{(x-\tilde{x}_1)^2 + (y-\tilde{y}_1)^2} - \hat{d}_1 \\ v_2 = \sqrt{(x-\tilde{x}_2)^2 + (y-\tilde{y}_2)^2} - \hat{d}_2 \\ \vdots \\ v_\varepsilon = \sqrt{(x-\tilde{x}_\varepsilon)^2 + (y-\tilde{y}_\varepsilon)^2} - \hat{d}_\varepsilon \end{cases} \tag{11.27}$$

where, the parameter vector is defined as $\boldsymbol{u} = [x, y, \hat{d}_1, \hat{d}_2, \ldots, \hat{d}_\varepsilon]^T$, and $\hat{\boldsymbol{d}} = [\hat{d}_1, \hat{d}_2, \ldots, \hat{d}_\varepsilon]$ is the estimate of the true distance vector $\underset{\sim}{\boldsymbol{d}}$. Then the objective function $F(\mathbf{u})$ is defined as:

$$F(\boldsymbol{u}) = \sum_{i=1}^{\varepsilon} v_i^2 = \sum_{i=1}^{\varepsilon} \left(\sqrt{(x-\tilde{x}_i)^2 + (y-\tilde{y}_i)^2} - \hat{d}_i \right)^2 \tag{11.28}$$

Therefore, the position estimate is produced by:

$$\begin{aligned} &\hat{\mathbf{p}} = arg \underset{mathbfu}{min} F(\mathbf{u}) \\ &s.t. g_i \le 0, i = 1, 2, \ldots, 2\varepsilon + 4 \end{aligned} \tag{11.29}$$

Here $\{g_i \le 0\}$ are the constraints, and $\{g_i\}$ are defined as:

$$g_i = \begin{cases} x_{\min} - x, & i = 1 \\ y_{\min} - y, & i = 2 \\ x - x_{\max}, & i = 3 \\ y - y_{\max}, & i = 4 \\ -\hat{d}_{i-4}, & 5 \le i \le \varepsilon + 4 \\ \hat{d}_{i-(\varepsilon+4)} - \theta_d, & \varepsilon + 5 \le i \le 2\varepsilon + 4 \end{cases} \tag{11.30}$$

where $(x_{\max}, y_{\max})$ represents the maximum coordinate values in the independent coordinate system, while $(x_{\min}, y_{\min})$ represents the minimum coordinate values. Subsequently, the initial position estimate and the distance estimates obtained from the GTBPD ranging model are employed as the initial values of the parameter vector:

$$\boldsymbol{u}_0 = [\hat{x}_0, \hat{y}_0, \hat{\xi}_1, \hat{\xi}_2, \ldots, \hat{\xi}_\varepsilon]^T \tag{11.31}$$

Next, the Taylor expansion is used to simplify the objective function $F(\mathbf{u})$ of the nonlinear constrained problem at the kth iteration into a quadratic function:

$$\begin{aligned} &F(u) = \frac{1}{2}[u - u_k]^T \nabla^2 F(u_k)[u - u_k] + \nabla F(u_k)[u - u_k] \\ &s.t.\ \nabla g_i(u_k)^T [u - u_k] + g_i(u_k) \leq 0 \quad i = 1, 2, \ldots, 2\varepsilon + 4 \end{aligned} \tag{11.32}$$

Here ∇ is the gradient (derivative) operator. $\mathbf{u}_k$ is the value of the parameter vector $\mathbf{u}$ at the kth iteration. Define

$$\begin{cases} \mathbf{S}_k = u - \mathbf{u}_k \\ \mathbf{H}_k = \nabla^2 F(\mathbf{u}_k) \\ \mathbf{C}_k = \nabla F(\mathbf{u}_k) \\ \mathbf{A}_k = \left[\nabla g_1(\mathbf{u}_k), \nabla g_2(\mathbf{u}_k), \ldots, \nabla g_{2\varepsilon+4}(\mathbf{u}_k)\right]^T \\ \mathbf{B}_k = -\left[g_1(\mathbf{u}_k), g_2(\mathbf{u}_k), \ldots, g_{2\varepsilon+4}(\mathbf{u}_k)\right]^T \end{cases} \tag{11.33}$$

Here the second-order derivative matrix $\mathbf{H}_k$ can be calculated approximately using Newton's method by the BFGS formula, i.e. Hessian matrix. Then $F(\mathbf{u})$ can be described as the general format of sequential quadratic programming:

$$\begin{aligned} &F(\mathbf{u}) = \frac{1}{2}\mathbf{S}_k^T \mathbf{H}_k \mathbf{S}_k + \mathbf{C}_k^T \mathbf{S}_k \\ &s.t.\ \mathbf{A}_k \mathbf{S}_k \leq \mathbf{B}_k \end{aligned} \tag{11.34}$$

According to the active set method, take the case where the constraint inequalities in Eq. (11.34) are equal:

$$\mathbf{A'}_k \mathbf{S}_k = \mathbf{B'}_k (\mathbf{A'}_k \in \mathbf{A}_k, \mathbf{B'}_k \in \mathbf{B}_k) \tag{11.35}$$

then the Lagrangian function $\Gamma(\mathbf{S}_k, \boldsymbol{\lambda})$ becomes:

$$\min\ \Gamma(\mathbf{S}_k, \boldsymbol{\lambda}) = \frac{1}{2}\mathbf{S}_k^T \mathbf{H}_k \mathbf{S}_k + \mathbf{C}_k^T \mathbf{S}_k + \boldsymbol{\lambda}^T (\mathbf{A}_{k'} \mathbf{S}_k - \mathbf{B}_{k'}) \tag{11.36}$$

Here $\boldsymbol{\lambda} = [\lambda_1, \lambda_2, \ldots, \lambda_e]^T$ is the Lagrangian multiplier estimated from the Kuhn-Tucker equation, and e is the number of constraints in Eq. (11.30) that satisfy the equation condition. By setting $\nabla\Gamma(\mathbf{S}_k, \boldsymbol{\lambda})$ to zero, the parameters $\mathbf{S}_k$ and $\boldsymbol{\lambda}$ are determined and the function is minimized. Thus, we have

$$\begin{aligned} &\mathbf{S}_k^T \mathbf{H}_k + \mathbf{C}_k + \mathbf{A}_k^{'T} \boldsymbol{\lambda} = \mathbf{O}_1 \\ &\mathbf{A}_{k'} \mathbf{S}_k - \mathbf{B}_{k'} = \mathbf{O}_2 \end{aligned} \tag{11.37}$$

Here $\mathbf{O}_1$ is a zero vector of size $(\varepsilon + 2) \times 1$ and $\mathbf{O}_2$ is a zero vector of size $e \times 1$. Equation (11.37) is equivalent to:

$$\begin{bmatrix} \mathbf{S}_k^T & \mathbf{A'}_k^T \\ \mathbf{A'}_k & \mathbf{O'}_2 \end{bmatrix} \begin{bmatrix} \mathbf{H}_k \\ \lambda \end{bmatrix} = \begin{bmatrix} -\mathbf{C}_k \\ \mathbf{B'}_k \end{bmatrix} \tag{11.38}$$

which is actually a system of linear equations with $(\mathbf{S}_k, \boldsymbol{\lambda})$ as variables, and the solution of this equation is obtained for $\mathbf{S}_k$, which yields the optimized estimation of $\mathbf{u}$ for the kth iteration, and the process repeats until (11.36) satisfies a certain condition or the predefined number of iterations is reached.

In summary, the functional block diagram depicted in Fig. 11.12 outlines the proposed GTBPD-LSQP localization algorithm, which involves three main phases:

1. Offline RSSI Transformation: Initially, the extract stable feature of RSSI vector is obtained by transformation.
2. Offline Ranging Model Construction: Subsequently, a ranging model is constructed using a BP neural network. The initial values of the network's weights and biases are optimized through GA.
3. Online Phase: The RSSI vector obtained at the target point undergoes transformation. Utilizing the established ranging model, distances from the target point to each reference point are computed. Position determination is then executed based on the least-squares estimator and SQP algorithm [28].

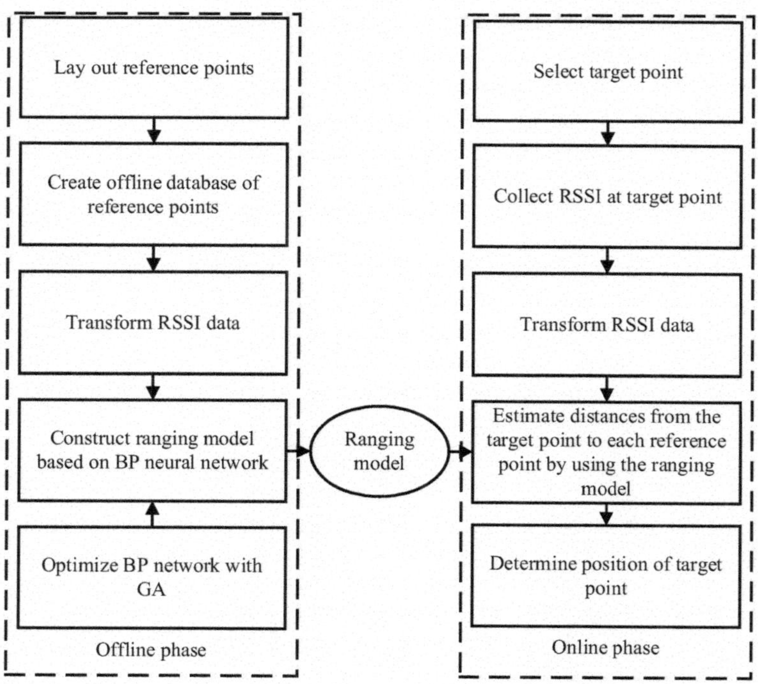

Fig. 11.12 Framework of the proposed GTBPD-LSQP localization algorithm

Expanding upon these steps, the algorithm nables robust and accurate indoor localization by effectively addressing signal fluctuations and leveraging the predictive capabilities of neural networks. Through the integration of optimization techniques and iterative algorithms, it ensures precise positioning even in challenging indoor environments.

11.6 Experimental Analyses

We conducted experiments in the academic office buildings No. 4 and No. 5 at the China University of Mining and Technology (CUMT), to evaluate the localization accuracy of our proposed algorithm. Specifically, we established experimental fields within these buildings: two fields measuring 7 m * 10 m were arranged on the first and second floors of the No. 5 academic office building. Notably, these fields were situated in the lobby areas adjacent to the stairs, characterized by complex building structures and a high volume of foot traffic. Additionally, we set up a 9 m * 15 m experimental field in a discussion hall located in the No. 4 academic office building. In both office buildings, APs were evenly distributed to facilitate data services for the campus network.

Figure 11.13 provides visual representations of the field conditions during data collection. The left image depicts the first experimental field, the middle image showcases the second experimental field, and the right image illustrates the third experimental field. It's worth noting that these experimental areas are frequently traversed by students, leading to fluctuations in RSSI due to the movement of people. These experimental setups enable us to evaluate the performance of our algorithm in real-world scenarios characterized by complex indoor environments and dynamic human activities.

The size of every fingerprint database is $n * M$, with n denoting the number of reference points and M representing the total number of APs scanned in the indoor location area. An AP that has been scanned previously may not be scanned next time. The RSSI value corresponding to an un-scanned AP at a reference point is set as -100dBm, in order to make sure that the array length of the fingerprint data for

Fig. 11.13 Actual experimental environments

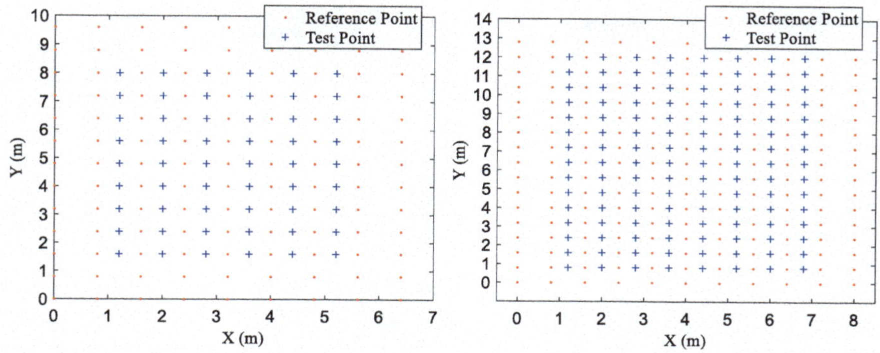

Fig. 11.14 Layout of reference and test points at three experimental fields

each reference point in the fingerprint database is the same. We conduct the scanning and collection of Wi-Fi data via the MI 5 smartphone once per second. Repeat the scanning for 10 times at each reference point, and then take the average of the RSSI values of the 10 collections for each AP as the RSSI value. In this way, the fingerprint of each reference point is obtained, which is corresponding to the vector $\mathbf{RSSI}_i$.

We set up independent coordinate systems in each of the three experimental fields. The left image of Fig. 11.14 shows the arrangement of reference and test points in the first and second indoor experimental fields. Each of these two experimental fields contains 117 reference points and 54 online test points which are evenly distributed. The right image of Fig. 11.14 shows the layout of reference and test points in the third experimental field, where 187 reference points and 120 test points were laid out.

In the first experimental field, a total of 163 AP signals were received, and when establishing the BP neural network, the number of hidden layer nodes was set to 22. In the second experimental field, a total of 210 AP signals were received, and the number of hidden layer nodes was set to 24. And in the third experimental field, a total of 200 AP signals were received, and then the number of hidden layer nodes was also set to 24. As mentioned before, the determination of the number of hidden layer nodes is based on (11.10) and the analysis of the experimental results.

11.6.1 Performance of RSSI Vector Translation and Scaling

We conducted a comparative analysis by collecting RSSI data twice at three randomly selected locations that were not part of offline database, and randomly time interval between the two data collections was set as 5–20 min. Figure 11.15 illustrates the results of this analysis, presenting the original RSSI vectors collected at the three location points and the corresponding vectors after translation and scaling. In the visualization, the left image showcases the original RSSI vectors collected during the two sessions, while the right image displays the Transformed RSSI (TR) vectors obtained after applying translation and scaling. In Fig. 11.15, the horizontal axis represents the IDs of the various APs, with a total of 40 APs scanned at each location point. If an AP scanned during the first collection is not detected during the second collection, the corresponding RSSI value for that AP is set to -100 dBm. This analysis allows us to assess the consistency and stability of the RSSI data across different collection sessions and ascertain the effectiveness of the translation and scaling process in mitigating signal fluctuations and enhancing the reliability of the collected data.

The length of the dashed line in Fig. 11.15 indicates the discrepancy between the two RSSI values collected at the same location point with the same AP. From the graphs (a), (c), and (e) in Fig. 11.15, it is evident that the two RSSI values for the same AP at the same location point can differ by approximately 10 dBm. An AP that can be scanned the first time might not be able to be scanned the second time. At any given time, as long as other conditions such as the smartphone hardware conditions remain unchanged, the RSSI values received from the same AP at the same location will be equal without being influenced by the environment. However, the environment typically changes over time, and the RSSI values received from the same AP at the same point may vary for different phones due to the heterogeneity of the phones. Hence, the variation of RSSI is unavoidable, and the deviation of its value can have a significant impact on the accuracy of fingerprint-based Wi-Fi localization. Therefore, the RSSI vector is translated and scaled to minimize the impact. As can be observed from the three plots (b), (d), and (f) in Fig. 11.15, the RSSI vectors collected twice at the three points have a better match after translation and scaling.

Here, we use the mean relative absolute deviation (*MRAD*) to evaluate the degree of match between two vectors, which is calculated by:

$$MRAD = \frac{1}{V}\sum_{i=1}^{V}\frac{|R_{1,i} - R_{2,i}|}{|R_{1,i} + R_{2,i}|/2} \tag{11.39}$$

Here $R_{1,i}$ denotes the value of the ith element of vector in the first collection, and similarly $R_{2,i}$ denotes the ith element of vector in the second collection, and V is the length of the vector.

At the same location, when comparing the *MRAD* between original RSSI vectors collected at two different times and the *MRAD* between the TR vectors obtained after z-score normalization of the corresponding RSSI vectors, significant reductions in *MRAD* are observed. Table 11.2 illustrates these *MRAD* values between the original RSSI vectors and the *MRAD* between the TR vectors at three distinct positions. The results demonstrate a substantial decrease in MRAD when utilizing TR vectors. Following the translation and scaling of the RSSI vectors, the *MRAD* of the TR vectors at all three points is notably reduced, typically ranging from one eighth to one tenth of the MRAD before transformation. This indicates a remarkable improvement in matching the TR vectors derived from translated and scaled RSSI vectors collected at different times. Consequently, this transformation effectively

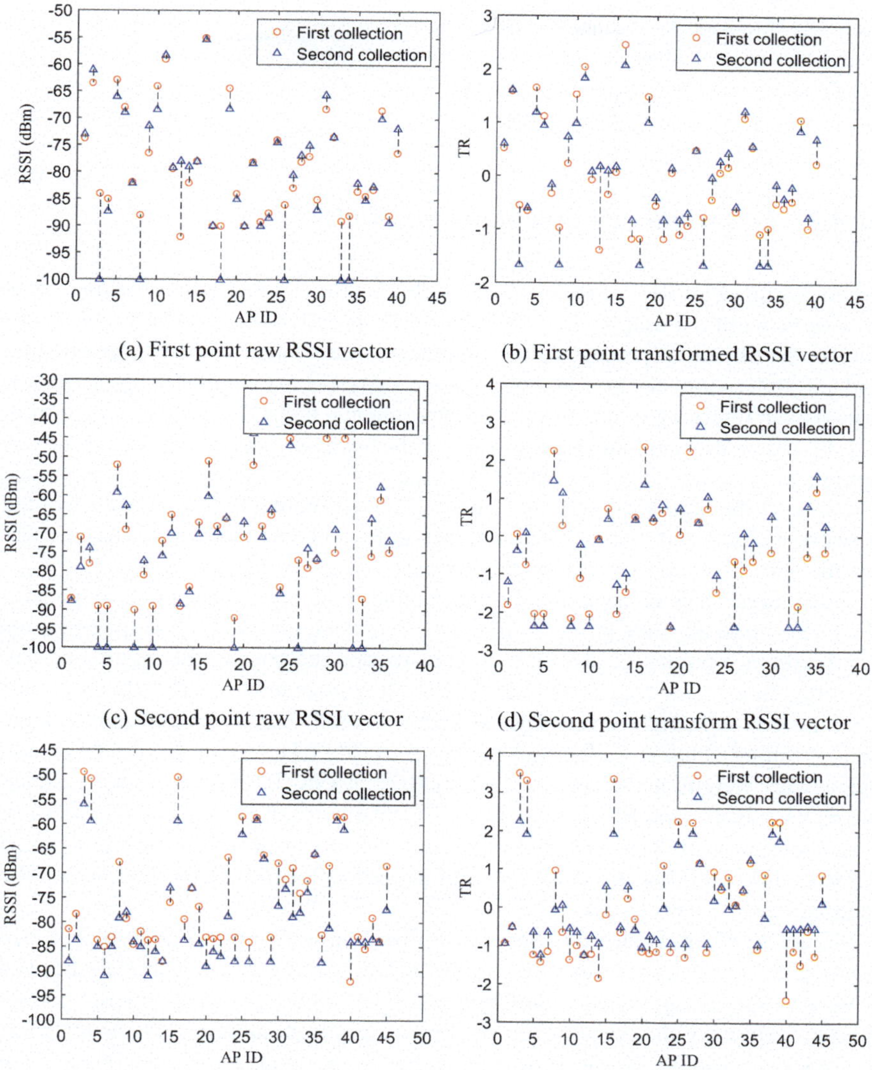

(a) First point raw RSSI vector

(b) First point transformed RSSI vector

(c) Second point raw RSSI vector

(d) Second point transform RSSI vector

(e) Third point raw RSSI vector

(f) Third point transform RSSI vector

Fig. 11.15 Original RSSI vector and the vector after translation and scaling

mitigates the fluctuation effects of RSSI, thereby preserving relatively stable RSSI vector feature information while minimizing redundancy.

Table 11.2 Mean relative absolute deviation of the two collections

	Point	1	2	3
MRAD (%)	RSSI	4.161	9.952	6.073
	TR	0.433	0.961	0.725

11.6.2 Performance of the Ranging Model

Firstly, we make use of the measured data to check if the GA that we have designed can provide better initial weights and biases for our proposed ranging model. Figure 11.16 depicts the alteration process of the minimum fitness value within the population under 100 iterations for the GA optimization by respectively employing the fingerprint data of reference points from three experimental fields. It can be observed from Fig. 11.16 that the minimum fitness values in the three experimental fields via the GA optimization network decrease along with the increase of the number of iterations. The smaller the fitness value is, the better the corresponding initialized values of the network weights and biases. The outcomes in Fig. 11.16 demonstrate that the GA has the ability to optimize our ranging model.

Then three ways of constructing the ranging model are considered. Table 11.3 shows the mean absolute error (MAE) of the estimated distances from all test points to all reference points for the different construction methods of the ranging model. Note that Site i ($i = 1, 2, 3$) denotes for the ith experimental field. BPD represents the model in which the original RSSI vector is used directly as the input sample data, meaning $\{Input_1, ..., Input_M\} = \{ [r_{i,1}-r_{j,1}]^2, ..., [r_{i,1}-r_{j,1}]^2 \}$ in (11.17), and the network weights and biases are not initialized by GA. TBPD denotes the model in which the transformed RSSI vector is used as the input sample data, but the network is not initialized by GA. GTBPD represents the model in which the transformed RSSI vector is used as the input data and the network is initialized by GA, i.e. the ranging model we proposed.

Figure 11.17 illustrates the cumulative distribution function (CDF) of distance estimation errors across three experimental fields using ranging models derived from three distinct approaches. Notably, the GTBPD model exhibits superior performance, surpassing the other two models by a significant margin. This underscores the efficacy of the RSSI transformation and network initialization. To illustrate the ranging performance of the GTBPD, let's examine the percentage of distance estimation errors within 3 m: For the first and second experimental fields, the GTBPD model achieves approximately 76% accuracy within 3 m, representing a notable 10% enhancement over the alternative ranging models. In the third experimental field, the GTBPD model maintains a distance estimation error within 3 m at about 71.8%, while the GBPD model and BPD model report 62.2 and 57.5% respectively, indicating the clear superiority of the GTBPD approach.

Table 11.3 provides further insight into the performance discrepancies, specifically through the MAE of distance estimation. Through translation and scaling of RSSI, the MAE experiences reductions of 0.165, 0.401, and 0.318 m across the

Fig. 11.16 Changes in the minimum fitness value of the GA-optimized ranging model

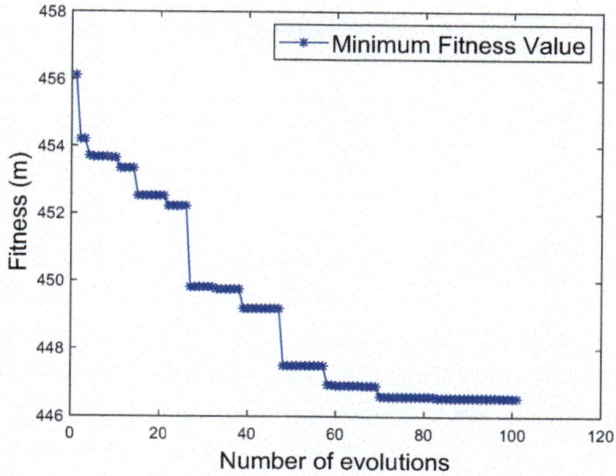

(a) First experimental field

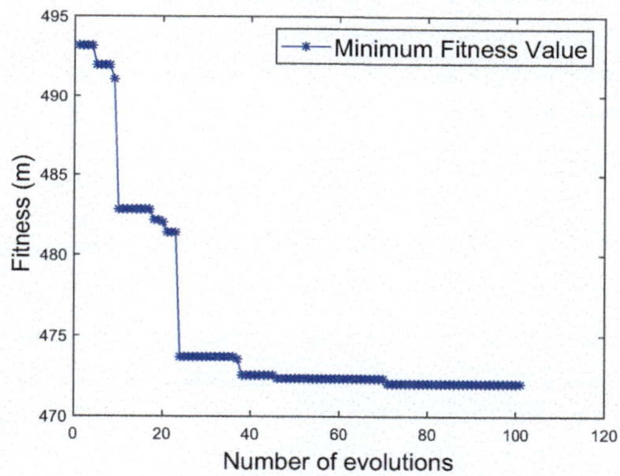

(b) Second experimental field

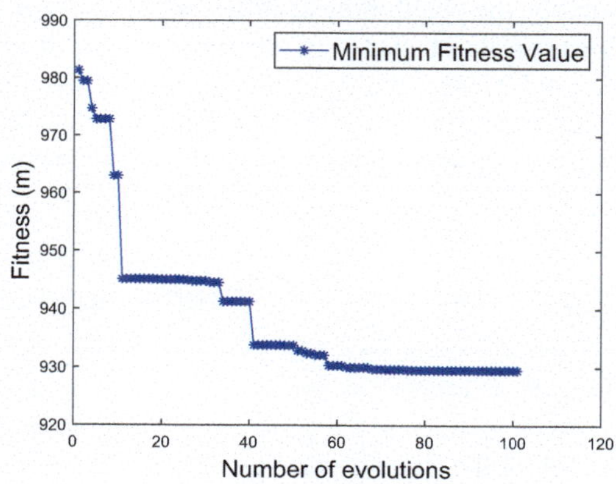

(c) Third experimental field

Table 11.3 Mean absolute error of distance estimation under different ranging models

	Models	GTBPD	TBPD	BPD
MAE (m)	Site 1	2.070	2.726	2.891
	Site 2	2.000	2.605	3.006
	Site 3	2.285	2.647	2.965

three experimental fields respectively. This indicates the efficacy of the RSSI vector transformation method in extracting pertinent feature information from RSSI signals, thereby mitigating their inherent fluctuations. Moreover, the utilization of GA optimization significantly enhances the accuracy of the ranging model. Table 11.3 illustrates that compared to the TBPD model, the MAE of the GTBPD model experiences reductions of 0.656, 0.605, and 0.362 m in the three experimental fields respectively. This underscores the effectiveness of network initialization facilitated by GA. Additionally, the MAE of distance estimation using the proposed GTBPD model consistently hovers around 2 m across diverse indoor environments. This indicates robust performance and the model's adaptability to varied indoor settings.

11.6.3 The Effect of Distance Threshold

Based on the experimental field's dimensions, the distance threshold θ_d is set between 3 and 7 m for experimental analysis. Table 11.4 presents the statistics of positioning errors for three experimental sites with different distance thresholds. The results indicate that optimal positioning accuracy is achieved with $\theta_d = 4$ m for the first experimental site, $\theta_d = 5$ m for the second site, and $\theta_d = 4$ m for the third site. Notably, the positioning accuracy tends to diminish as the θ_d increases or decreases. However, the disparity in accuracy among different thresholds within the range is minimal. Hence, the algorithm's sensitivity to the selection of θ_d is low, suggesting a preference for $\theta_d = 4$ m in similar scenarios. Nonetheless, selecting an appropriate threshold θ_d should be based on the location area's size and experimental findings.

11.6.4 Analysis of the Positioning Accuracy

Once the GTBPD ranging model for the location area of interest is established, it becomes applicable for position determination using RSSI data at the target point. In each of the three experimental fields, a random test point was selected. Different ranging models established for these fields were then utilized to predict distances from the selected test points to all reference points within their respective experimental fields. Scatter plots of predicted versus true distances for these test points are depicted in Fig. 11.18. Ideally, when the predicted distances match the true distances from the

Fig. 11.17 Distance error CDF curves for different ways of constructing the ranging model

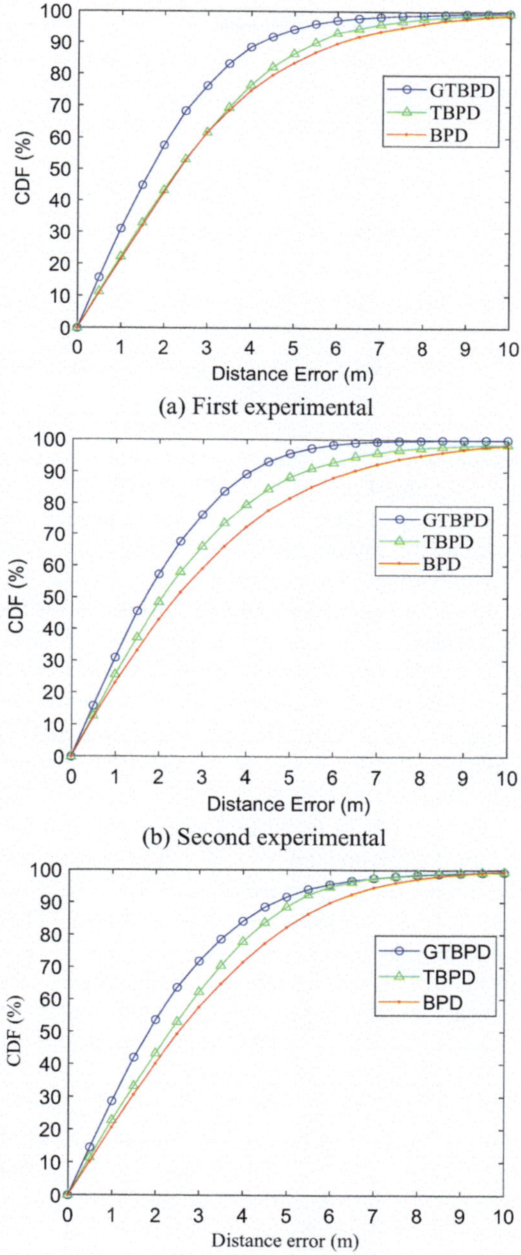

(a) First experimental

(b) Second experimental

(c) Third experimental

Table 11.4 Positioning error at different distance thresholds θ_d

	θ_d (m)	3	4	5	6	7
Site 1	Mean (m)	2.154	2.099	2.143	2.196	2.28
	STD (m)	1.014	0.961	0.867	0.823	0.835
	RMSE (m)	2.381	2.308	2.311	2.345	2.428
Site 2	Mean (m)	2.198	2.112	2.084	2.129	2.12
	STD (m)	1.166	0.967	0.877	0.999	0.786
	RMSE (m)	2.488	2.323	2.261	2.351	2.261
Site 3	Mean (m)	2.855	2.635	2.705	2.795	2.845
	STD (m)	1.202	1.219	1.202	1.146	1.106
	RMSE (m)	3.098	2.903	2.960	3.021	3.053

test point to each reference point, scatter points in Fig. 11.18 align precisely along the solid blue line.

From Fig. 11.18, it is apparent that scatter points generally cluster near the blue line. By eliminating distances exceeding the distance threshold θ_d along with their corresponding reference points, the positions of test points are determined using the least squares estimator and SQP. Additionally, the positions of all test points within the localization area were estimated using four algorithms introduced in Sect. 11.2: WKNN algorithm, Bayesian localization algorithm, EWKNN algorithm, and the GA-ANN-based localization algorithm proposed by [13], denoted as GA-ANN. Comparisons were made with the proposed GTBPD-LSQP algorithm. The least-squares estimator based on the GTBPD ranging model is denoted as GTBPD-LS algorithm.

Tables 11.5, 11.6 and 11.7 present the positioning errors of six algorithms across the three experimental fields. k values ranging from 6 to 14 were explored to analyze the impact on the performance of the WKNN algorithm. The findings underscore the superior performance of our proposed GTBPD-LSQP algorithm, consistently exhibiting the lowest mean error and RMSE across all experimental fields. Firstly, the GTBPD-LS algorithm marginally outperforms the GA-ANN algorithm, yielding accuracy improvements of 0.126, 0.046, and 0.085 m for the respective experimental fields. Secondly, the GTBPD-LSQP algorithm surpasses the GTBPD-LS algorithm in all experimental fields, showcasing accuracy enhancements of 0.239, 0.139, and 0.361 m respectively. This improvement is achieved through iterative optimization, employing the estimate derived from the GTBPD-LS algorithm as the initial estimate, albeit at the expense of increased computational complexity. Furthermore, when compared against the other four algorithms, the proposed GTBPD-LSQP algorithm significantly outperforms them. For instance, compared to the best-case scenario of the WKNN algorithm, the GTBPD-LSQP algorithm achieves improvements of 0.92, 1.28, and 1.038 m across the three experimental fields, highlighting its robust performance.

Fig. 11.18 Scatter plot of predicted and true distances

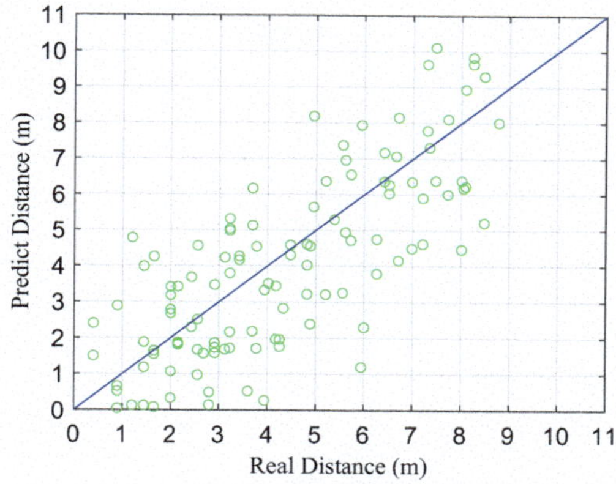

(a) First experimental

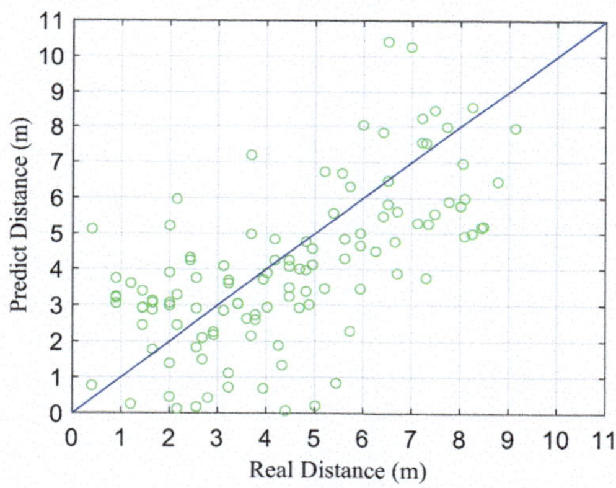

(b) Second experimental

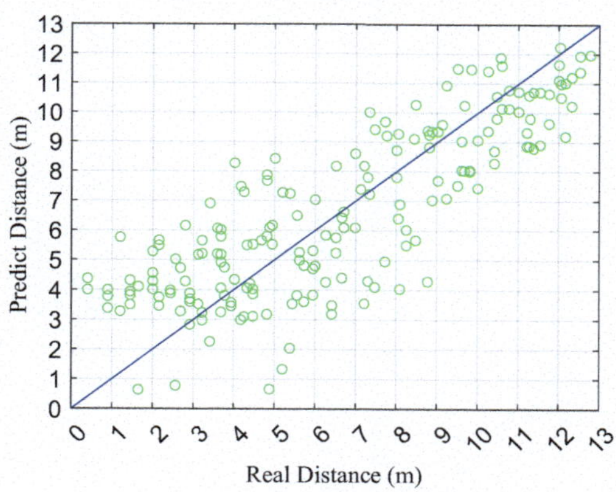

(c) Third experimental

Table 11.5 Comparison of positioning errors at first experimental field

Algorithms	GTBPD-LSQP	GTBPD-LS	GA-ANN	EWKNN	WKNN ($k = 6$)	WKNN ($k = 8$)	WKNN ($k = 10$)	WKNN ($k = 12$)	WKNN ($k = 14$)	Bayesian
Mean (m)	2.099	2.324	2.45	2.685	3.24	3.143	3.063	2.978	3.02	3.098
STD (m)	0.961	1.042	1.436	1.74	1.375	1.391	1.282	1.247	1.167	1.059
RMSE (m)	2.308	2.547	2.84	3.199	3.52	3.437	3.321	3.229	3.236	3.274

Table 11.6 Comparison of positioning errors at second experimental field

Algorithms	GTBPD-LSQP	GTBPD-LS	GA-ANN	EWKNN	WKNN ($k = 6$)	WKNN ($k = 8$)	WKNN ($k = 10$)	WKNN ($k = 12$)	WKNN ($k = 14$)	Bayesian
Mean (m)	2.112	2.227	2.199	2.628	3.341	3.319	3.126	3.194	3.202	3.257
STD (m)	0.967	1.051	1.206	1.314	1.601	1.706	1.462	1.473	1.468	1.235
RMSE (m)	2.323	2.462	2.508	2.938	3.705	3.731	3.451	3.517	3.523	3.483

Table 11.7 Comparison of positioning errors at third experimental field

Algorithms	GTBPD-LSQP	GTBPD-LS	GA-ANN	EWKNN	WKNN ($k = 6$)	WKNN ($k = 8$)	WKNN ($k = 10$)	WKNN ($k = 12$)	WKNN ($k = 14$)	Bayesian
Mean (m)	2.635	2.894	3.045	3.27	3.534	3.463	3.379	3.374	3.437	3.503
STD (m)	1.219	1.508	1.395	2.17	2.357	2.267	2.029	2.038	2.04	1.467
RMSE (m)	2.903	3.264	3.349	3.925	4.248	4.139	3.942	3.941	3.997	3.798

In Fig. 11.19, the cumulative distribution of positioning errors across three distinct experimental fields is depicted for six algorithms. Notably, the WKNN algorithm achieves its minimal RMSE value for $k = 12$ in the first field, $k = 10$ in the second field, and $k = 12$ in the third field. Examining Fig. 11.19 reveals that when the positioning error threshold is set at 3 m, the CDF of positioning errors for each algorithm varies across the experimental fields. For the GTBPD-LSQP algorithm, the CDF of positioning errors is approximately 85.5, 83.5, and 61.7% for the three fields respectively. The GTBPD-LS algorithm demonstrates CDF values of about 72.7, 74.5, and 52.1%. The GA-ANN algorithm exhibits CDF values around 74.1, 68.5, and 50.0%. EWKNN shows CDF values of 63.6, 67.3, and 54.2%. WKNN demonstrates CDF values of 52.7, 61.8, and 54.2%. Lastly, the Bayesian algorithm presents CDF values of 52.7, 45.5, and 42.5%. The CDF analysis reveals that the performance of the GTBPD-LSQP algorithm surpasses that of the other five algorithms. This outcome aligns with the earlier assessment of RMSE, reinforcing the superior performance of our proposed approach across diverse indoor environments.

Our GTBPD-LSQP algorithm demonstrates remarkable adaptability to various indoor settings, consistently outperforming the four existing algorithms considered in the evaluation. This robust performance underscores its effectiveness in accurately localizing targets, even amidst challenging indoor conditions.

11.7 Conclusion

This chapter explores Wi-Fi localization techniques, focusing on our innovative approach utilizing a BP neural network. We begin by addressing the challenge of Wi-Fi signal fluctuation, translating and scaling the RSSI vectors captured by smartphones to stabilize the input data. This step is crucial for ensuring the consistency and reliability of the subsequent localization process.

Next, we introduce the GTBPD ranging model, which leverages these transformed RSSI vectors and a BP neural network optimized by GA to estimate the distances between various locations within indoor environments. The optimization of initial weights and biases in the neural network via GA enhances the model's accuracy and robustness. Our extensive experiments demonstrate the robustness of this ranging model, highlighting its adaptability across different indoor settings, such as varying room sizes and layouts.

Additionally, we present the GTBPD-LSQP localization algorithm, which combines the GTBPD ranging model with linear least squares and Sequential Quadratic Programming (SQP) algorithms. This hybrid approach utilizes SQP's nonlinear iterative optimization capabilities to reduce errors caused by signal fluctuations in distance estimation, thereby improving the overall positioning accuracy.

Fig. 11.19 Positioning error CDF curves of different localization algorithms for three different experimental fields

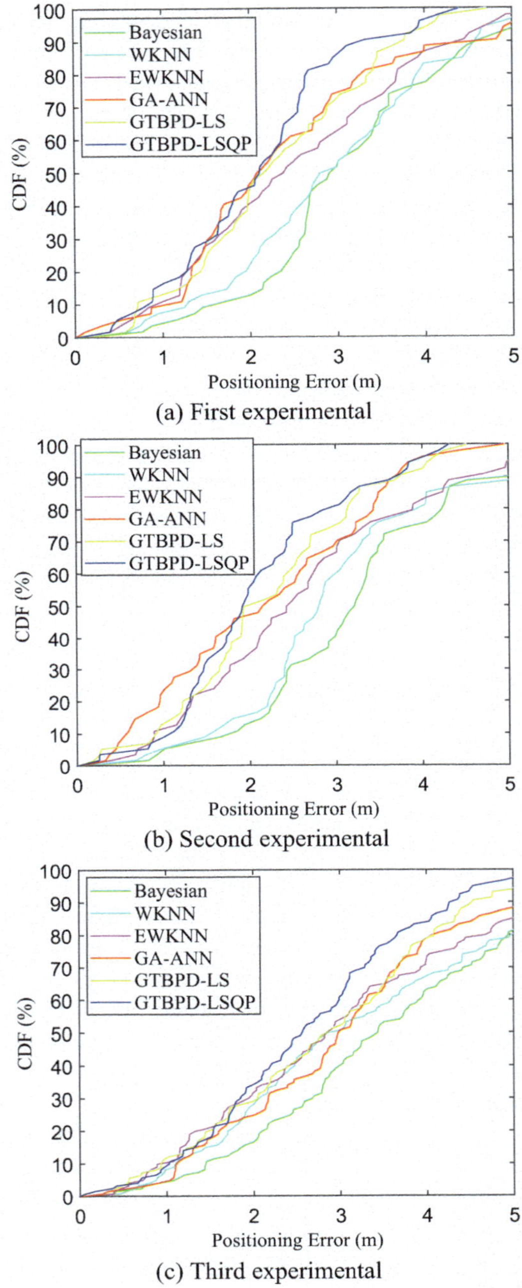

Experimental results validate the superior performance of our GTBPD-LSQP algorithm. When compared to four existing algorithms—WKNN, Bayesian localization, EWKNN, and the GA-ANN algorithm—our approach demonstrates significantly enhanced localization accuracy and robustness. This improved performance is particularly evident in diverse indoor environments, where the GTBPD-LSQP algorithm consistently outperforms the others, making it a reliable solution for precise indoor positioning.

References

1. Lin Y, Yu K, Hao L, Wang J, Bu J (2022) An indoor Wi-Fi localization algorithm using ranging model constructed with transformed RSSI and BP neural network. IEEE Trans Commun 70(3):2163–2177
2. Zhang H, Wang Z, Xia W, Ni Y, Zhao H (2022) Weighted adaptive KNN algorithm with historical information fusion for fingerprint positioning. IEEE Wirel Commun Lett 11(5):1002–1006
3. Xia S, Liu Y, Yuan G, Zhu M, Wang Z (2017) Indoor fingerprint positioning based on Wi-Fi: an overview. ISPRS Int J Geo Inf 6(5):135–160
4. Shin B, Lee JH, Lee T, Kim HS (2012) Enhanced weighted K-nearest neighbor algorithm for indoor Wi-Fi positioning systems. In: 2012 8th international conference on computing technology and information management (NCM and ICNIT), pp 574–577
5. Binghao L (2006) Indoor positioning techniques based on wireless LAN. In: 1st IEEE international conference on wireless broadband and ultra wideband communications, pp 1–6
6. Jawad HM, Jawad AM, Nordin R, Gharghan SK, Abdullah NF, Ismail M, Abu-AlShaeer MJ (2019) Accurate empirical path-loss model based on particle swarm optimization for wireless sensor networks in smart agriculture. IEEE Sens J 20(1):552–561
7. Zhou B, Wu Z, Chen Z, Liu X, Li Q (2023) Wi-Fi RTT/Encoder/INS-based robot indoor localization using smartphones. IEEE Trans Veh Technol 72(5):6683–6694
8. Zhou H, Liu J (2022) An enhanced RSSI-based framework for localization of bluetooth devices. In: 2022 ieee international conference on electro information technology (eIT), pp 296–304
9. Zhou T, Xu K, Shen Z, Xie W, Zhang D, Xu J (2022) AoA-based positioning for aerial intelligent reflecting surface-aided wireless communications: an angle-domain approach. IEEE Wirel Commun Lett 11(4):761–765
10. Huang C, Zhuang Y, Liu H, Li J, Wang W (2020) A performance evaluation framework for direction finding using BLE AoA/AoD receivers. IEEE Internet Things J 8(5):3331–3345
11. Song X, Fan X, Xiang C, Ye Q, Liu L, Wang Z, Fang G (2019) A novel convolutional neural network based indoor localization framework with WiFi fingerprinting. IEEE Access 7(1):110698–110709
12. Mehmood H, Tripathi NK (2013) Optimizing artificial neural network-based indoor positioning system using genetic algorithm. Int J Digital Earth 6(2):158–184
13. Ma L, Sun Y, Zhou M, Xu Y (2010) WLAN indoor GA-ANN positioning algorithm via regularity encoding optimization. In: 2010 international conference on communications and intelligence information security, pp 261–265
14. Zhu W, Zeng Z, Yang Q, Zhao X, Zhang J (2021) Research on indoor positioning algorithm based on BP neural network. J Phys Conf Ser 1–6
15. Zhang W, Liu K, Zhang W, Zhang Y, Gu J (2016) Deep neural networks for wireless localization in indoor and outdoor environments. Neurocomputing 194(1):279–287
16. Roy P, Chowdhury C (2021) A survey of machine learning techniques for indoor localization and navigation systems. J Intell Rob Syst 101(3):63–70

17. Shang S, Wang L (2022) Overview of WiFi fingerprinting-based indoor positioning. IET Commun 16(7):725–733
18. Feng X, Nguyen KA, Luo Z (2022) A survey of deep learning approaches for WiFi-based indoor positioning. J Inf Telecommun 6(2):163–216
19. Hu W, Liang J, Jin Y, Wu F, Wang X, Chen E (2018) Online evaluation method for low frequency oscillation stability in a power system based on improved XGboost. Energies 11(11):3238–3248
20. Shafi I, Ahmad J, Shah SI, Kashif FM (2006) Impact of varying neurons and hidden layers in neural network architecture for a time frequency application. In: 2006 IEEE international multitopic conference, pp 188–193
21. Shen H, Wang Z, Gao C, Qin J, Yao F, Xu W (2008) Determining the number of BP neural network hidden layer units. J Tianjin Univ Technol 24(5):13–20
22. Cui X, Yang J, Li J, Wu C (2020) Improved genetic algorithm to optimize the Wi-Fi indoor positioning based on artificial neural network. IEEE Access 8(1):74914–74921
23. Wang J, Wen Y, Gou Y, Ye Z, Chen H (2017) Fractional-order gradient descent learning of BP neural networks with Caputo derivative. Neural Netw 89(1):19–30
24. Wu C, Wang H (2016) BP neural network optimized by improved adaptive genetic algorithm. Electron Des Eng 24(24):29–33
25. Garg H (2016) A hybrid PSO-GA algorithm for constrained optimization problems. Appl Math Comput 274(1):292–305
26. Lipowski A, Lipowska D (2012) Roulette-wheel selection via stochastic acceptance. Physica A 391(6):2193–2196
27. Yu K, Guo YJ (2008) Improved positioning algorithms for nonline-of-sight environments. IEEE Trans Veh Technol 57(4):2342–2353
28. Yu K, Sharp I, Guo YJ (2009) Ground-based wireless positioning. Wiley

Chapter 12
Intelligent Indoor Positioning Based on Wireless Signals

Yu Han and Zan Li

Abstract Due to the rapid development of intelligent devices, higher requirements are put forward for Location-Based Services (LBS), and indoor positioning based on wireless signals has become one of the important research areas. Because of its good performance, fingerprint positioning has become a mainstream technical solution for indoor positioning based on wireless signals. However, because this technology needs a rich fingerprint database as support, it still faces many technical challenges. In this chapter, we analyze and introduce indoor positioning technologies based on recent artificial intelligence solutions, mainly including traditional machine learning and deep learning methods, and sensor fusion approaches, which solve the problems of inadequate accuracy and efficiency in indoor positioning based on wireless signals. In addition, we introduce examples of intelligent indoor positioning using traditional machine learning, deep learning and crowdsensing.

12.1 Introduction

Nowadays, indoor positioning technology has penetrated into all aspects of people's lives and has a wide range of applications in medical care, logistics, manufacturing, retail, emergency rescue and other fields. However, due to the complexity and diversity of the indoor environment, there are still some challenges for indoor positioning, mainly including low positioning accuracy and low calibration efficiency. Therefore, researchers are constantly developing emerging indoor positioning technologies to solve the problems [1, 2].

Intelligent indoor positioning is the process of navigation and localization of users or objects in the indoor environment. Which is mainly divided into a number of stages as shown in Fig. 12.1. First, users need to collect data through smart devices, and the

Y. Han · Z. Li (✉)
The College of Communication Engineering, Jilin University, Changchun, China
e-mail: zanli@jlu.edu.cn

Y. Han
e-mail: hyu22@mails.jlu.edu.cn

K. Yu (ed.), *Positioning and Navigation Using Machine Learning Methods*, Navigation: Science and Technology 14, https://doi.org/10.1007/978-981-97-6199-9_12

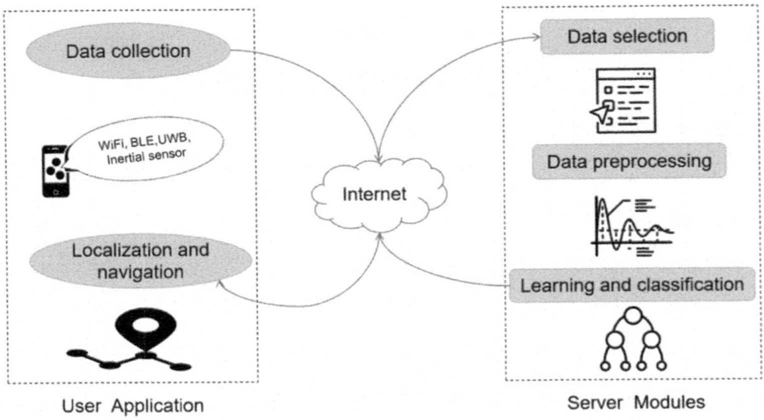

Fig. 12.1 The process of smartphone based on intelligent indoor positioning

collected data may include wireless signal data [3] such as Wireless Fidelity (WiFi), Bluetooth Low Energy (BLE), Ultra-WideBand (UWB), and built-in sensor data [4] such as acceleration, heading angle, and geomagnetic data. Then, according to the requirements, the server needs to select the required data set for preprocessing and apply the appropriate positioning algorithm to learn and train the data. Finally, the system applies the trained model to realize the position estimation and navigation of the user.

Data collection can be carried out mainly in two different ways: professional surveyor-based and crowdsourcing-based methods. In the first method, professional surveyors need to collect the required data at predefined reference points to get labeled data with location information. Crowdsourcing-based methods distribute the workload of data acquisition to massive ordinary mobile device users. This can not only save the measurement cost but also adaptively update the database according to environmental changes. Crowdsourcing data collection relies on the feedback of people but people may intentionally or unintentionally provide the wrong location labels into the database or refuse to provide location information due to privacy security issues, so one of the biggest challenges of this method is the labeling problem.

Indoor positioning algorithms can be mainly divided into sensor-based and wireless signal-based positioning algorithms. The sensor-based positioning method is mainly based on the data of the built-in sensors of smart devices to estimate the trajectory through Simultaneous Localization And Mapping (SLAM) [22] or Pedestrian Dead Reckoning (PDR) [23] algorithms. The wireless signal-based positioning algorithms can be divided into ranging-based and fingerprinting-based methods. The ranging-based methods mainly use Time of Arrival (ToA) [5], Time Difference of Arrival (TDoA) [6] and Angle of Arrival [7] to realize location estimation by trilateral positioning algorithm, which need to know the location of at least three base stations, and all of them have high requirements for time synchronization. The fingerprinting-based methods are mainly based on signal strength. Due to the multipath effect

[8], it is difficult to build a suitable propagation model to estimate the relationship between distance and signal strength. However, the fingerprinting-based methods do not require knowledge of the location of the base station, nor do they require time and angle measurements, so as to increase its feasibility for indoor deployment. Fingerprint-based positioning method [9] is mainly to obtain the location estimate of the user by matching the characteristics of the signal strength through pattern-matching algorithms.

Traditional pattern-matching methods based on Machine Learning (ML) [10] such as K-Nearest Neighbor (KNN) [11], Random Forest (RF) [24] and Support Vector Machine (SVM) [12] can only use the original features of data, and their performance is limited. The Deep Learning (DL) [13] methods, such as the Convolutional Neural Network (CNN) [14] and Long Short Term Memory (LSTM) [30], can stack multiple independent network structures together in a certain order to form a deeper network model to improve the feature extraction ability of data, which helps to further improve the positioning accuracy.

A key factor affecting positioning accuracy is the density of fingerprints, i.e., the number of reference points per unit area. In general, increasing the granularity of reference points can potentially enhance localization accuracy. However, constructing a high-density fingerprint database in a large indoor environment demands a lot of time cost and labor resources. Moreover, any changes in the indoor setting necessitate additional site surveying to update the fingerprint database. To alleviate this burden, crowdsourcing [15] has emerged as an alternative approach for collecting fingerprints. The fundamental idea of crowdsourcing is to implicitly assign the redundant and heavy fingerprint data collection task to multiple ordinary users, which can not only help save costs but also adaptively update the database based on environmental changes. However, due to privacy concerns or limited access to buildings, the crowdsourced data collected is mostly unlabeled data that does not carry location information. Therefore, in practice, labeled data is often scarce, which leads to insufficient data to accurately train the positioning model. A semi-supervised method of indoor positioning based on autoencoder is proposed to solve the problem of scarcity of labeled data and improve positioning performance [17].

12.2 Challenges of Indoor Positioning

Based on the above research, researchers have provided a variety of technical solutions for indoor positioning, but due to the complexity of indoor positioning tasks, there are still many challenges to be solved. We discuss the major challenges of indoor positioning as follows.

Impact of Interference and Large Noise

Due to the complexity of indoor scenes, there are many factors affecting signal transmission and sensor measurement. Wireless signals like WiFi, BLE, and RFID are significantly susceptible to multipath fading, obstacles, human motion, and interfering devices, which will accumulate noise in free-space propagation, resulting in a deviation between the received signal and the transmitted signal. In addition, there may be cumulative errors in the sensors built into smart devices. These factors may have a great impact on the positioning accuracy.

In most of the existing work, experiments are carried out with specific mobile devices within a specific and small experimental area. In this case, the algorithm can often achieve good positioning performance. However, in some complex large buildings, the algorithm may be subject to technical limitations and signal interference, which makes the positioning accuracy often difficult to reach the expectation, and the stability and robustness are poor.

Calibration Efforts

In the actual positioning scene, the collection and labeling of fingerprint data are very time-consuming and laborious, so it is often very difficult to construct a fine-grained fingerprint database. At the same time, once the indoor environment changes and the fingerprint data set is not updated immediately, there will be differences between the radio map and the testing data, which will significantly affect the performance of positioning algorithms.

In the process of crowdsourcing data collection, users may not provide the location label of the data due to privacy and security issues, so the scarcity of labeled data sets often occurs in fingerprint positioning tasks.

12.3 Indoor Positioning Based on Machine Learning

The fingerprint positioning method is easy to implement and has low cost, so it has gradually become the mainstream trend of indoor positioning technology. The fingerprint positioning method is composed of the offline phase and the online phase, and its process is shown in Fig. 12.2. In the offline phase, the surveyor needs to collect the signal strength from different Access Points (APs) at each Reference Point (RP) to represent the fingerprint information of the RP, and the fingerprint information of all RPs constitutes the fingerprint database of the region. In the online phase, according to the fingerprint information provided by the user, the appropriate pattern-matching algorithm is selected to match with the fingerprint database information to estimate the target user's position. In this section, we mainly introduce and summarize the fingerprint positioning technology based on the framework of machine learning.

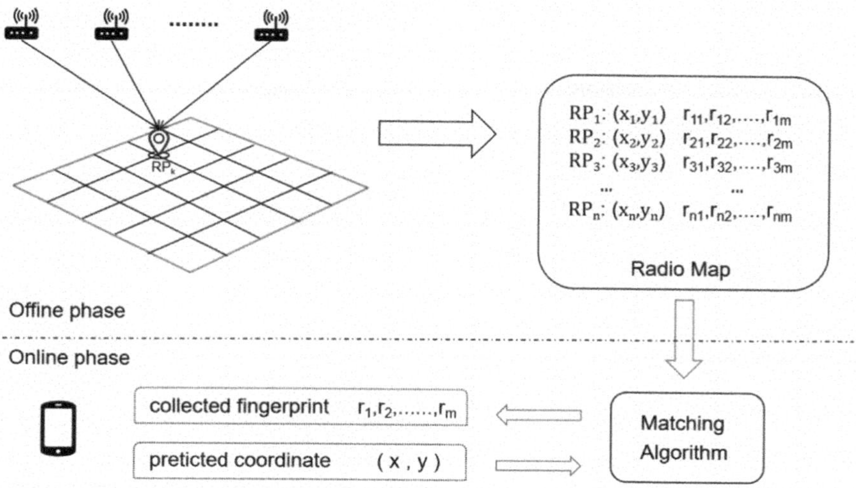

Fig. 12.2 The process of fingerprinting positioning

According to the type of data provided, we can use three types of training methods to build the location model: supervised learning, unsupervised learning and semi-supervised learning.

12.3.1 Supervised Learning

Supervised learning requires a fingerprint database with rich labels, which requires intensive data collection efforts. If professional surveyor-based methods are used to collect data, supervised learning techniques can be used to build a localization model.

Traditional Machine Learning Methods

Traditional machine learning-based positioning algorithms [10] are mainly classified into deterministic and probabilistic positioning algorithms [26]. In the deterministic positioning algorithms, the environment is divided into cells to build the radio map, and the estimated position is obtained by finding the best match between the new measurements and the fingerprints in the radio map. In probabilistic methods, also known as distribution-based methods, the radio map is constructed with the signal strength distribution obtained at the RPs, and the probability distribution function is used to estimate the location of the user.

The probabilistic positioning algorithms are generally based on the probabilistic model of Bayes theorem to estimate the location. In the offline phase, by making an independence assumption among signals from different APs, we multiply the probabilities of all APs to obtain the conditional probability $Pr\,(o|l_i)$ of receiving a particular observation o at location l_i as given by Eq. 12.1, which is exactly the content stored in a radio map.

$$Pr\,(o|l_i) = \prod_{j=1}^{k} Pr(RSS_j|AP_j, l_i), \tag{12.1}$$

where $Pr(RSS_j|AP_j, l_i)$ is the probability that AP_j has the signal strength measurement RSS_j at location l_i. In the online phase, a posterior distribution over all the locations is computed using the Bayes rule:

$$Pr\,\left(l_i|o^*\right) = \frac{Pr\,(o^*|l_i)\,Pr\,(l_i)}{\sum_{i=1}^{n} Pr\,(o^*|l_i)\,Pr\,(l_i)}, \tag{12.2}$$

where o^* is a new observation obtained. $Pr\,(l_i|o^*)$ encodes prior knowledge about where a user may be. In [28], the authors first use Bayesian estimation to calculate the conditional a posteriori probability, and then use the maximum a posteriori probability criterion to obtain the target location. Researchers in [27] have proposed a Bayesian-based particle filter method for indoor localization.

The deterministic positioning algorithm mostly uses similarity criteria such as minimum Euclidean distance criterion and cosine similarity to match the user's fingerprint information with the fingerprint database, The user's location is then determined by the locations of the RPs corresponding to the fingerprint that best matches the target user's fingerprint, and the overall process is shown in Fig. 12.3. The KNN [11] adopts the Euclidean distance to determine the k nearest fingerprints of an unknown fingerprint. The SVM algorithm realizes the classification of reference points by finding a set of hyperplanes in the fingerprint data set of N reference points, and the authors in [12] use SVM to determine the area of the object. The RF constructs an ensemble learning model by constructing a large number of decision trees. Each decision tree predicts a class, and the most common class is considered as the final prediction result. In [25], the author designs an enhanced fingerprint pattern-matching algorithm using an RF regression model.

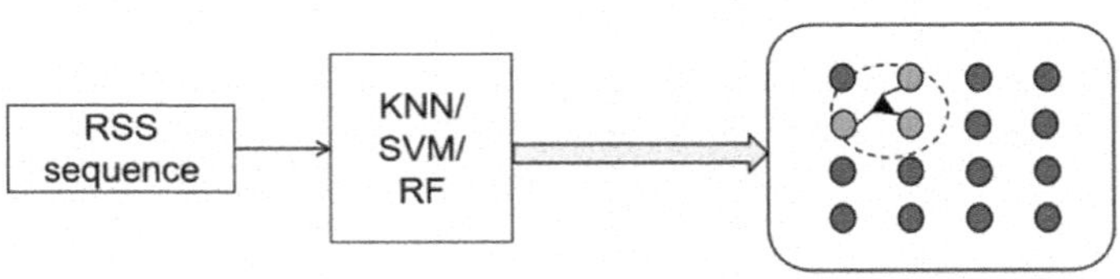

Fig. 12.3 Traditional machine learning for indoor location

Deep Learning Methods

Compared with traditional machine learning methods, deep learning algorithms have more efficient data analysis and processing capabilities, and once the models are trained with enough labeled data, the learned features can be used to make predictions on unknown data.

Typical deep learning algorithms such as Convolutional Neural Networks (CNN) are generally composed of multiple convolutional layers and fully connected layers, which are usually used to process grid-structured data. As shown in Fig. 12.4, a set of RSS fingerprint sequences are first input into the neural network for convolution operations to extract fingerprint feature information, and then they are input into the fully connected layer for flattening, and finally the position coordinates (X, Y) are obtained [14]. Wireless signals vary not only with the distance from the target but also with the change of time, so the Recurrent Neural Network (RNN) can be introduced to process time series data, and Fig. 12.5 shows the localization process based on the RNN model. The input to the RNN is an RSS vector r_t and the hidden state vector h_{t-1} from the previous time unit, and after RNN training, it outputs a new hidden state vector h_t that contains information from the previous and current input data. The hidden state feature vector h_t is then fed into a multiple regression function to obtain the current position estimation p_t [29]. However, a single CNN or RNN model may have the problem of gradient disappearance and gradient explosion during the training process. Therefore, researchers in [31] propose a spatial-temporal positioning algorithm combining Residual Network (ResNet) and LSTM, in which the ResNet extracts the spatial state information of the signal and the LSTM extracts the time state information of the signal. By combining the two improved deep learning networks, positioning accuracy is greatly improved.

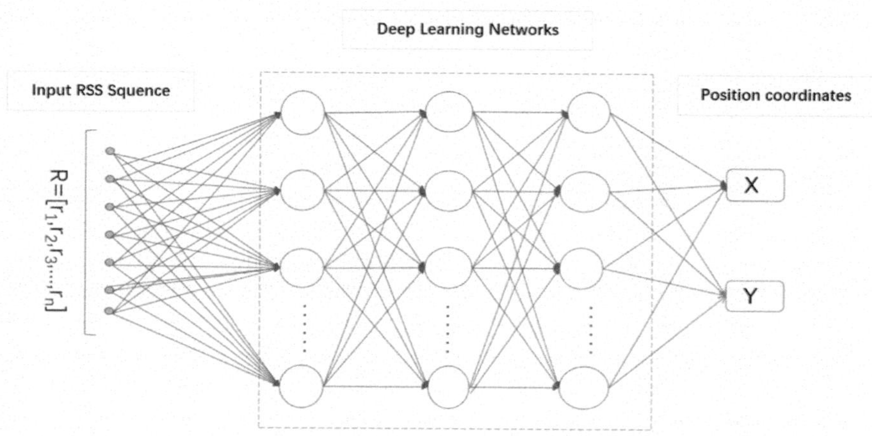

Fig. 12.4 Indoor location based on CNN

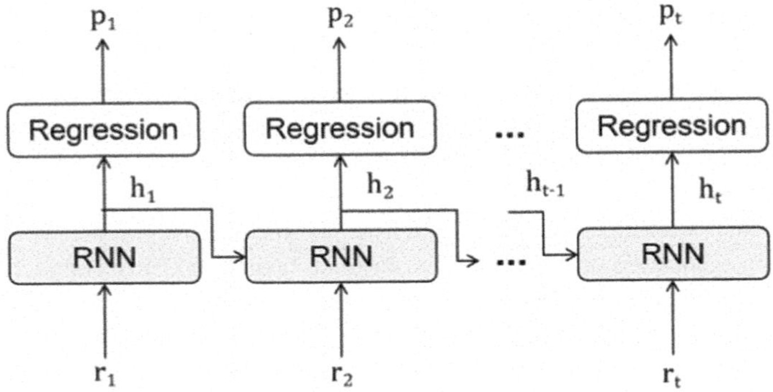

Fig. 12.5 Indoor location based on RNN

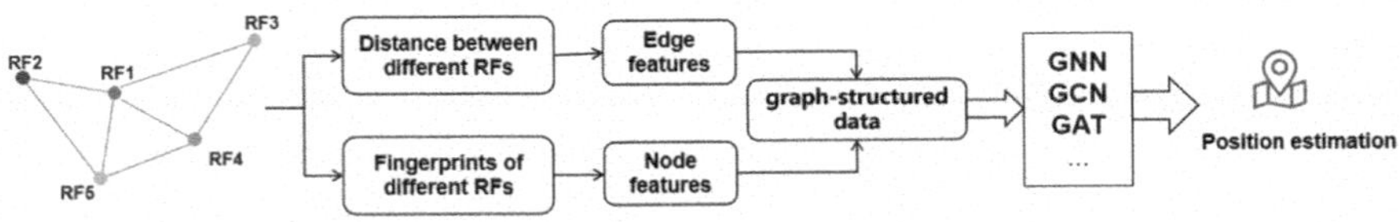

Fig. 12.6 A graph-based framework for indoor localization

Graph Neural Network (GNN) is a model that uses graph-structured data as input to make predictions or solve classification tasks. We can represent the signal strength and position information of beacons as a graph structure, where the positions represent the nodes in the graph and the links between them represent the edges, and then transform the fingerprint data into new features and input them into the GNN for positioning, as shown in Fig. 12.6. In [32], the authors use a graph regression approach to predict the location coordinates, which is compared with several existing GNN models. The authors of [34] propose a scheme to convert fingerprints into graphs by geometric methods and then use a Graph Sample and Aggregate (GraphSAGE) estimator for localization, which allows multiple wireless signals to be used as fingerprint features at the same time. Fingerprint-based indoor positioning methods usually have the problem of low feature discrimination of discrete signal fingerprints or high time cost of continuous signal fingerprint acquisition. To solve this problem, the authors in [33] propose a collaborative indoor positioning framework based on Graph Attention (GAT). The framework first constructs an adaptive graph representation using multiple discrete signal fingerprints collected by several users as inputs to effectively model the relationship between collaborative fingerprints, and then using GAT as the basic unit, a deep network with residual structure and hierarchical attention mechanism is designed to extract and aggregate features from the constructed graphs for collaborative localization.

In recent years, Federated Learning (FL) has been widely used in the field of deep learning, which is a way to train models without exchanging raw data. FL models can be trained locally on the respective data, and then the model parameters

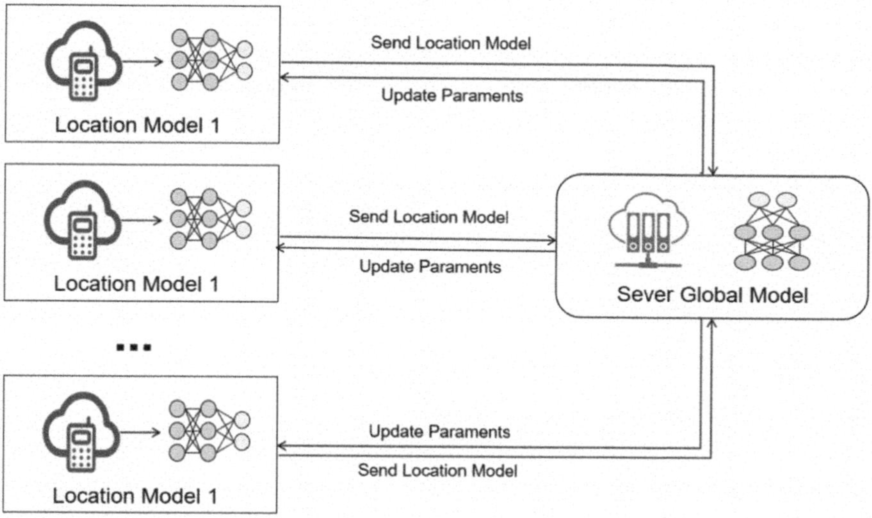

Fig. 12.7 Federated learning for indoor location

are uploaded to the server for aggregation, which helps to solve the privacy and security issues of indoor positioning [37]. Figure 12.7 illustrates a basic model of FL in indoor localization, First, each client uses the private fingerprint database to train the local positioning model and uploads the model to the central server, then the central server aggregates the uploaded local model to generate a new global model and sends the new model parameters to the client to update the local model, and iteratively optimizes to generate the final positioning model [38, 39]. The authors in [40] propose Monte Carlo (MC) dropout to reduce the communication overhead and improve the computational efficiency of FL in localization.

12.3.2 Unsupervised Learning

In large and complex indoor environments, offline fingerprint collection is a very time-consuming and laborious process. Crowdsourcing data can avoid complex data collection works, but it is often unlabeled. Unsupervised learning algorithms can use clustering to group unlabeled data, so unsupervised learning can be used to train crowdsourced unlabeled data to predict user location.

The k-means is a typical unsupervised learning algorithm, which mainly classifies the data by clustering. The K-means localization algorithm first needs to determine K reference points, and then attaches the collected fingerprint data to the nearest reference points and calculates the positions of k reference points again until they do not change anymore. The authors in [19] design a wireless indoor localization using k-means clustering combined with a logical floor plan mapping method. Researchers in [20] propose an unsupervised learning indoor localization model based on the memetic algorithm, which first uses a global search algorithm to obtain an initial model, and then uses a local optimization algorithm based on k-means to estimate the user location. The method builds an accurate indoor localization model using only unlabeled fingerprints by integrating global search and local optimization algorithms into MA, avoiding the necessity of position labels. In [21], researchers propose a method to automatically construct and optimize the unlabeled fingerprint data acquired by random walks to construct radio maps based on unsupervised learning, which does not require a complex site survey.

12.3.3 Semi-Supervised Learning

Traditional supervised learning requires a large amount of labeled data for training, but it is very difficult to obtain labeled data samples covering all indoor positioning scenes. Unlabeled crowdsourced data is easy to obtain and has a wider coverage, while unsupervised learning can extract unlabeled data features. Therefore, it is a good choice to mix labeled and unlabeled data for localization in a semi-supervised learning way, in this way, we can acquire features of large amounts of unlabeled data through unsupervised training, and then make supervised fine-tuning with a small amount of labeled data.

AutoEncoder (AE) [18] is a kind of neural network using an unsupervised learning method, which is composed of two main components: encoder and decoder. The encoder converts the input data into low-dimensional vectors, and then the decoder maps the low-dimensional vectors to the original input space for data reconstruction. The AE is often used for tasks such as dimensionality reduction, anomaly detection, and generative modeling, where the goal is to learn a compressed representation of input data while preserving its salient features. Therefore, the AE can be used as a feature extractor of the semi-supervised model to extract the features of unlabeled data X_u, so as to obtain its feature representation r, and then fine-tune the network model with a small amount of labeled data X_L in downstream tasks to obtain the localization target Y. The process of the AE-based semi-supervised positioning system is shown in Fig. 12.8.

There are many types of AE, such as Stacking AutoEncoders (SAE) [35], Denoising AutoEncoder (DAE) and Variational AutoEncoder (VAE) [36], etc., and researchers have also proposed many different AE-based semi-supervised indoor positioning solutions to deal with various problems of indoor positioning tasks. In [43], the authors utilize the SAE to extract high-level features from unlabeled crowdsourced data to improve localization performance during classification. RSS-based indoor positioning technology is particularly vulnerable to security threats. The authors in [42] propose a semi-supervised deep learning security enhancement framework combining DAE and CNN, aiming to achieve security enhancement without affecting the effectiveness and efficiency of indoor positioning. The model first

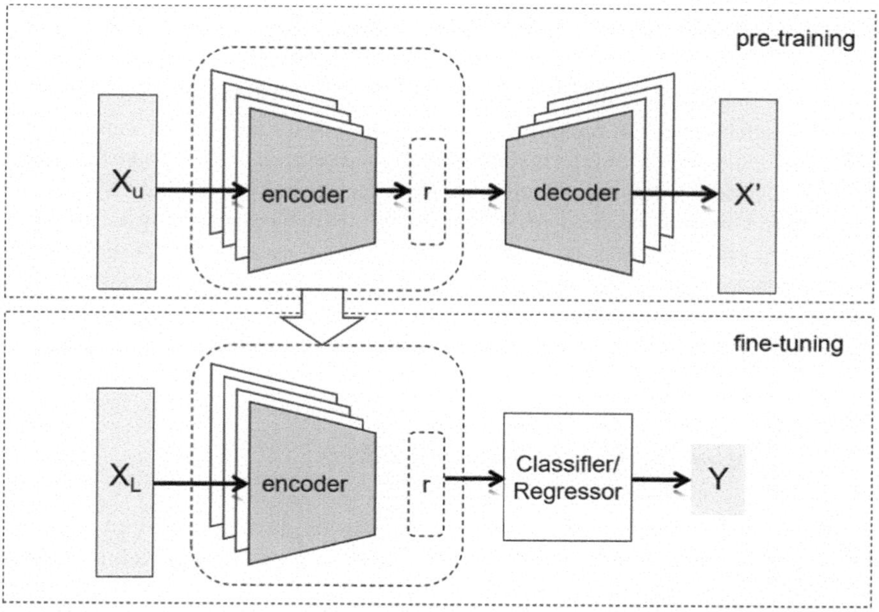

Fig. 12.8 The process of AE-based semi-supervised positioning system

uses DAE components to denoise the abrupt RSS values to reduce the impact of malicious attacks and then uses CNN to perform fingerprint matching. Researchers in [44] propose a semi-supervised learning model based on the Variational AutoEncoder (VAE). During unsupervised learning, the VAE is used as the feature extractor to learn the latent distribution of the original input, and then the labeled data is used to train the predictor.

In order to alleviate the scarcity of labeled data, researchers in [41] propose a centralized indoor localization method based on pseudo-labels. This scheme uses unlabeled fingerprint data collected by mobile crowdsourcing to generate pseudo-labels and combines it with a small amount of labeled data to reduce the burden of labeling fingerprints. The AP locations in the environment may change over time, causing deviations in the fingerprint database and affecting the location performance. In [16], the authors propose a crowdsourced indoor location method based on ensemble learning, which can automatically identify altered APs in the crowdsourced data and update the database.

12.4 Sensor-Fusion Positioning Techniques

In complex large-scale indoor environments, a single positioning technology may not be enough to meet the challenges faced by indoor positioning, so we can consider combining multiple sensor information sources, and select the appropriate algorithm to build a hybrid positioning system to improve the accuracy and reliability of positioning. We can perform dead reckoning on the IMU data according to the PDR algorithm to obtain the relative motion trajectory of the user, and then use the absolute position coordinates obtained by the fingerprint localization algorithm based on wireless signals to correct the trajectory, so as to obtain the fusion localization result. The overall framework of the hybrid positioning system is shown in Fig. 12.9.

12.4.1 Sensor Fusion via Bayesian Filters

In order to realize target positioning by sensor fusion, it is necessary to integrate and synchronize data from multiple sources. Bayesian filters such as Kalman Filters (KF) and Particle Filters (PF) can effectively integrate multi-sensor information to improve the accuracy and stability of the positioning system.

In [45], the authors combine the two-dimensional PDR trajectory information based on PF correction with the information based on KF to detect floor changes to form a three-dimensional target tracking system. The researchers in [46] propose to use the Adaptive Feedback Extended Kalman Filter (AFEKF) algorithm to deeply integrate the results of BLE position, distance measurement and PDR, so as to adaptively adjust the position estimation at the next moment, making the position estimation more accurate and the algorithm more robust. The authors in [47] develop a light-weight Pedestrian Inertial Navigation System (PINS) using iterative Extended Kalman Filter (iEKF), and propose a crowdsourced WiFi fingerprint database generation framework based on deep learning. Finally, three different types of multi-source integration models are used to integrate PINS and WiFi fingerprint information to deal with different scenarios.

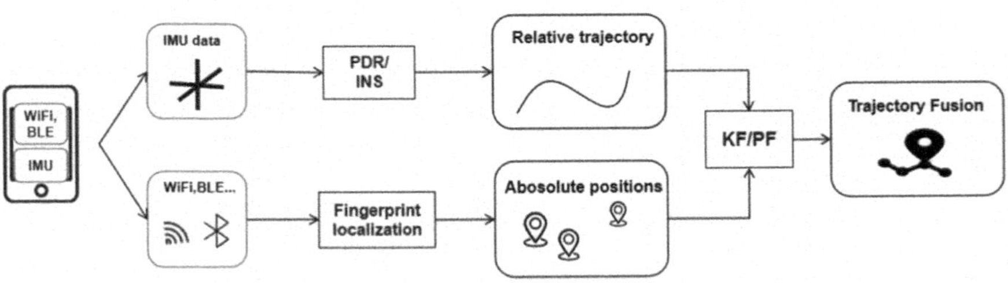

Fig. 12.9 The framework of the hybrid positioning system

12.4.2 Indoor Positioning via Trajectory Fusion

The inertial sensor-based positioning method can produce cumulative errors or trajectory drift over time, and wireless signals-based positioning is prone to environmental interference. By fusing different positioning trajectory information, the advantages can be complementary.

The authors in [48] use an enhanced particle filter to fuse PDR, GPS and WiFi fingerprints, and propose a three-step tracking and matching algorithm to obtain crowdsourced radio maps, achieving high-precision outdoor-indoor seamless localization. In order to solve the problem of trajectory fragmentation caused by crowdsourcing data, the authors in [49] propose a robust iterative trace merging algorithm based on WiFi access points as signal markers to merge a large number of trajectories, and further improve the accuracy of crowdsourcing indoor positioning by removing trajectory outliers through the enhanced matching algorithm. In [50], the researchers propose a graph-based SLAM framework that adaptatively establishes WiFi-based edges and GPS-based edges as prior information for corresponding positions in IMU trajectory estimation to achieve trajectory fusion and generate crowdsourced radio maps. After fusion, the generated crowdsourced radio map can be used for other users to conduct WiFi-based fingerprint positioning, reducing the cost of building a fingerprint database.

12.5 Experimental Results and Performance Comparison

In this section, we provide examples from traditional machine learning, deep learning and crowdsourcing for indoor positioning, which are conducted by our laboratory.

12.5.1 Traditional Machine Learning

Fingerprinting approaches typically use traditional machine learning algorithms, such as KNN, SVM, and RF, to solve the classification or regression problem. Ensemble learning combines multiple weak models to improve positioning accuracy in which even if one weak learner gets a wrong prediction, other weak learners can correct the error to some extent. In practice, ensemble learning can mitigate overfitting to obtain higher predictive accuracy and higher generalization ability. As an often used ensemble learning approach with high prediction accuracy and the ability of processing nonlinear data, RF regression uses decision trees as basic learners. The decision tree is easy to understand and implement, which has a fast training rate and high efficiency.

In our previous work of [25], we apply an RF model to fuse the RSS and timestamp fingerprints to locate indoor users. We design a positioning testbed for narrow-band

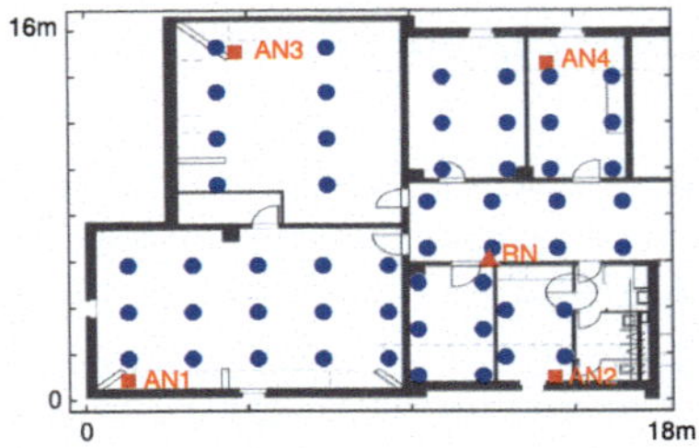

(a) The distribution map of reference points and APs in the first scenario

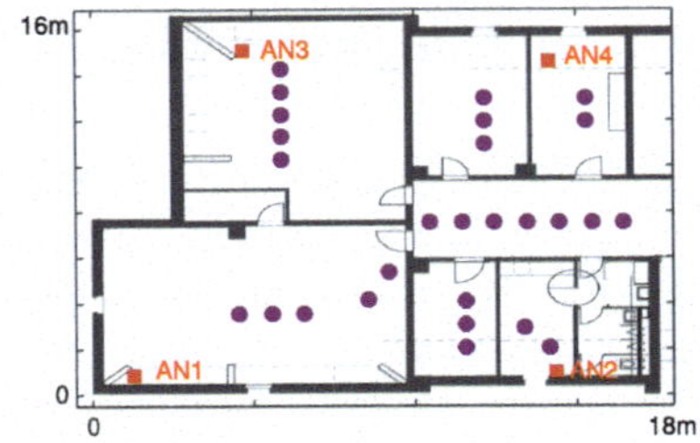

(b) Testing positions in the first scenario

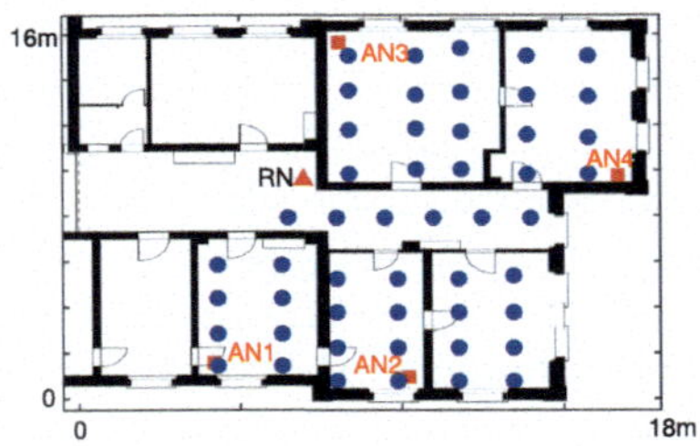

(c) The distribution map of reference points and APs in the second scenario

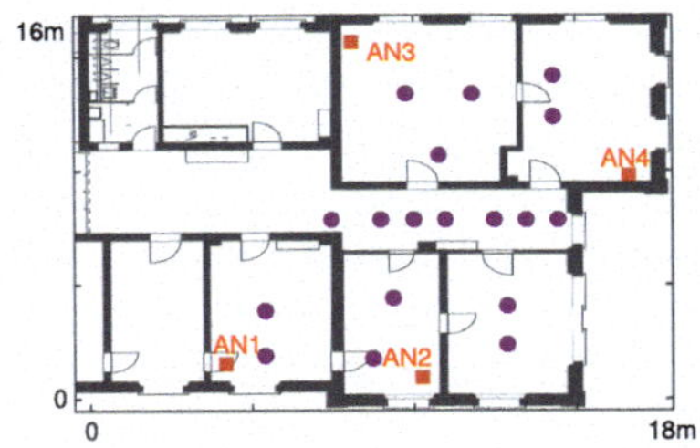

(d) Testing positions in the second scenario

Fig. 12.10 Measurement setup

Table 12.1 Measurement notation

Notation	Scenario	Conducted time
M1	First	1 day after constructing radio map
M2	First	5 days after constructing radio map
M3	Second	1 day after constructing radio map
M4	Second	5 days after constructing radio map

signals (ZigBee) based on software defined radio techniques to extract RSS and timestamps. We deploy the system in two indoor scenarios in the INF building at the University of Bern, in which the distribution maps of reference points and APs as shown in Fig. 12.10a and Fig. 12.10c respectively. We compare the performance of the Fusion KNN-RF, Fusion WKNN, RSS WKNN, and DTDOA WKNN algorithms in the four experiments shown in Table 12.1. KNN-RF is the proposed algorithm integrating the KNN and RF model, and *Fusion* indicates the fused RSS and DTDOA, in which DTDOA is the differential time difference of arrival information. We refer to our previous work [25] for more details.

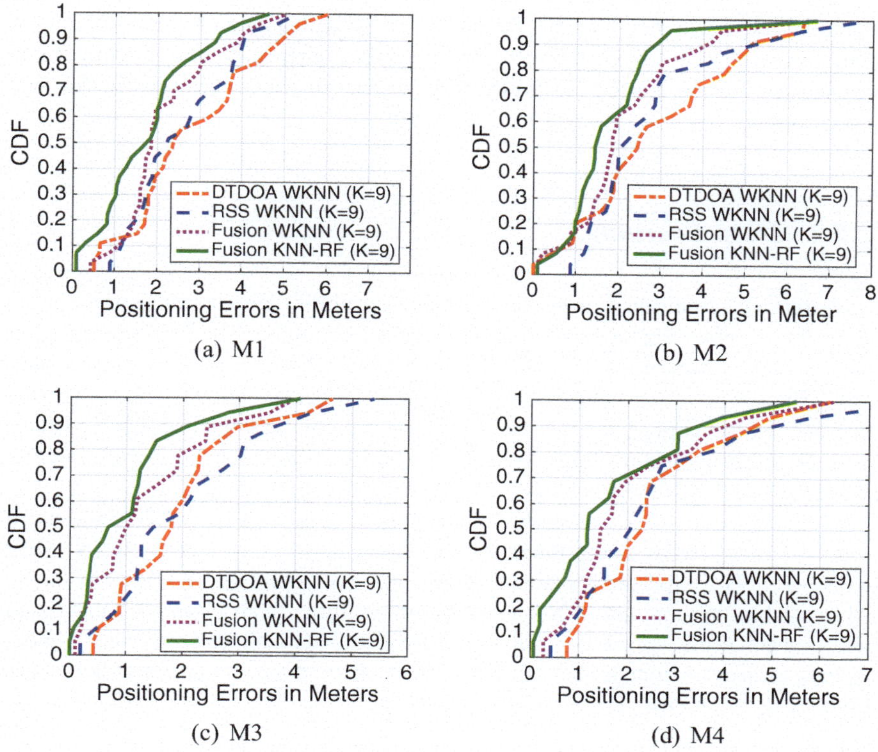

Fig. 12.11 Positioning accuracy

Table 12.2 Mean positioning errors

EXP	DTDOA WKNN (m)	RSS WKNN (m)	Fusion WKNN (m)	Fusion KNN-RF (m)
M1	2.88	2.61	2.24	1.8
M2	2.89	2.77	2.22	1.86
M3	1.95	2.08	1.38	1.09
M4	2.59	2.64	2.1	1.7
Overall	2.58	2.52	1.99	1.61

CDFs of positioning accuracy are shown in Fig. 12.11a–d and the mean positioning errors are summarized in Table 12.2. KNN-RF outperforms traditional KNN. As shown in Table 12.2, Fusion KNN-RF achieves an accuracy of 1.61 m outperforming Fusion WKNN with an improvement of 19.1%. With the same input information, the performance of Fusion KNN-RF is significantly better than Fusion WKNN thanks to the RF for regression. the proposed Fusion KNN-RF significantly outperforms RSS WKNN by fusing the DTDOA-RSS features and adopting KNN-RF for pattern matching. As shown in Table 12.2, Fusion KNN-RF with a mean accuracy of 1.61 m outperforms the RSS-based fingerprinting by 36.1%. The results in Fig. 12.11a–d also indicate that Fusion KNN-RF significantly outperforms the traditional RSS-based fingerprinting.

12.5.2 Deep Learning

Compared with traditional machine learning algorithms, deep learning-based algorithms have a more abstract and complex stacked hierarchy. The two most popular types of deep neural networks are CNN and RNN. RNN is a kind of neural network with a recurrent structure, which processes sequence data through a recurrent structure and can capture the time series information in the sequence. CNN extracts data features through convolution operation, which can capture local spatial structure information.

In our research work, we conduct experiments on a set of public datasets of Jaume I University by using three deep learning network structures, CNN, MLP and ResNet. The experimental scenario and reference point distribution are shown in Fig. 12.12. We compare KNN with the above three deep learning localization algorithms in terms of average localization error and median localization error under this dataset, as shown in Table 12.3. For ease of observation, we plot the CDF curve of the localization error as shown in Fig. 12.13.

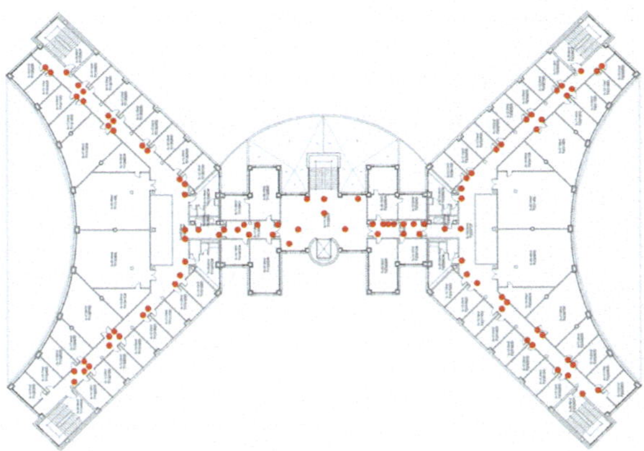

Fig. 12.12 Map of experimental scenario

Table 12.3 Positioning errors

Positioning error	KNN (m)	MLP (m)	CNN (m)	ResNet (m)
Mean	7.12	5.90	5.06	4.65
Median	7.04	5.33	4.73	3.78

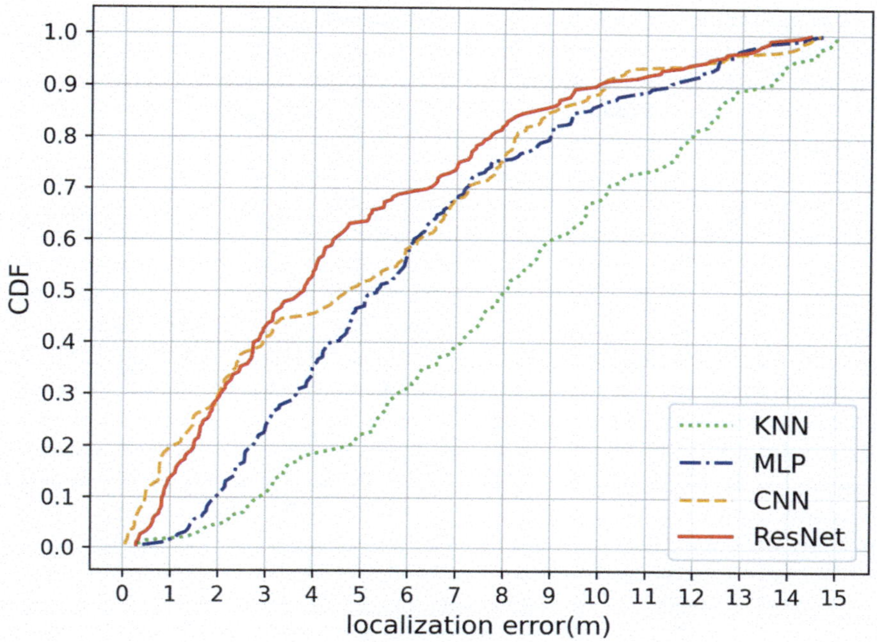

Fig. 12.13 CDF of positioning error

According to the experimental results, it can be seen that the deep learning localization algorithm is significantly better than the traditional machine learning localization algorithm represented by KNN. This is because the KNN algorithm can only use the original features of the data, while the deep learning algorithm can process and analyze fingerprint data more efficiently through the neural network. Because CNN can effectively capture the spatial characteristics of the input data through convolution, the positioning accuracy of CNN is improved by 14% compared with MLP. At the same time, it can be observed that on the basis of CNN, ResNet reduces the positioning error by 0.41 m. This is because the input information of the network structure of CNN is propagated forward through each layer, which may produce the problem of vanishing gradients. However, by adding the residual connection, ResNet enables the information to be transmitted directly. Therefore, the model can train a deeper network and improve positioning accuracy.

12.5.3 Crowdsourcing Solutions

Indoor positioning based on crowdsourcing data labels RSS values with the locations on the merged crowdsourcing traces to generate a radio map. Compared with the traditional radio maps created by site surveying, crowdsourcing radio maps are normally noisy. Hence, it is still challenging to locate users with high accuracy based on crowdsourcing indoor positioning. Moreover, the accuracy of positioning is further limited due to the sparse coverage of the crowdsourcing radio map.

In our previous work of [49], we have designed a crowdsourcing indoor positioning system namely WiFi-RITA. In WiFi-RITA, we merge massive noisy user traces to recover indoor walking paths, in which massive noisy user traces are large quantities of short traces with uncertain rotation errors. In WiFi-RITA, traces are merged by iteratively translating and rotating relying on the ubiquitous signal-marks of WiFi access points. WiFi-RITA positioning further improves positioning accuracy based on noisy crowdsourcing radio maps with limited coverage, in which we design a multivariate Gaussian model to generate a grid-based radio map and an enhanced matching algorithm to improve our fingerprinting accuracy. Then, we fuse PDR and the fingerprinting based on a particle filter to further improve performance, especially in the uncovered areas of the radio map.

To evaluate WiFi-RITA positioning, we conduct a set of comprehensive experiments in Fig. 12.14 (6656 m^2) and Fig. 12.15 (8372 m^2), in which the experiments are conducted for trace merging and user positioning. For trace merging, 10 users randomly walk along the predefined paths in two scenarios as shown in Figs. 12.14a and 12.15a. We use Huawei Mate8, Mate9 and P10 for data collection with different phone placements, i.e., in coat pockets, in bags, in trouser pockets, and freely holding in hands. The data collection duration is about 10 h with around 1 h for each user. For locating users, the users walk on five predefined paths as the green lines in Fig. 12.14b to f and Figure 12.15b to f.

Figure 12.16 shows the accuracy of the merged traces, in which the median accuracy achieves 1.3 m for Scenario 1 and 1.0 m for Scenario 2.

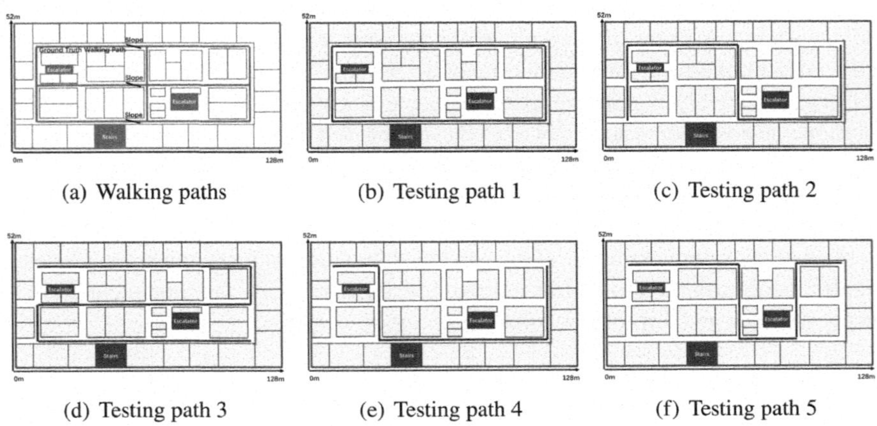

(a) Walking paths (b) Testing path 1 (c) Testing path 2

(d) Testing path 3 (e) Testing path 4 (f) Testing path 5

Fig. 12.14 Walking and testing paths in Scenario 1

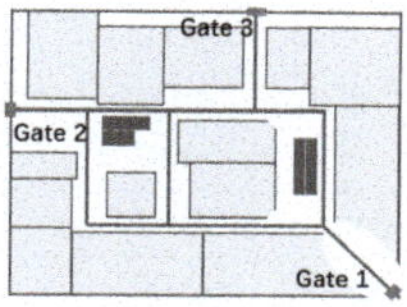

(a) Walking paths

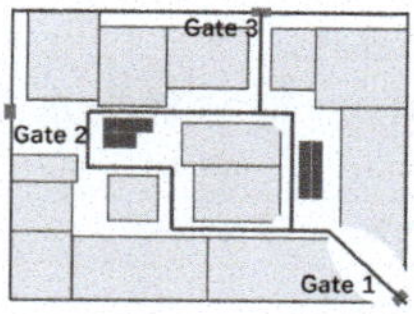

(b) Testing path 1

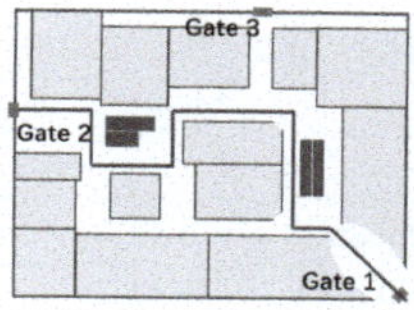

(c) Testing path 2

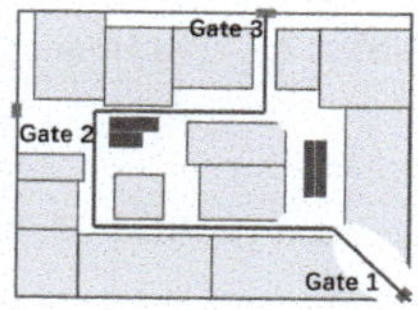

(d) Testing path 3

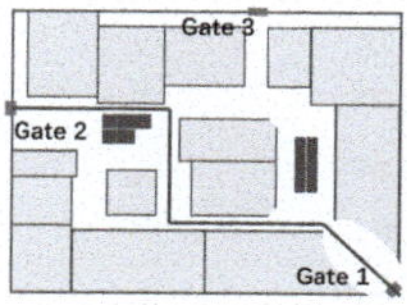

(e) Testing path 4

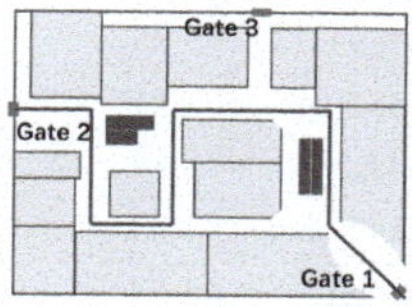

(f) Testing path 5

Fig. 12.15 Walking and testing paths in Scenario 2

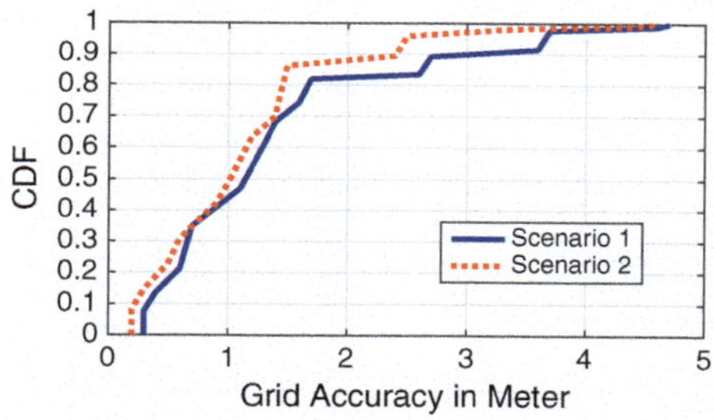

Fig. 12.16 Accuracy of grids

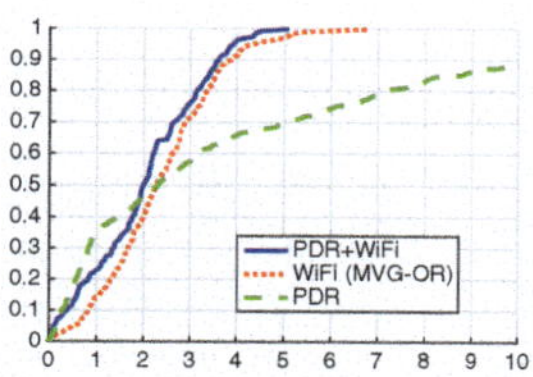

(a) EXP 1 - 3

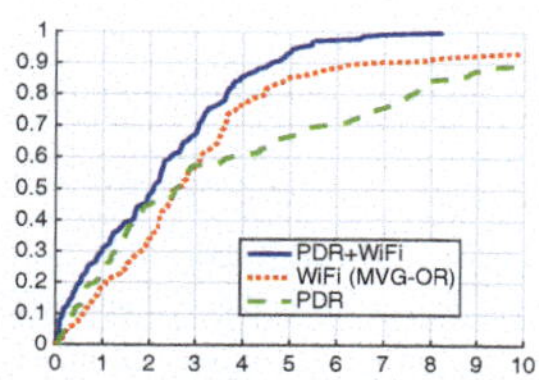

(b) EXP 4 - 5

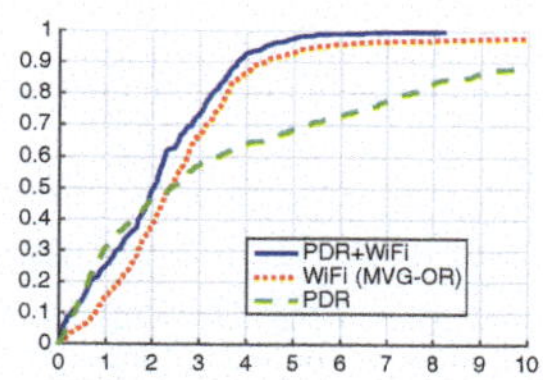

(c) Overall

Fig. 12.17 Positioning errors of different positioning algorithms in Scenario 1

The limited coverage of radio maps degrades the performance of fingerprinting. To solve this problem, we evaluate the enhanced particle filter by fusing PDR and fingerprinting. Figure 12.17 shows the CDF of positioning errors and Table 12.4 summarizes the mean and median accuracy. Figure 12.18 shows the estimated locations of WiFi-RITA-Positioning in two representative moving paths. According to the evaluation results, the enhanced particle filter by fusing PDR and fingerprinting sig-

Table 12.4 Positioning errors in the two testing scenarios

			KNN	MVG (m)	MVG-OR (m)	PDR (m)	PDR+WiFi (m)
Scenario 1	EXP 1–3	Mean	3.35	2.76	2.38	4.04	2.03
		Median	3.32	2.82	2.29	2.32	2.00
	EXP 4–5	Mean	4.62	4.29	3.68	4.25	2.24
		Median	3.65	3.08	2.61	2.59	2.08
	Overall	Mean	3.75	3.24	2.79	4.11	2.10
		Median	3.41	2.93	2.39	2.41	2.02
Scenario 2	EXP 1–3	Mean	3.45	2.84	2.51	3.62	2.11
		Median	3.43	2.91	2.59	2.21	2.15
	EXP 4–5	Mean	4.74	4.37	3.73	3.78	2.37
		Median	3.71	3.16	2.77	2.38	2.19
	Overall	Mean	3.97	3.45	3.00	3.68	2.21
		Median	3.52	3.12	2.65	2.32	2.16

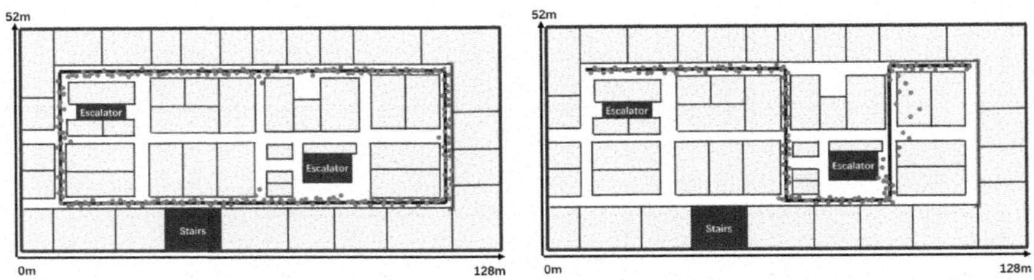

(a) EXP 1 (b) EXP 5

Fig. 12.18 Estimated points based on WiFi-RITA positioning in Scenario 1

nificantly outperforms PDR and WiFi fingerprinting of MVG-OR, especially on trace 4 and trace 5. Fingerprinting can not locate users in the uncovered areas of the radio map, in which the particle filter improves the performance by fusing PDR. Hence, in Fig. 12.18b, the fusion approach outperforms MVG-OR by 20% and 39% respectively with a median accuracy of 2.08 m and a mean accuracy of 2.24 m. PDR is accurate at the beginning but significantly degrades with time. The enhanced particle filter by fusing PDR and fingerprinting achieves a median accuracy of 2.02 m and mean accuracy of 2.10 m, which significantly outperforms PDR and fingerprinting with MVG-OR.

12.6 Conclusion

In this chapter, we give an overview of the current challenges of intelligent indoor positioning based on wireless signals and various indoor positioning technical solutions that have emerged to solve these challenges.

When the fingerprint data is sufficient, the existing positioning algorithms based on deep learning can achieve high positioning accuracy. However, in most practical cases, it is often difficult to obtain enough labeled data to build a fine-grained fingerprint database. Therefore, a semi-supervised localization method based on crowdsourcing data is proposed, which can alleviate the problem of scarcity of labeled data while ensuring positioning accuracy. Wireless signals are susceptible to interference in complex indoor environments, fusion positioning technology based on multi-source sensors can overcome the shortcomings of single localization technology and achieve complementary advantages, which has become the key research direction in the field of indoor localization. Finally, we take our previous work as examples to briefly introduce the current several mainstream positioning approaches based on WiFi and PDR.

References

1. Farahsari PS, Farahzadi A, Rezazadeh J, Bagheri A (2022) A survey on indoor positioning systems for IoT-based applications. IEEE Internet Things J 9(10):7680–7699. https://doi.org/10.1109/JIOT.2022.3149048
2. Zafari F, Gkelias A, Leung KK (2019) A survey of indoor localization systems and technologies. IEEE Commun Surv Tutor 21(3):2568–2599. https://doi.org/10.1109/COMST.2019.2911558
3. Wen Q, Liang Y, Wu C, Tavares A, Han X (2018) Indoor localization algorithm based on artificial neural network and radio-frequency identification reference tags. Adv Mech Eng 10(12):1687814018808682
4. Liu GX, Shi LF, Xun JH, Chen S, Liu H, Shi YF (2018) Hierarchical calibration architecture based on Inertial/magnetic sensors for indoor positioning. Indoor Navigation and Location-Based Services (UPINLBS), pp 1–9. https://doi.org/10.1109/UPINLBS.2018.8559914
5. Pan G, Wang T, Zhang S, Xu S (2020) High accurate time-of-arrival estimation with fine-grained feature generation for Internet-of-Things applications. IEEE Wirel Commun Lett 9(11):1980–1984. https://doi.org/10.1109/LWC.2020.3010251
6. Chen X, Gao Z (2017) Indoor ultrasonic positioning system of mobile robot based on TDOA ranging and improved trilateral algorithm. Vision and computing (ICIVC), pp 923–927
7. Liu J, Wang T, Li Y, Li C, Wang Y, Shen Y (2022) A transformer-based signal denoising network for AOA estimation in NLOS environments. IEEE Commun Lett 26(10):2336–2339. https://doi.org/10.1109/LCOMM.2022.3187661
8. Tuta J, Juric MB (2016) A self-adaptive model-based Wi-Fi indoor localization method. Sensors 16(12):2074
9. Luo J, Zhang Z, Wang C et al (2019) Indoor multifloor localization method based on WiFi fingerprints and LDA. IEEE Trans Ind Inf 15(9):5225–5234
10. Ouameur MA, Caza-Szoka M, Massicotte D (2020) Machine learning enabled tools and methods for indoor localization using low power wireless network. Int Things 12:100300
11. Li D, Zhang B, Li C (2015) A feature-scaling-based k-nearest neighbor algorithm for indoor positioning systems. IEEE Int Things J 3(4):590–597

12. Abbas HA, Boskany NW, Ghafoor KZ, et al(2021) Wi-Fi based accurate indoor localization system using SVM and LSTM algorithms. In: 2021 IEEE 22nd international conference on information reuse and integration for data science (IRI), pp 416–422
13. Seok KY, Lee J H (2018) Deep learning based fingerprinting scheme for wireless positioning. In: International conference on artificial intelligence in information and communication (ICAIIC), pp 312–314. https://doi.org/10.1109/ICAIIC48513.2020.9065054
14. Ibrahim M, Torki M, ElNainay M (2018) CNN based indoor localization using RSS time-series. In: 2018 IEEE symposium on computers and communications (ISCC), pp 01044–01049. https://doi.org/10.1109/ISCC.2018.8538530
15. Wang B, Chen Q, Yang LT et al (2016) Indoor smartphone localization via fingerprint crowdsourcing: challenges and approaches. IEEE Wirel Commun 23(3):82–89. https://doi.org/10.1109/MWC.2016.7498078
16. Yang J, Zhao X, Li Z (2019) Crowdsourcing indoor positioning by light-weight automatic fingerprint updating via ensemble learning. IEEE Access 7:26255–26267. https://doi.org/10.1109/ACCESS.2019.2901736
17. Ouyang RW, Wong AKS, Lea CT et al (2013) Indoor location estimation with reduced calibration exploiting unlabeled data via hybrid generative/discriminative learning. IEEE Trans Mobil Comput 11(11):1613–1626. https://doi.org/10.1109/TMC.2011.193
18. Fontaine J, Ridolfi M, Van Herbruggen B et al (2020) Edge inference for UWB ranging error correction using autoencoders. IEEE Access 8:139143–139155. https://doi.org/10.1109/ACCESS.2020.3012822
19. Wu C, Yang Z, Liu Y, Xi W (2013) WILL: wireless indoor localization without site survey. IEEE Trans Parallel Distrib Syst 24(4):839–848. https://doi.org/10.1109/TPDS.2012.179
20. Jung S, Moon B, Han D (2015) Unsupervised learning for crowdsourced indoor localization in wireless networks. IEEE Trans Mobil Comput 15(11):2892–2906
21. Trogh J, Joseph W, Martens L, Plets D (2019) An unsupervised learning technique to optimize radio maps for indoor localization. Sensors 19(4):752. https://doi.org/10.3390/s19040752
22. Dong Y, Yan D, Li T et al (2022) Pedestrian gait information aided visual inertial SLAM for indoor positioning using handheld smartphones. IEEE Sens J 22(20):19845–19857. https://doi.org/10.1109/JSEN.2022.3203319
23. Wu C, Yang Z, Liu Y (2014) Smartphones based crowdsourcing for indoor localization. IEEE Trans Mobil Comput 14(2):444–457. https://doi.org/10.1109/TMC.2014.2320254
24. Calderoni L, Ferrara M, Franco A et al (2015) Indoor localization in a hospital environment using random forest classifiers. Expert Syst Appl 42(1):125–134
25. Li Z, Braun T, Zhao X et al (2018) A narrow-band indoor positioning system by fusing time and received signal strength via ensemble learning. IEEE Access 6:9936–9950. https://doi.org/10.1109/ACCESS.2018.2794337
26. Bozkurt S, Elibol G, Gunal S, Yayan U (2015) A comparative study on machine learning algorithms for indoor positioning. In: International symposium on innovations in intelligent systems and applications (INISTA)
27. Seshadri V, Zaruba GV, Huber M (2005) A bayesian sampling approach to in-door localization of wireless devices using received signal strength indication. In: Third IEEE international conference on pervasive computing and communications, pp 75–84
28. Chai X, Yang Q (2007) Reducing the calibration effort for probabilistic indoor location estimation. IEEE Trans Mobil Comput 6:649–662. https://doi.org/10.1109/TMC.2007.1025
29. Khassanov Y, Nurpeiissov M, Sarkytbayev A et al (2021) Finer-level sequential wifi-based indoor localization. IEEE/SICE Int Symp Syst Int (SII) 2021:163–169
30. Poulose A, Han DS (2021) Feature-based deep LSTM network for indoor localization using UWB measurements. In: International conference on artificial intelligence in information and communication (ICAIIC), pp 298–301. https://doi.org/10.1109/ICAIIC51459.2021.9415277
31. Wang R, Luo H, Wang Q et al (2020) A spatial-temporal positioning algorithm using residual network and LSTM. IEEE Trans Instrum Meas 69(11):9251–9261. https://doi.org/10.1109/TIM.2020.2998645

32. Fu Y, Xiong X, Liu Z, Chen X, Liu Y, Fu Z (2022) A GNN-based indoor localization method using mobile RFID platform. In: International conference on smart and sustainable technologies (SpliTech), pp 1–6. https://doi.org/10.23919/SpliTech55088.2022.9854370
33. He T, Niu Q, Liu N (2023) GC-LOC: a graph attention based framework for collaborative indoor localization using infrastructure-free signals. Proc ACM Interact Mobil Wearable and Ubiquitous Technol 6(4):1–27
34. Luo X, Meratnia N (2022) A geometric deep learning framework for accurate indoor localization. In: IEEE 12th international conference on indoor positioning and indoor navigation (IPIN), pp 1–8. https://doi.org/10.1109/IPIN54987.2022.9918118
35. Song X (2019) A novel convolutional neural network based indoor localization framework With WiFi fingerprinting. IEEE Access 7:110698–110709. https://doi.org/10.1109/ACCESS.2019.2933921
36. Chidlovskii B, Antsfeld L (2019) Semi-supervised variational autoencoder for WiFi indoor localization. In: International conference on indoor positioning and indoor navigation (IPIN), pp 1–8. https://doi.org/10.1109/IPIN.2019.8911825
37. Nagia N, Rahman MT, Valaee S (2022) Federated learning for WiFi fingerprinting. In: IEEE international conference on communications, pp 4968–4973. https://doi.org/10.1109/ICC45855.2022.9838945
38. Liu Y, Li H, Xiao J, Jin H (2019) FLoc: fingerprint-based indoor localization system under a federated learning updating framework. In: International conference on mobile Ad-Hoc and sensor networks (MSN), pp 113–118. https://doi.org/10.1109/MSN48538.2019.00033
39. Wu P, Imbiriba T, Park J, Kim S, Closas P (2021) Personalized federated learning over non-IID data for indoor localization. In: International workshop on signal processing advances in wireless communications (SPAWC), pp 421–425
40. Park J et al (2022) Federated learning for indoor localization via model reliability with dropout. IEEE Commun Lett 26(7):1553–1557. https://doi.org/10.1109/LCOMM.2022.3170878
41. Li W, Zhang C, Tanaka Y (2020) Pseudo label-driven federated learning-based decentralized indoor localization via mobile crowdsourcing. IEEE Sens J 20(19):11556–11565. https://doi.org/10.1109/JSEN.2020.2998116
42. Ye Q, Fan X, Bie H, Puthal D, Wu T, Song X, Fang G (2023) SE-LOC: security-enhanced indoor localization with semi-supervised deep learning. IEEE Trans Netw Sci Eng 10(5):2964–2977. https://doi.org/10.1109/TNSE.2022.3174674
43. Khatab ZE, Gazestani AH, Ghorashi SA et al (2021) A fingerprint technique for indoor localization using autoencoder based semi-supervised deep extreme learning machine. Signal Process 181:107915
44. Qian W, Lauri F, Gechter F (2021) Supervised and semi-supervised deep probabilistic models for indoor positioning problems. Neurocomputing 435:228–238
45. Luo J, Zhang C, Wang C (2020) Indoor multi-floor 3D target tracking based on the multi-sensor fusion. IEEE Access 8:36836–36846. https://doi.org/10.1109/ACCESS.2020.2972962
46. Kong X, Wu C, You Y, Yuan Y (2023) Hybrid indoor positioning method of BLE and PDR based on adaptive feedback EKF with low BLE deployment density. IEEE Trans Instrum Meas 72:1–12. https://doi.org/10.1109/TIM.2022.3227957
47. Yu Y, Chen R, Chen L et al (2021) H-WPS: hybrid wireless positioning system using an enhanced wi-fi FTM/RSSI/MEMS sensors integration approach. IEEE Int Things J 9(14):11827–11842
48. Li Z, Zhao XHu, F, Zhao Z, Villacrés JLC, Braun T, (2019) SoiCP: a seamless outdoor-indoor crowdsensing positioning system. IEEE Int Things J 6(5):8626–8644
49. Li Z, Zhao X, Zhao Z, Braun T (2021) WiFi-RITA positioning: enhanced crowdsourcing positioning based on massive noisy user traces. IEEE Trans Wirel Commun 20(6):3785–3799. https://doi.org/10.1109/TWC.2021.3053582
50. Gu Y, Zhou C, Wieser A, Zhou Z (2018) Trajectory estimation and crowdsourced radio map establishment from foot-mounted IMUS, wi-fi fingerprints, and GPS positions. IEEE Sens J 19(3):1104–1113. https://doi.org/10.1109/JSEN.2018.2877804

Chapter 13
High Precision Positioning Algorithms Based on Improved Sparse Bayesian Learning in MmWave MIMO Systems

Jiancun Fan, Wei Zou, and Xiaoyuan Dou

Abstract Sparse Bayesian learning (SBL) is a millimeter-wave (mmWave) positioning method that leverages the sparsity of channels to estimate parameters such as angle of arrival (AOA) and time delay for positioning. Compared to other parameter estimation algorithms, such as the Multi-signal classification (MUSIC) algorithm, Expectation–Maximization (EM) algorithm, and Space-alternating Generalized Expectation–Maximization (SAGE) algorithm, SBL demonstrates superior performance and robustness in millimeter wave scenarios. However, most existing SBL solutions only account for angle sparsity. In this chapter, we address the joint sparsity of both the angle domain and time delay domain, and propose a new two-dimensional adaptive grid refinement method to enhance the existing SBL framework. To address the grid mismatch problem common in all sparse estimation algorithms, we have also introduced a low-complexity grid evolution algorithm. Additionally, we derive the Cramer-Rao bound (CRB) for AOA, time delay, and position estimation based on the mmWave multipath signals from base stations (BS), and subsequently analyze estimation errors. Simulation results indicate that the proposed algorithm outperforms existing algorithms and approaches the CRB. Simulations using real-world datasets also confirm these findings.

13.1 Introduction

Currently, mainstream high-precision positioning technology is divided into two main categories: ranging and non-ranging positioning techniques. Ranging positioning technology calculates the distance between a known position node and the

J. Fan (✉) · W. Zou · X. Dou
School of Information and Communications Engineering, Xi'an Jiaotong University, Xi'an, China
e-mail: fanjc0114@gmail.com

W. Zou
e-mail: zouweixjtu@163.com

X. Dou
e-mail: douz6334@stu.xjtu.edu.cn

K. Yu (ed.), *Positioning and Navigation Using Machine Learning Methods*, Navigation: Science and Technology 14, https://doi.org/10.1007/978-981-97-6199-9_13

target node by measuring various parameters and then determines the target node's position through geometric relationships. Non-ranging positioning, on the other hand, utilizes various relationships between known location nodes and target nodes, such as node coverage, signal transmission characteristics, wireless fingerprints, and other related features to ascertain the position of unknown nodes. Research has demonstrated that range-based positioning generally achieves higher accuracy, particularly when based on propagation delay measurements. In range-based positioning, the target device observes signals from one or more reference transmitters, then parameter estimation methods are used to determine position-related parameters such as distance and angle, and finally, the device's position is calculated based on these estimated parameters.

In communication and positioning systems, millimeter-wave (mmWave) technology offers significant advantages due to its inherent characteristics. First, the high frequency of mmWave leads to significant losses when encountering obstacles, resulting in limited scattering and making line-of-sight (LOS) propagation dominant, thereby creating sparse channels. Second, the short wavelength of mmWave allows for the integration of a large number of antennas into a small space, providing high angular resolution. Additionally, mmWave's larger available bandwidth offers higher delay resolution [1]. Consequently, mmWave can be employed for high-precision positioning. Accurate estimation of positioning parameters is crucial for precise target location determination. Typically, positioning parameters include the angle of arrival (AOA) and time delay. Joint estimation of AOA and delay enables a single receiver to determine the target position, reducing system overhead and enhancing efficiency. Traditional subspace methods, such as multi-signal classification (MUSIC) [2, 3] and estimation of signal parameters via rotational invariant techniques (ESPRIT) algorithm [4], are widely used for joint AOA and delay estimation.

In recent years, compressed sensing (CS) techniques using sparse representation have emerged as a novel approach to address parameter estimation problems [5, 6]. Sparse Bayesian learning (SBL) is a recent parameter estimation method based on the CS concept. The traditional SBL method uses a fixed grid search to estimate parameters for signal reconstruction, but this approach often results in a mismatch between parameters and the grid, leading to larger estimation errors. References [7, 8] propose the off-grid SBL (OGSBL) method to address grid mismatch, replacing grid points with off-grid interval values to reduce estimation errors of the traditional SBL method. However, the OGSBL method requires first-order Taylor expansion to approximate off-grid interval values, leading to greater approximation errors and higher algorithm complexity. Additionally, some works improve the SBL method for different scenarios.

To further enhance SBL positioning methods in mmWave systems, we propose a new two-dimensional adaptive grid refinement method and a joint AOA and time delay estimation method based on an improved SBL algorithm. The contributions of the proposed method are summarized as follows: (1) To mitigate the performance degradation of traditional subspace methods under low SNR and a low number of snapshots, we formulate AOA and time delay estimation as an SBL problem. (2) To address grid mismatch in traditional SBL solutions, we propose a two-dimensional adaptive grid refinement method, treating fixed grid points as parameters that can be adaptively adjusted within a given range. (3) By integrating the new two-dimensional adaptive grid refinement method with the SBL framework, we propose an improved SBL algorithm that reduces estimation errors and algorithm complexity.

13.2 System Model

Consider a mmWave *multiple-input multiple-output* (MIMO) system, as shown in Fig. 13.1. In this positioning system, there is a *base station* (BS) with a known location and R mobile stations (MSs) with unknown locations, and several scatterers are randomly distributed between the BS and MSs. Assume that each MS is equipped with a *uniform linear array* (ULA) composed of M antennas, and the BS is equipped with a ULA composed of M_t antennas. Define d, θ, f_c, Δf and λ as antenna array spacing, AOA, carrier frequency, frequency interval and wavelength, and define $\varphi = \frac{d \sin\theta}{\lambda}$. The positions of BS and MS are denoted as $\mathbf{s} = \left[s_x, s_y\right]^T \in \mathbb{R}^2$ and $\mathbf{p}_r = \left[p_x, p_y\right]^T \in \mathbb{R}^2$ respectively; $\mathbf{s}$ is known, while $\mathbf{p}_r$ is to be determined.

Due to the strong directivity of mmWave transmission, line-of-sight (LOS) propagation is its primary mode. When the signal travels along a non-line-of-sight (NLOS) path, it undergoes significant attenuation after one or two reflections. Consequently,

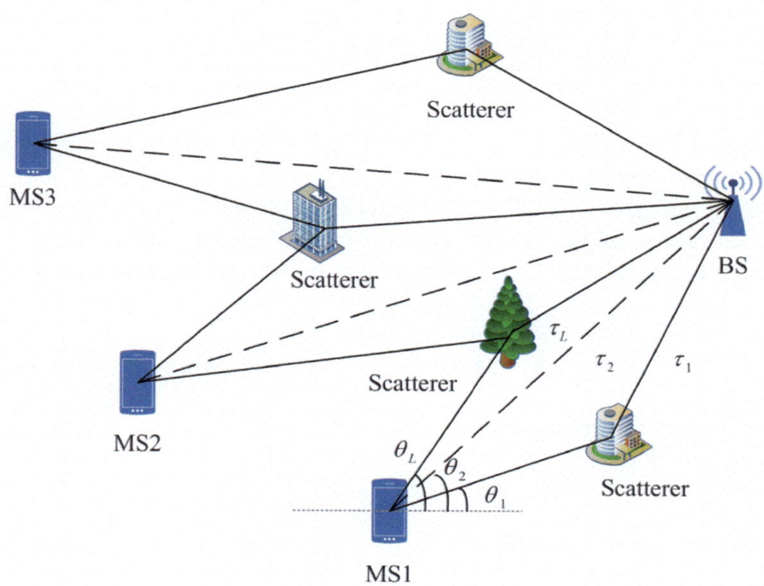

Fig. 13.1 System model with channel parameters

the chapter considers a location environment characterized by a LOS setting, which includes one LOS path alongside multiple NLOS paths. It is assumed that the signal encounters only one reflection upon interacting with a scatterer.

Figure 13.1 shows the position-related parameters in the multipath channel. These parameters include θ_l, τ_l and $d_l = c \cdot \tau_l$, which represent AOA, delay, and path length of the l th path (c represents the speed of light), where the range of AOA is $[-180°, 180°)$. In this case, the baseband continuous signal received at the mth antenna can be written as

$$y_m(t) = \left(\sum_{l=1}^{L} \beta_l e^{-j2\pi m \varphi_l} \delta(t - \tau_l) \right) * x(t) + w_m(t), \tag{13.1}$$

where L denotes the number of paths, $*$ represents convolution, $x(t)$ represents the transmitted signal, β_l is the equivalent channel gain of the l th path, and $w_m(t)$ is noise.

After Fourier transform, the frequency domain signal received at the m th antenna can be expressed as

$$y_m(f) = \left(\sum_{l=1}^{L} \beta_l e^{-j2\pi m \varphi_l} e^{-j2\pi f \tau_l} \right) \cdot x(f) + w_m(f). \tag{13.2}$$

Stacking $\{y_m(f)\}$ of M antennas in a vector form produces

$$\mathbf{y}(f) = \left(\sum_{l=1}^{L} \beta_l \mathbf{a}(\varphi_l) e^{-j2\pi f \tau_l} \right) \cdot x(f) + \mathbf{w}(f), \tag{13.3}$$

where $\mathbf{a}(\varphi_l) = \left[1, e^{-j2\pi\varphi_l}, \ldots, e^{-j2\pi(M-1)\varphi_l}\right]^T \in \mathbb{C}^{M \times 1}$ represents the antenna array response corresponding to the angle θ_l.

We assume that *orthogonal frequency division multiplexing* (OFDM) modulation with N subcarriers is used. Define Δf as the subcarrier interval, and then the nth subcarrier operates at $n\Delta f Hz$. Therefore, the received signal on each subcarrier can be expressed as

$$\mathbf{y}(n\Delta f) = \mathbf{g}(n\Delta f) \cdot x(n\Delta f) + \mathbf{w}(n\Delta f). \tag{13.4}$$

Stack all the received signals on N subcarriers to obtain a $MN \times 1$ dimensional vector, expressed as

$$\mathbf{y} = \sum_{l=1}^{L} \beta_l \mathbf{q}(\varphi_l, \tau_l) + \mathbf{w}, \tag{13.5}$$

where $\mathbf{y} \in \mathbb{C}^{MN\times 1}$, $\mathbf{q}(\varphi_l, \tau_l) = vec\{\mathbf{a}(\varphi_l)\mathbf{b}^T(\tau_l)\} \in \mathbb{C}^{MN\times 1}$, $vec\{\cdot\}$ refers to vectorization, that is, transforming a matrix into a one-dimensional column vector, $\mathbf{b}(\tau_l) = [1 \cdot x(0), e^{-j2\pi\Delta f\tau_l} \cdot x(\Delta f), \cdots, e^{-j2\pi(N-1)\Delta f\tau_l} \cdot x((N-1)\Delta f)]^T \in \mathbb{C}^{N\times 1}$ represents the frequency-domain steering vector pointing to the delay τ_l, $\mathbf{w} = [w_1, w_2, \ldots, w_{MN}]^T \in \mathbb{C}^{MN\times 1}$ represents the additive zero-mean complex Gaussian noise with covariance $\sigma^2 I \in \mathbb{C}^{MN\times 1}$. Our goal is first to estimate φ_l, τ_l and β_l from $\mathbf{y}$, and then the user's position $\mathbf{p}$ is estimated by the obtained φ_l and τ_l.

13.3 Conventional Parameter Estimation Algorithms

13.3.1 Subspace Algorithms, MUSIC and ESPRIT

As one of two important subspace algorithms, the MUSIC algorithm was proposed by Schmidt in [1]. The algorithm can perform high-resolution AOA estimation. By exploiting the orthogonality between the signal subspace and noise subspace, the spatial spectrum function is constructed and searched for peak value where the direction is regarded as the estimated AOA.

The array response or steering vector corresponding to the input signal forms a signal subspace, and to eliminate the noise vector, the signal subspace should be orthogonal to the noise subspace. The orthogonality satisfies $\alpha(\theta)^H Q_n = 0$, where $\alpha(\theta)$ is the steering vector, Q_n is the noise projection vector. Therefore, the pseudo spectrum provided by MUSIC is given by the following formula,

$$P_{MUSIC} = \frac{1}{\alpha^H(\theta)Q_n Q_n^H \alpha(\theta)} \tag{13.6}$$

Due to the complexity of global search, several improved MUSIC algorithms have been developed, such as Root-MUSIC. Another notable subspace method, ESPRIT, was introduced by Roy et al. in [2]. This algorithm estimates the angle of departure (AOD) based on shift invariance between subarrays with identical structures, including [3, 4] (TLS-ESPRIT). In these algorithms, the signal of one element exhibits a constant phase shift relative to the previous element of the steering vector. The signal parameters are derived as eigenvalues of a nonlinear function, mapping one set of vectors to another. Eigenvalue decomposition is performed on the two subarrays to estimate the basis of the signal subspace.

Since ESPRIT is limited to arrays with specific structures, several extended algorithms have been proposed, such as multiple invariant ESPRIT and unitary ESPRIT. Although MUSIC and ESPRIT were originally designed to estimate AOA, they can also be applied to estimate other parameters, such as TOA.

However, the estimation accuracy of MUSIC and ESPRIT decreases significantly at low SNR. Additionally, the subspace method involves substantial computation due to eigenvalue decomposition, especially for multidimensional parameter estimation.

These algorithms also require a large number of snapshots to accurately capture the signal or noise subspace. Therefore, their performance degrades significantly when the number of snapshots is small or the SNR is low.

13.3.2 Iterative Algorithm

In iterative algorithm, the constraint ensures that the estimate $\hat{s}$ is K-sparse. The constraint restricts $\hat{s}$ to be consistent with the measurement matrix y to solve the optimization problem [5]:

$$\hat{s} = \underset{\tilde{s}}{argmin}\|A\tilde{s} - x\|_2^2 s.t. \|\tilde{s}\|_0 \leq K \tag{13.7}$$

Each iteration consists of two operations, i.e., hard thresholding and gradient descent. Here, $\hat{s}^{(i)}$ is at the ith iteration and $T(.)$ is the thresholding operator. Several algorithms belonging to this category are the Iterative Hard Thresholding (IHT) [9] and the Fast Iterative Shrinkage Thresholding Algorithm (FISTA) and Basis Pursuit Denoising (BPDN). the performance of BPDN and IHT increases with the increase of the number of antenna elements.

13.3.3 Statistical Sparse Recovery

The Bayesian frame of reference is utilized for statistical sparse recovery which approaches the signal vector in a probabilistic manner. In the Maximum-a-Posteriori (MAP) procedure, the estimate of s is given as

$$\hat{s} = \arg\max_{s} \ln f(s|y) = \arg\max_{s} \ln f(y|s) + \ln f(s) \tag{13.8}$$

where $f(s)$ is the prior distribution of s and it is modeled such that the $f(s)$ reduces with magnitude of s. For instance, the Expectation–Maximization (EM) algorithm [10] is an iterative method used to solve Maximum Likelihood (ML) estimation problems where some information is missing or unknown. EM estimates unobserved data from incomplete observed data through expectation and maximization steps. Given that the EM algorithm provides a general method rather than a specific solution, there is ongoing debate about whether it qualifies as a true algorithm. The EM algorithm is based on the concept of having complete but unobserved data and incomplete but observed data. The space-alternating generalized expectation–maximization (SAGE) algorithm was first introduced in [11]. Each iteration of the SAGE algorithm essentially corresponds to an iteration of the EM algorithm. SAGE significantly reduces complexity as the number of parameters increases by decomposing the optimization problem into several simpler sub-problems. The EM algorithm struggles to optimize the likelihood function given the high dimensionality required for the maximization step due to the large number of model parameters. SAGE uses an initial rough estimation for the zeroth iteration, and its performance heavily depends on the quality of these initial estimates.

Bayesian Compressed Sensing (BCS) and Sparse Bayesian Learning (SBL) are part of the Gaussian prior model, where the prior is modeled as Gaussian with variance parameterized by a hyper-parameter, and can be estimated from data using the EM algorithm or ML estimation.

13.3.4 Discussion

There are many techniques and methods to estimate DOA, but there is a compromise between calculation time and resolution. Two conventional methods, namely MUSIC and ESPRIT, provide high resolution, which requires highly complicated mathematical calculations, especially at a low signal-to-noise ratio. Statistical sparse recovery techniques similar to SBL also provide high resolution at high SNR and false peaks at low SNR. At this stage, deep learning, as a subset of machine learning, can also be used to learn the conversion between received signals and channels by training neural networks instead of heavy calculation, but it is also limited by training data and calculation overhead. The important problem of sparse statistical recovery algorithms is that there is a grid mismatch between the estimated values that are not on the grid points, while the existing off-grid algorithms need to pay a large system complexity. Therefore, this chapter will introduce sparse Bayesian learning and its improved algorithm in detail.

13.4 Improved Sparse Bayesian Learning Algorithm

The traditional SBL framework is grid-based, and conventional grid compression sensing technology can simplify Eq. (13.5) to

$$\mathbf{y} = \mathbf{Q}(\varphi, \boldsymbol{\tau})\boldsymbol{\beta} + \mathbf{w}, \tag{13.9}$$

where $\mathbf{Q}(\varphi, \boldsymbol{\tau})$ is a dictionary matrix and $\boldsymbol{\beta}$ is an unknown sparse weight vector to be estimated.

For single user,

$$\mathbf{Q}(\varphi, \boldsymbol{\tau}) = \left[\mathbf{q}(\overline{\varphi}_1, \overline{\tau}_1), \ldots, \mathbf{q}(\overline{\varphi}_K, \overline{\tau}_K)\right] \in \mathbb{C}^{MN\times K} \tag{13.10}$$

$$\boldsymbol{\beta} = [\beta_1, \beta_2, \ldots \beta_K]^T \in \mathbb{C}^{K\times 1} \tag{13.11}$$

For R users,

$$\mathbf{Q}(\varphi, \boldsymbol{\tau}) = \left[\mathbf{q}(\overline{\varphi}_1, \overline{\tau}_1), \ldots, \mathbf{q}(\overline{\varphi}_{RK}, \overline{\tau}_{RK})\right] \in \mathbb{C}^{MN \times RK} \tag{13.12}$$

$$\boldsymbol{\beta} = [\beta_1, \beta_2, \ldots \beta_{RK}]^T \in \mathbb{C}^{RK \times 1} \tag{13.13}$$

$(\varphi, \boldsymbol{\tau}) = \{(\overline{\varphi}_k, \overline{\tau}_k), k = 1, \ldots, K\}$ is a fixed grid that non-uniformly divides the entire angular delay domain. We do not use a uniform grid to avoid greatly increasing the computational complexity. And $K \gg L$ is the number of discrete grid points.

13.4.1 Sparse Bayesian Learning Formulation

We first obtain the prior distribution of $\boldsymbol{\beta}$ with promotion of sparsity. Specifically, we apply a two-layer hierarchical prior model [12]. The first layer is zero-mean Gaussian prior distribution, i.e. $p(\boldsymbol{\beta}|\boldsymbol{\alpha}) = \mathcal{CN}(\boldsymbol{\beta}|\mathbf{0}, \mathbf{A})$, where $\boldsymbol{\alpha} = [\alpha_1, \ldots, \alpha_{RK}]$ is the hyperparameter matrix, $\mathbf{A} \triangleq \text{diag}(\boldsymbol{\alpha}^{-1})$. The second layer is modeled as a Gamma prior distribution [13], i.e. $p(\boldsymbol{\alpha}) = \prod_{k=1}^{RK} \Gamma(\alpha_k; \varepsilon, \rho)$, where $\Gamma(\cdot)$ is the Gamma function, and generally we set ε and ρ to 0 as in [12] to obtain a broad hyperprior.

Therefore, the prior distribution of $\boldsymbol{\beta}$ is

$$p(\boldsymbol{\beta}) = \int \mathcal{CN}\left(\boldsymbol{\beta}|\mathbf{0}, diag\left(\boldsymbol{\alpha}^{-1}\right)\right)p(\boldsymbol{\alpha})d\boldsymbol{\alpha} \propto \prod_{k=1}^{K} \left(\rho + |\alpha_k|^2\right)^{-(\varepsilon+1/2)}, \tag{13.14}$$

where diag $(\cdot)$ refers to the diagonal matrix.

The reason for using the two-layer distribution is that it is difficult to directly find the MAP estimation of $\boldsymbol{\beta}$. This problem has been solved in the SBL before, especially using the *relevance vector machine* (RVM) [14]. Instead of imposing a Laplace prior on RVM, it uses a hierarchical prior, which has similar properties to the Laplace prior but allows convenient conjugate index analysis.

Then, we express the *probability density function* (PDF) of $\mathbf{y}$ as

$$p(\mathbf{y}|\varphi, \boldsymbol{\tau}, \boldsymbol{\beta}, \xi) = \mathcal{CN}\left(\mathbf{Q}(\varphi, \boldsymbol{\tau})\boldsymbol{\beta}, \xi^{-1}\mathbf{I}\right), \tag{13.15}$$

where $\xi = \sigma^{-2}$ refers to the reciprocal of noise variance.

Finally, the posterior distribution of $\boldsymbol{\beta}$ can be expressed as a Gaussian distribution, i.e.

$$p(\boldsymbol{\beta}|\mathbf{y}, \xi, \boldsymbol{\alpha}, \varphi, \boldsymbol{\tau}) = \mathcal{CN}(\boldsymbol{\beta}|\boldsymbol{\mu}, \boldsymbol{\Sigma}), \tag{13.16}$$

where

$$\boldsymbol{\Sigma} = \left(\mathbf{Q}^H(\varphi, \tau)\mathbf{Q}(\varphi, \tau) + \mathbf{A}^{-1}\right)^{-1}, \tag{13.17}$$

$$\boldsymbol{\mu} = \Sigma\mathbf{Q}^H(\varphi, \boldsymbol{\tau})\mathbf{y}, \tag{13.18}$$

and the sparse solution of $\boldsymbol{\mu}$ corresponds to the sparse solution of $\boldsymbol{\beta}$.

In order to estimate the hyperparameter $\hat{\xi}, \hat{\boldsymbol{\alpha}}, \hat{\boldsymbol{\psi}}, \hat{\boldsymbol{\tau}}$, we must maximize the posterior $p(\xi, \boldsymbol{\alpha}, \varphi, \boldsymbol{\tau}|\mathbf{y})$, or equivalently maximize the posterior $\ln p(\mathbf{y}, \xi, \boldsymbol{\alpha}, \varphi, \boldsymbol{\tau})$, i.e.

$$\left\{\hat{\xi}, \hat{\boldsymbol{\alpha}}, \hat{\boldsymbol{\varphi}}, \hat{\boldsymbol{\tau}}\right\} = \underset{\xi,\boldsymbol{\alpha},\boldsymbol{\varphi},\boldsymbol{\tau}}{\arg\max} \ln p(\mathbf{y}, \boldsymbol{\xi}, \boldsymbol{\alpha}, \boldsymbol{\varphi}, \boldsymbol{\tau}). \tag{13.19}$$

Since deriving the closed-form expression of (13.17) is challenging, we employ the block Minorize-Maximization (MM) algorithm from [8] to address this issue. The block MM algorithm is an iterative optimization method that leverages the convexity of functions to find their optimal solutions. When the objective function is difficult to optimize directly, the algorithm does not aim to find the optimal solution of the original objective function immediately. Instead, it uses an alternative surrogate objective function to replace the initial one, simplifying the optimization process. With each iteration, the optimal solution of the surrogate function gradually approaches the optimal solution of the original objective function. It can be demonstrated that the algorithm converges.

Specifically, we construct the surrogate function at any fixed point $(\dot{\xi}, \dot{\boldsymbol{\alpha}}, \dot{\boldsymbol{\varphi}}, \dot{\boldsymbol{\tau}})$ as

$$\begin{aligned} &\mathcal{L}(\xi, \boldsymbol{\alpha}, \boldsymbol{\varphi}, \boldsymbol{\tau}|\dot{\xi}, \dot{\boldsymbol{\alpha}}, \dot{\boldsymbol{\varphi}}, \dot{\boldsymbol{\tau}}) \\ &= \int p(\boldsymbol{\beta}|\mathbf{y}, \dot{\xi}, \dot{\boldsymbol{\alpha}}, \dot{\boldsymbol{\varphi}}, \dot{\boldsymbol{\tau}}) \ln \frac{p(\boldsymbol{\beta}, \mathbf{y}, \xi, \boldsymbol{\alpha}, \boldsymbol{\varphi}, \boldsymbol{\tau})}{p(\boldsymbol{\beta}|\mathbf{y}, \dot{\xi}, \dot{\boldsymbol{\alpha}}, \dot{\boldsymbol{\varphi}}, \dot{\boldsymbol{\tau}})} d\boldsymbol{\beta}, \end{aligned} \tag{13.20}$$

then iteratively update the hyperparameters. In the jth iteration, we update $\xi, \boldsymbol{\alpha}, \varphi, \boldsymbol{\tau}$ as

$$\xi^{(j+1)} = \underset{\xi}{\arg\max}\, \mathcal{L}\left(\xi, \boldsymbol{\alpha}^{(j)}, \varphi^{(j)}, \boldsymbol{\tau}^{(j)}|\xi^{(j)}, \boldsymbol{\alpha}^{(j)}, \varphi^{(j)}, \boldsymbol{\tau}^{(j)}\right), \tag{13.21}$$

$$\alpha^{(j+1)} = \underset{\alpha}{\arg\max}\, \mathcal{L}\left(\xi^{(j+1)}, \boldsymbol{\alpha}, \varphi^{(j)}, \boldsymbol{\tau}^{(j)}|\xi^{(j+1)}, \boldsymbol{\alpha}^{(j)}, \varphi^{(j)}, \boldsymbol{\tau}^{(j)}\right), \tag{13.22}$$

$$\begin{aligned} \varphi(j+1) = \underset{\tau}{\arg\max} \\ \mathcal{L}\left(\xi^{(j+1)}, \boldsymbol{\alpha}^{(j+1)}, \varphi, \boldsymbol{\tau}^{(j)}\middle|\xi^{(j+1)}, \boldsymbol{\alpha}^{(j+1)}, \varphi, \boldsymbol{\tau}^{(j)}\right) \end{aligned} \tag{13.23}$$

$$\begin{aligned} &\boldsymbol{\tau}^{(j+1)} = \underset{\tau}{\arg\max} \\ &\mathcal{L}\left(\xi^{(j+1)}, \boldsymbol{\alpha}^{(j+1)}, \varphi^{(j+1)}, \boldsymbol{\tau}\,|\xi^{(j+1)}, \boldsymbol{\alpha}^{(j+1)}, \varphi^{(j+1)}, \boldsymbol{\tau}^{(j)}\right). \end{aligned} \tag{13.24}$$

Then we substitute (13.16) and (13.20) into the iterative formula to get a closed-form solution of ξ, $\boldsymbol{\alpha}$, φ, $\boldsymbol{\tau}$ through updating.

For ξ and α_k, the surrogate function can be simplified to a convex function, so the closed solution can be obtained, then ξ is updated as

$$\xi^{(j+1)} = \frac{MN + v}{\chi + \Phi\big(\xi^{(j)}, \boldsymbol{\alpha}^{(j)}, \varphi^{(j)}, \boldsymbol{\tau}^{(j)}\big)}, \tag{13.25}$$

where $\Phi(\xi, \boldsymbol{\alpha}, \varphi, \boldsymbol{\tau}) = \mathbf{tr}\big(\mathbf{Q}(\varphi, \boldsymbol{\tau})\Sigma\mathbf{Q}^H(\varphi, \boldsymbol{\tau})\big) + \|\mathbf{y} - \mathbf{Q}(\varphi, \boldsymbol{\tau})\boldsymbol{\mu}\|_2^2$.

For α_k, it is updated as

$$\alpha_k^{(j+1)} = \frac{\varepsilon}{\rho + \big[\mathrm{diag}\big(\Xi\big(\xi^{(j+1)}, \boldsymbol{\alpha}^{(j)}, \varphi^{(j)}, \boldsymbol{\tau}^{(j)}\big)\big)\big]_k}, \forall k, \tag{13.26}$$

where $\Xi(\xi, \boldsymbol{\alpha}, \varphi, \boldsymbol{\tau}) = \boldsymbol{\Sigma} + \boldsymbol{\mu} \cdot \boldsymbol{\mu}^H$.

For φ_k and τ_k, since the surrogate function is non-convex and it is difficult to find the global optimal solution, we use the exact block MM algorithm in [8] to update φ and $\boldsymbol{\tau}$, that is, we apply the gradient update to the surrogate function as

$$\varphi^{(j+1)} = \varphi^{(j)} + \eta_\varphi \cdot \mathcal{L}'^{(j)}(\varphi_k), \tag{13.27}$$

$$\boldsymbol{\tau}^{(j+1)} = \boldsymbol{\tau}^{(j)} + \eta_\tau \cdot \mathcal{L}'^{(j)}(\tau_k), \tag{13.28}$$

where η is the step size of the backtracking search, $\mathcal{L}'(\varphi_k)$ and $\mathcal{L}'(\tau_k)$ are the derivative of the objective function to differentiate φ_k and τ_k respectively.

13.4.2 Grid Refinement

The traditional SBL method operates on a fixed grid. However, the actual AOA and delay may not necessarily align with the given grid points, resulting in larger estimation errors. To address this issue, we propose an enhanced SBL algorithm that incorporates a novel two-dimensional adaptive grid refinement technique into the traditional SBL framework. The key aspects of this method are twofold:

Firstly, we address the fixed-grid problem in sparse estimation. For the fixed parameter grid in system modeling, we transform the fixed grid into an adjustable grid, where the grid points are treated as adjustable parameters. This allows the mesh fineness to be adjusted based on different accuracy requirements.

Secondly, to distinguish from general high-density grids, we selectively refine only the areas around grid points that may contain actual values, significantly improving the algorithm's efficiency.

Figure 13.2 illustrates the grid refinement process. The specific refinement method is as follows:

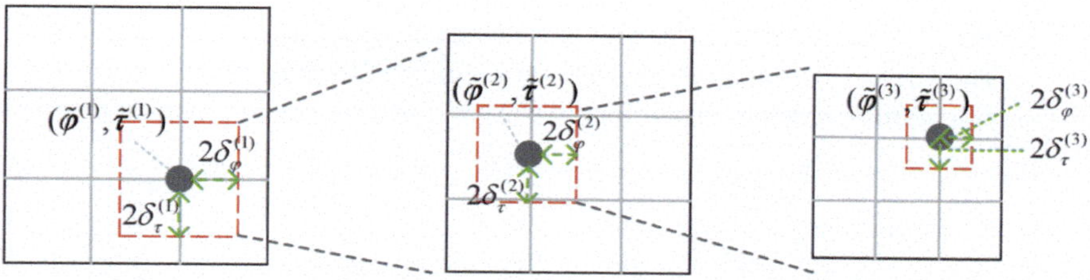

Fig. 13.2 Grid refinement method

(1) We build a two-dimensional grid and assume that the abscissa of the grid represents the angle domain, and the ordinate of the grid represents the delay domain. Meanwhile, let δ_φ and δ_τ denote the grid interval of the angle domain and the delay domain respectively.
(2) At the jth iteration, the grid where there may be real values of the AOA and TOA is updated to a new grid with $\left\{\begin{array}{l}\left[\hat{\varphi}_k^{(j)} - 2\delta_\varphi^{(j)}, \hat{\varphi}_k^{(j)} + 2\delta_\varphi^{(j)}\right] \\ \left[\hat{\tau}_k^{(j)} - 2\delta_\tau^{(j)}, \hat{\tau}_k^{(j)} + 2\delta_\tau^{(j)}\right]\end{array}\right.$ as the grid size, where $\hat{\varphi}_k^{(j)}$ and $\hat{\tau}_k^{(j)}$ represent the estimated angle and delay values at the jth iteration respectively.
(3) Then, in the next iteration, let $\left\{\begin{array}{l}\delta_\varphi^{(j+1)} = \delta_\varphi^{(j)}/\zeta \\ \delta_\tau^{(j+1)} = \delta_\tau^{(j)}/\zeta\end{array}\right.$, where ζ is the refinement interval. We hope to obtain a fine grid by slowly narrowing the grid interval using more refinement levels. A too large ζ will result in the new grid not containing the true values, while too small ζ will make the mesh refinement time too long, resulting in a large number of iterations. Therefore, ζ needs to be set to a reasonable value.

13.4.3 SBL with Grid Refinement

We combine the grid refinement method with the SBL framework to get the improved SBL algorithm. The specific idea of the proposed algorithm is as follows:

(1) For the first iteration, i.e.$j = 1$, create a rough two-dimensional grid $(\varphi, \boldsymbol{\tau}) = \{(\overline{\varphi}_k, \overline{\tau}_k), k = 1, \ldots, K\}$ around the possible true values. However, the original grid should not be too rough to avoid large errors.
(2) Use grid points at this time to obtain $\mathbf{Q}(\boldsymbol{\psi}, \boldsymbol{\tau})$ in (13.9), then the hyperparameters: $\hat{\xi}, \boldsymbol{\alpha}, \hat{\psi}, \hat{\tau}$ are updated with the MM algorithm.
(3) Recalculate $\mathbf{Q}(\boldsymbol{\psi}, \boldsymbol{\tau})$ using the estimated $\hat{\xi}, \boldsymbol{\alpha}, \hat{\psi}, \hat{\tau}$, and obtain $\boldsymbol{\mu}$. Then calculate the average power of each grid point in line k of $\boldsymbol{\beta}$, as

$$P(k) = |\mu_k|, k = 1, \cdots, K, \tag{13.29}$$

where μ_k is the element of $\boldsymbol{\mu}$, and $\boldsymbol{\mu}$ is calculated by (13.18).

The larger $P(k)$ is, the higher the probability of true angle and delay in the corresponding direction. Therefore, we use the grid refinement method to refine the grid around the grid point where $P(k)$ is the largest, namely $\left(\overset{\sim(j)}{\varphi}, \overset{\sim(j)}{\tau}\right)$, where j represents the number of iterations.

(4) Return to step (2), until the grid is fine enough, that is, the grid interval of the angle domain $\delta_\varphi \leq 10^{-5}$ degree and the grid interval of the delay domain $\delta_\tau \leq 10^{-15}$ s, the iteration stops.

At this time, $\hat{\varphi}$ and $\hat{\tau}$ are the estimated results, and then the estimated AOA is obtained by $\hat{\theta} = \arcsin(\hat{\varphi} \cdot \lambda / d)$.

(5) Finally, using the estimated AOA and delay, the user's position can be obtained by the following formula:

$$\mathbf{p} = \sum_{l=1}^{L} \left(\mathbf{s} + c \cdot \tau_l \begin{bmatrix} \cos(\theta_l) \\ \sin(\theta_l) \end{bmatrix} \right) / L. \tag{13.30}$$

13.4.4 Comparison of Algorithm Complexity

In this section, we analyze and compare the algorithm complexity of ESPRIT algorithm in [4], the OGSBL algorithm in [8] and the improved SBL algorithm.

For the case of a single snapshot, the algorithm complexity of the improved SBL algorithm proposed in this chapter is calculated as follows:

(1) For each iteration, the complexity of calculating the expectation $\boldsymbol{\mu}$ and variance $\boldsymbol{\Sigma}$ is $\mathcal{O}(K^2)$ and $\mathcal{O}(MNK^2)$ respectively.
(2) For each iteration, the complexity of updating the parameters ξ and $\boldsymbol{\alpha}$ is $\mathcal{O}(K)$ and $\mathcal{O}(K)$ respectively.
(3) Because the parameters $\boldsymbol{\psi}$ and $\boldsymbol{\tau}$ and $\boldsymbol{\beta}$ are sparse together, the effective size of $\boldsymbol{\psi}$ and $\boldsymbol{\tau}$ is small. Therefore, the complexity of the parameters $\boldsymbol{\psi}$ and $\boldsymbol{\tau}$ can be ignored.

In summary, the final algorithm complexity of the proposed algorithm is $\mathcal{O}(MNK^2 \cdot T_S)$, where T_S is the number of iterations.

Table 13.1 shows the algorithm complexity comparison of ESPRIT algorithm, OGSBL algorithm and the improved SBL algorithm. As can be seen, although the complexity formula of the OGSBL algorithm and the improved SBL algorithm are the same, the number of iterations of the OGSBL algorithm is much higher than that of the improved SBL algorithm, to achieve a relatively ideal positioning effect. In general, the number of iterations of the OGSBL algorithm T_O is about 20, while that of the improved SBL algorithm T_S is only about 10. Therefore, the complexity of the improved SBL algorithm is significantly less than that of the OGSBL algorithm. According to the parameter configuration during the simulation in this chapter, that

Table 13.1 Comparison of Complexity of Different Algorithms

Algorithm	Complexity
ESPRIT	$\mathcal{O}(M^3N^3)$
OGSBL	$\mathcal{O}(MNK^2 \cdot T_O)$
Proposed method	$\mathcal{O}(MNK^2 \cdot T_S)$

is, $M = 16$, $N = 256$, $K = 10$, the complexity of the improved SBL algorithm is also less than that of the ESPRIT algorithm. In summary, the proposed improved SBL algorithm also has certain advantages in algorithm complexity.

13.5 Numerical Result

We conduct several simulations to evaluate the performance of our proposed method. All results are generated from 1000 Monte Carlo trials. We assume that the coordinates of MS and BSs are given in a plane coordinate system. We use a ULA composed of M antennas at the MS, and the number of antennas $M = 16$. Similarly, we use a ULA composed of M_t antennas on the BS, and the number of antennas $M_t = 16$. We use a single snapshot, that is, the number of snapshots $T = 1$, the carrier frequency $f_c = 60$ GHz, the bandwidth $B = 100$ MHz, the frequency interval $\Delta f = 240$ KHz, the number of subcarriers $N = 256$, the number of users $R = 3$, the number of paths between each BS and user $L = 3$. For the grid, we use a two-dimensional grid and assume that the abscissa of the grid represents the angle, the ordinate of the grid represents the delay domain, and the number of grids $K = 10$. For the initial grid settings, usually, 1–2° is the sampling range in the angle domain, and 10–20 ns is the sampling range in the delay domain. Generally, for a specific grid, grid refinement usually needs about 10 iterations.

There is no clear theoretical answer for the exact value of the grid refinement interval ζ. Therefore, it must be set according to the experiment. The basic idea is that the setting of this value can balance the speed of the grid refinement process and the number of SBL iterations, and make the final position estimate accurate enough. It is observed that when the value of ζ is set to 2 and 3, the final positioning error, in terms of root mean square (RMSE), is almost the same, but when ζ is set at 3, the number of simulation iterations is 8, and when ζ is set at 2, the number of iterations is 15. Therefore, we finally choose to set ζ to be 3.

13.5.1 Verification of the Adaptive Grid Refinement Method

In this section, we assess the effectiveness of the proposed two-dimensional adaptive grid refinement method through simulation results. The conventional MUSIC algorithm employs a fixed grid for angle and delay in spectral space search, leading to a

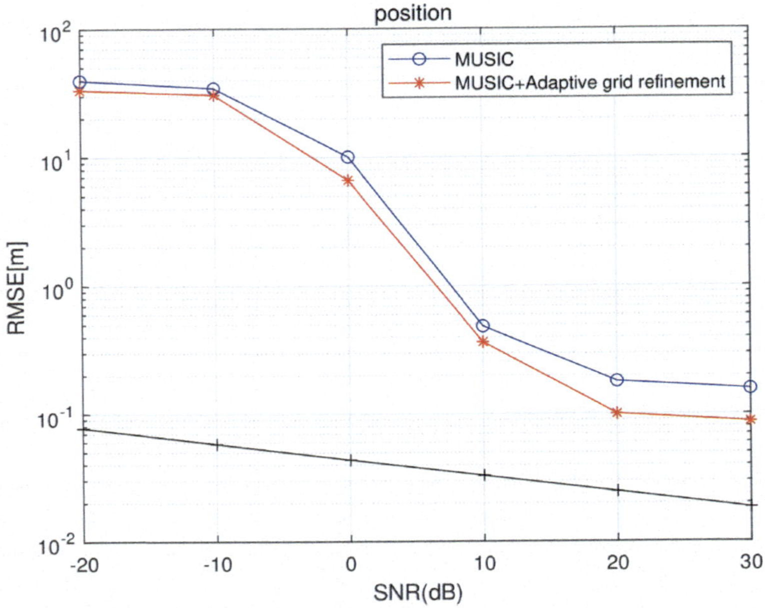

Fig. 13.3 Positioning accuracy comparison of the conventional MUSIC algorithm and the MUSIC algorithm combined with the adaptive grid refinement method

grid mismatch problem. To verify the effectiveness of the two-dimensional adaptive mesh refinement method proposed in this chapter, we integrate it with the MUSIC algorithm and then perform positioning. We compare its performance with that of the conventional MUSIC algorithm described in [2].

Figure 13.3 illustrates the comparison of positioning results between the conventional MUSIC algorithm and the MUSIC algorithm combined with the adaptive grid refinement method. It is evident that the MUSIC algorithm combined with the adaptive grid refinement method achieves better positioning accuracy than the conventional MUSIC algorithm. This demonstrates the effectiveness of the adaptive grid refinement method proposed in this chapter for performance enhancement.

13.5.2 Comparison of Positioning Results

We conduct several simulations to evaluate the performance of our proposed method, and compare with the MUSIC algorithm in [2], Root-MUSIC algorithm in [3], ESPRIT algorithm in [4], proposed method in [8] and OGSBL algorithm in [7].

Figures 13.4 and 13.5 illustrate how the RMSE of AOA and delay estimation varies with the SNR. Under the condition of a single snapshot, the estimation accuracy of the proposed algorithm for both parameters is significantly better than that of the

MUSIC algorithm, Root-MUSIC algorithm, and ESPRIT algorithm, regardless of whether the SNR is low or high. This indicates that the proposed algorithm effectively addresses the shortcomings of traditional subspace algorithms in low SNR and low snapshot scenarios. Compared with the OGSBL algorithm and the method described in [7], the proposed algorithm also demonstrates superior accuracy, highlighting the effectiveness of combining the two-dimensional adaptive grid refinement method with the SBL framework. Furthermore, the figures show that after several iterations, the proposed algorithm approaches the corresponding CRB bounds more closely than other algorithms, further proving the superiority of the proposed algorithm's estimation performance.

Figure 13.6 depicts the RMSE of position estimation in the LOS environment as a function of SNR. As SNR increases, the positioning error of the proposed algorithm consistently remains lower than that of other algorithms, indicating higher positioning accuracy. Furthermore, compared to other algorithms, the RMSE of the proposed algorithm approaches the corresponding CRB more closely as SNR increases, demonstrating the algorithm's superiority. At 30 dB, the RMSE of position estimation is 0.0234 m, achieving centimeter-level accuracy, indicating the proposed algorithm's capability for high-precision positioning.

Since there is a clock offset between the base station (BS) and the mobile station (MS), meaning they are not strictly synchronized, this introduces a certain error in delay measurement, thereby affecting positioning accuracy. Although this chapter does not delve deeply into clock synchronization, many scholars have extensively

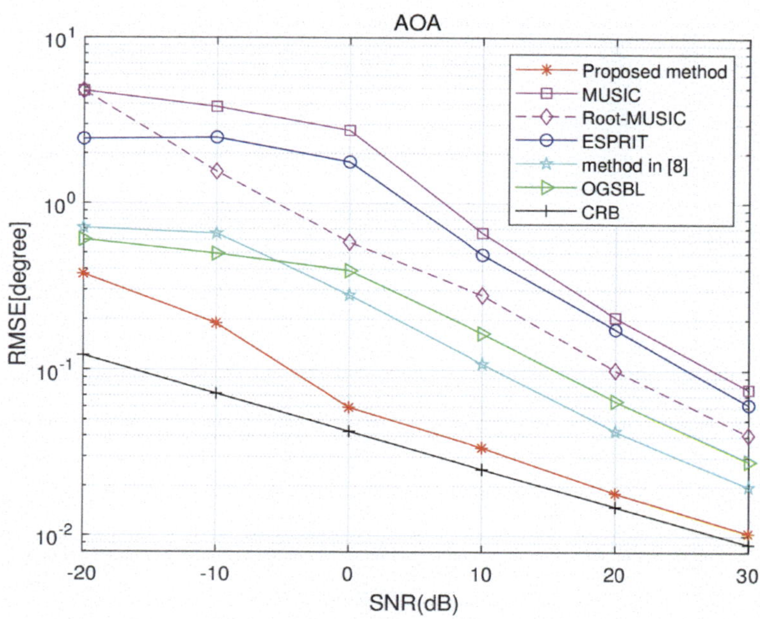

Fig. 13.4 The RMSE of AOA in dB varies with SNR

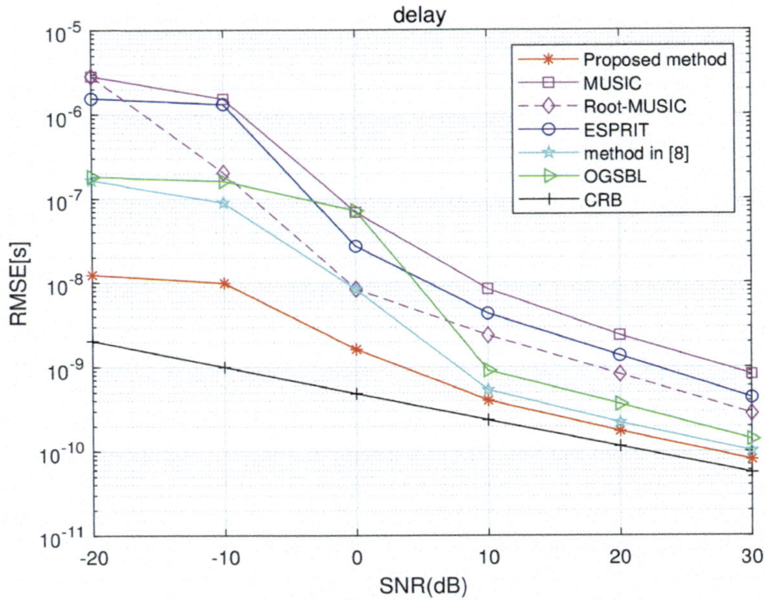

Fig. 13.5 The RMSE of delay in dB varies with SNR

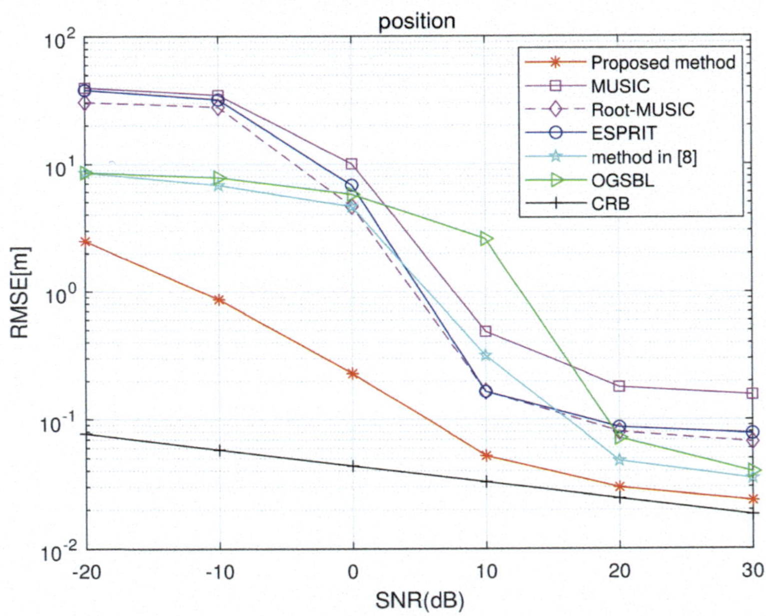

Fig. 13.6 The RMSE of position estimation varies with SNR in LOS condition

researched clock synchronization and calibration in positioning systems [15]. It is generally found that through specific methods or equipment, the clock offset can be controlled within an acceptable range, typically nanoseconds or smaller. Consequently, the impact of clock deviation on position estimation is relatively minor. Therefore, the results presented earlier in the chapter are based on the assumption that the base station and the mobile terminal are clock-synchronized.

In order to verify the influence of clock offset on the positioning result, the chapter adds a certain clock offset to the estimated delay. When the frequency interval is: $\Delta f = 240$ KHz and the number of snapshots is: $T = 1$, it can be seen that the sampling time interval is $\Delta t = \frac{1}{T \cdot \Delta f} = 32$ ns. Therefore, we set the clock offset to be about 3% relative to the sampling interval, that is, $e = 1$ns, and then obtains the position estimation, and compares it with the positioning result without clock offset. We respectively use ESPRIT algorithm in [4], OGSBL algorithm in [8] and the proposed algorithm in the chapter for simulation verification.

Figure 13.7 presents a comparison of positioning results with and without clock offset. When the SNR is low, the noise-induced estimation error of the delay is substantial, so the clock offset has minimal impact on the positioning result. However, at high SNR, where the noise-induced estimation error of the delay is at the nanosecond level, the introduction of clock offset increases the delay estimation error, significantly reducing positioning accuracy. Thus, employing a method or device to further minimize the clock offset would greatly enhance positioning accuracy. Additionally, it is evident that, even with the clock offset, the proposed algorithm is less affected compared to the ESPRIT algorithm and the OGSBL algorithm.

13.6 The Result of Measured Data

In this section, we validate the proposed algorithm and other classic algorithms using a localization test dataset [16] for 5G large-scale MIMO collected by K. Gao and H. Wang, comparing the performance of different algorithms in two typical scenarios. A server equipped with an Intel Xeon E5-2626 v2 CPU and a GeForce GTX 1080 GPU was used to run the algorithms with the dataset. A Nikon DTM-352C Total Station was utilized for site surveying and positioning-point labeling. Two typical 5G NR positioning scenarios were selected for dataset experiments, as shown in Figs. 13.8 and 13.9.

Scenario 1 is an indoor office hall in a new building at the Chinese Academy of Sciences (CAS) in Beijing, China. Five ISAC gNBs operate at 3.5 GHz with 100 MHz bandwidth and 40 W power. These five gNBs are suspended on plastic holders 2.4 m above the ground. For simulation, there is a random floating height of 0.1 m to prevent coplanarity. The UE, acting as a receiver, is mounted on a marked liftable cart 1.2 m above the ground to simulate a 1.8 m tall person holding a mobile phone.

Scenario 2 represents a typical urban canyon environment, where it is difficult for a UE to access sufficient satellites for positioning. The UEs are located on a

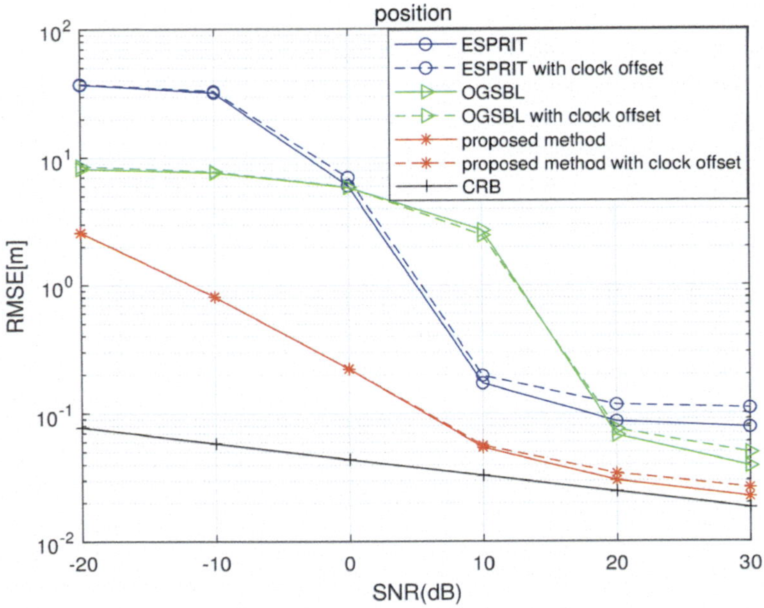

Fig. 13.7 Comparison of positioning results with clock offset and without clock offset

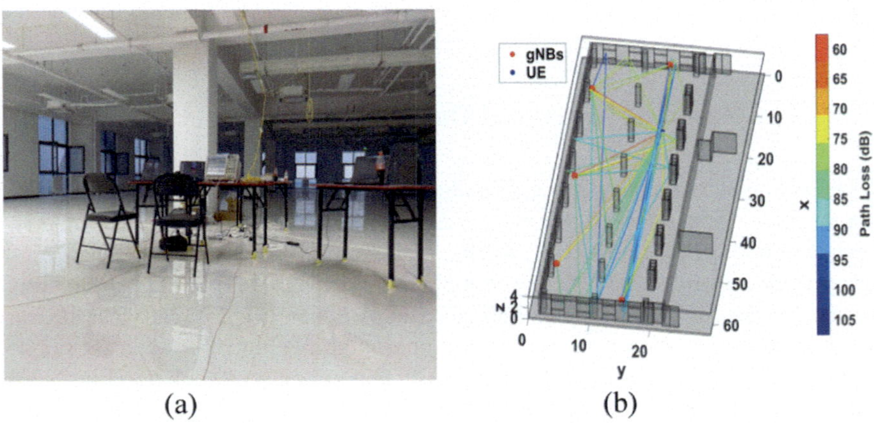

Fig. 13.8 A typical indoor positioning scenario in an office hall: **a** photograph and **b** electromagnetic simulation results

low-rise platform between two high-rise buildings. The dataset can be downloaded from IEEE DataPort at https://dx.doi.org/https://doi.org/10.21227/jsat-pb50.

In scenario 1, we face a challenging indoor environment characterized by the presence of numerous obstacles, walls, and numerous signal interference sources,

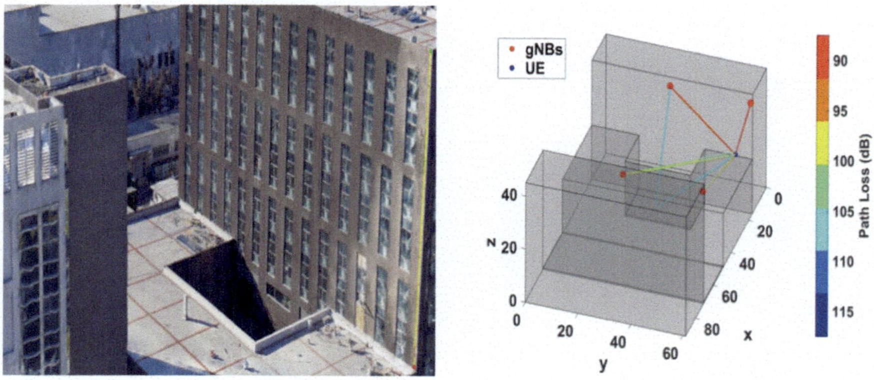

(a) (b)

Fig. 13.9 A typical outdoor positioning scenario in an urban canyon: **a** photograph and **b** electromagnetic simulation results

or areas where signal reflection and attenuation are significant due to various structures. The complexity of this environment leads to an increase in positioning errors, posing greater challenges for algorithms to accurately estimate positions. In this environment, positioning algorithms need to overcome signal obstruction and reflection problems caused by multiple obstacles, as well as interference from other electronic devices. These factors work together, resulting in the cumulative error distribution function (CDF) chart (as shown in Fig. 13.10) showing higher error rates and larger error fluctuations. Compared to the complexity of scenario 1, scenario 2 describes a simpler and more controllable outdoor environment. This is usually an open space between buildings or a setting with minimal obstacles and fewer signal interference sources. In such an environment, signal propagation is more direct and predictable, which enables positioning algorithms to estimate positions more accurately, as the number of error sources is greatly reduced. Due to the simplicity of scenario two, the localization algorithm can better utilize signal strength and quality, thereby improving the accuracy of localization. Therefore, the error CDF chart of different algorithms in this indoor environment (as shown in Fig. 13.11) shows lower error rates and more stable performance.

In scenario 1, the indoor environment, the localization error distribution of all algorithms shows larger error values, but in contrast to the original analysis, the performance in the outdoor environment, scenario 2, is better than in scenario 1. This indicates that the proposed method, followed by OGSBL, ESPRIT, and MUSIC, shows improved performance in the outdoor scenario. The CDF curve of the proposed method reaches a higher CDF value, meaning that it can reach high positioning accuracy even in the presence of significant outdoor environmental challenges.

Compared with the indoor scenario, the positioning error of all algorithms in the outdoor scenario generally decreases, indicating that the outdoor scenario provides

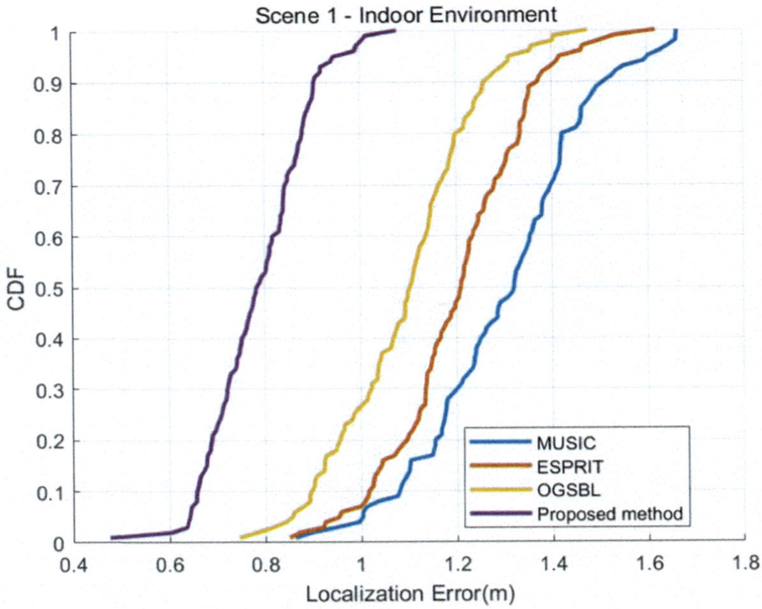

Fig. 13.10 Comparison of positioning results in the indoor environment

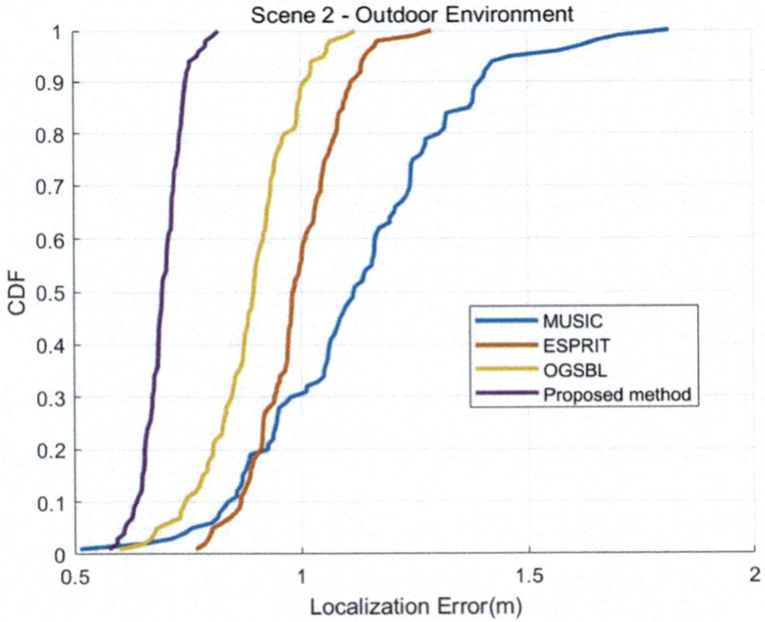

Fig. 13.11 Comparison of positioning results in the outdoor environment

conditions that are favorable for accurate localization. The proposed method maintains the best performance in both environments, but its advantages are more pronounced in the outdoor scenario. Through this analysis, we can understand the performance of each algorithm in two different localization scenarios more accurately. The proposed method has shown excellent performance in both scenarios, especially in the outdoor scenario, where its advantages are more pronounced.

13.7 Summary

In this chapter, we first explore the applications of the classical MUSIC and ESPRIT algorithms in mmWave MIMO systems. The MUSIC algorithm is renowned for its excellent resolution of signal subspaces, performing particularly well under high SNR conditions. However, its performance may be limited when SNR is low and there are few snapshots. In contrast, the ESPRIT algorithm, by recursively decomposing the signal subspace, reduces dependence on the noise subspace, demonstrating robustness in the case of low SNR and a limited number of snapshots.

Although the MUSIC and ESPRIT algorithms exhibit remarkable performance under certain conditions, they may experience performance degradation in extreme scenarios such as low SNR and a limited number of snapshots. To address this issue, we introduce a novel approach that leverages the joint sparsity of angle and delay in mmWave systems, formulating the estimation of angle and delay as a Sparse Bayesian Learning (SBL) problem. To tackle the challenges posed by traditional SBL algorithms, including grid mismatch and the high complexity of the OGSBL algorithm, we design a new two-dimensional adaptive grid refinement method. This method treats fixed grid points as adjustable parameters within a given range, gradually reducing the grid area with each iteration. By integrating the adaptive grid refinement method into the SBL framework, we propose an improved SBL algorithm to enhance the accuracy of Angle of Arrival (AOA), time delay, and Mobile Station (MS) position estimation. Simulation results demonstrate that under the condition of a single snapshot, the proposed algorithm achieves higher positioning accuracy compared to other algorithms. Tests on the real dataset also show that even in the presence of complex environmental challenges, the algorithm achieves high positioning accuracy. This approach excels in complex environments, providing a reliable solution for high-precision positioning.

References

1. Garcia N, Wymeersch H, Slock DTM (2018) Optimal precoders for tracking the AoD and AoA of a mmWave tath. IEEE Trans Signal Process 66(21):5718–5729
2. Guo Z, Wang X, Heng W (2017) Millimeter-Wave channel estimation based on 2-D beamspace MUSIC method. IEEE Trans Wirel Commun 16(8):5384–5394

3. Yin D, Zhang F (2020) Uniform linear array MIMO radar unitary root MUSIC angle estimation. In: 2020 Chinese automation congress (CAC), pp 578–581
4. Zhang J, Haardt M (2017) Channel estimation and training design for hybrid multi-carrier mmWave massive MIMO systems: the beamspace ESPRIT approach. In: 2017 25th European signal processing conference (EUSIPCO), pp 385–389. https://doi.org/10.23919/EUSIPCO.2017.8081234
5. Talaei F, Dong X (2019) Hybrid mmWave MIMO-OFDM channel estimation based on the multi-band sparse structure of channel. IEEE Trans Commun 67(2):1018–1030. https://doi.org/10.1109/TCOMM.2018.2871448
6. Lee J, Gil G, Lee YH (2016) Channel estimation via orthogonal matching pursuit for hybrid MIMO systems in millimeter wave communications. IEEE Trans Commun 64(6):2370–2386
7. Yang Z, Xie L, Zhang C (2013) Off-grid direction of arrival estimation using sparse bayesian inference. IEEE Trans Signal Process 61(1):38–43
8. Dai J, Liu A, Lau VKN (2018) FDD massive MIMO channel estimation with arbitrary 2D-array geometry. IEEE Trans Signal Process 66(10):2584–2599
9. Stoeckle C et al (2015) DoA estimation performance and computational complexity of subspace and compressed sensing-based methods. In: WSA 2015; 19th international ITG workshop on smart antennas, pp 1–6
10. Moon TK (1996) The expectation-maximization algorithm. IEEE Signal Process Mag 13(6):47–60
11. Fleury BH et al (1999) Channel parameter estimation in mobile radio environments using the SAGE algorithm. IEEE J Sel Areas Commun 17(3):434–450
12. Ji S, Xue Y, Carin L (2008) Bayesian compressive sensing. IEEE Trans Signal Process 56(6):2346–2356
13. Wipf DP, Rao BD (2004) Sparse bayesian learning for basis selection. IEEE Trans Signal Process 52(8):2153–2164. https://doi.org/10.1109/TSP.2004.831016
14. Tipping ME (2001) Sparse bayesian learning and the relevance vector machine. J Mach Learn Res 1(3):211–244
15. Zhao S et al (2021) A new TOA localization and synchronization system with virtually synchronized periodic asymmetric ranging network. IEEE Internet Things J 8(11):9030–9044
16. Gao K, Wang H, Lv H, Liu W (2022) Toward 5G NR high-precision indoor positioning via channel frequency response: a new paradigm and dataset generation method. IEEE J Sel Areas Commun 40(7):2233–2247. https://doi.org/10.1109/JSAC.2022.3157397

Chapter 14
UWB Non-line-of-Sight Propagation Identification and Localization

Jin Wang and Kegen Yu

Abstract This chapter focuses on non-line-of-sight (NLOS) identification and mitigation for UWB localization. First, an NLOS identification method based on One-Dimensional Wavelet Packet Analysis (ODWPA) and Convolutional Neural Network (CNN) is proposed, which achieves an average identification accuracy of about 95%, significantly higher than that of other traditional methods. Then, two different error models in LOS/NLOS environments are established respectively, which improve the ranging accuracy by about 29% on average. Finally, an improved Chan-Kalman localization algorithm based on NLOS identification is proposed. The experimental results show that, compared with other algorithms, the proposed algorithm achieves the highest localization accuracy in static scenario with an average of 12.6 cm, which is an improvement of about 32.8%. In dynamic scenario, the average error of the proposed algorithm is about 10.7 and 8.9 cm in X-axis and Y-axis, respectively, significantly outperforming other algorithms.

14.1 Introduction

In outdoor environments (e.g., city streets, deserts, and sea surfaces), the positioning accuracy of the Global Navigation Satellite System (GNSS) can reach meter or even sub-decimeter levels depending on the specific system used. However, in indoor scenarios, GNSS signal loss and attenuation are significant due to blockage by building roofs, walls, and other obstacles, making it impossible to obtain continuous and accurate position information. Therefore, developing indoor positioning technology is essential to meet the demand for indoor location services in the era of

J. Wang
Eacon Mining Technology, Beijing 100000, P. R. China
e-mail: cumt_jinwang@163.com

K. Yu (✉)
School of Environment Science and Spatial Informatics, China University of Mining and Technology, Xuzhou 221116, P. R. China
e-mail: kegen.yu@cumt.edu.cn

K. Yu (ed.), *Positioning and Navigation Using Machine Learning Methods*, Navigation: Science and Technology 14, https://doi.org/10.1007/978-981-97-6199-9_14

the "Internet of Things" with the help of continuously innovating wireless communication technology. Among existing indoor wireless positioning technologies, ultra-wideband (UWB) stands out for its strong signal penetration, high anti-interference capability, and high time resolution, achieving centimeter-level positioning accuracy. However, UWB location accuracy can be significantly reduced by non-line-of-sight (NLOS) propagation in complex indoor environments. Thus, identifying and mitigating NLOS errors is crucial for improving UWB positioning accuracy in such harsh conditions.

According to several key models of wireless positioning technology, such as angle of arrival (AOA) [1], time of arrival (TOA), time difference of arrival (TDOA) [2], and received signal strength indication (RSSI) [3], various traditional indoor positioning algorithms have been developed. Wade introduced a Taylor-series algorithm based on TDOA as early as 1976 [4], which improves the position estimation of unknown tags by solving the local least squares solution of the TDOA measurement error and iterating towards the true position. In 1990, Fang proposed a straightforward localization algorithm that directly uses four TDOA measurements to estimate the position of an unknown tag [5]. Chan and Ho introduced a non-iterative localization algorithm in 1994, implemented using TDOA and maximum likelihood estimation, and demonstrated it could reach the Cramer-Rao lower bound in the small error region [6]. These classical localization algorithms are effective in specific applications and generally achieve good localization accuracy in LOS environments. However, they fail to account for the impact of NLOS conditions on TDOA measurements, which significantly reduces their accuracy. Therefore, in real indoor settings, it is crucial to suppress the measurement errors caused by NLOS propagation. NLOS error suppression techniques can be broadly classified into two categories. One approach involves first identifying NLOS propagation, processing the identified NLOS measurements, and then determining the tag's position based on the processed data. The other approach focuses on directly optimizing the measurement information or the algorithm itself. This chapter will concentrate on the first category of methods and the associated experiments.

14.2 NLOS Identification Methods Based on Machine Learning

14.2.1 Method Description

One-dimensional Wavelet Packet Analysis

Wavelet analysis decomposes original signals into low-frequency component a1 and high-frequency component d1. The information lost in the low-frequency signal a1 is captured by the high-frequency signal d1 during decomposition. In the next level of decomposition, the low-frequency signal a1 is decomposed into the low-frequency signal a2 and the high-frequency signal d2, and the information lost in the low-frequency signal a2 is captured by the high-frequency signal d2, and so on, so that the local characteristics of the parent signals are characterized in great detail. However, the wavelet function used in wavelet analysis is not unique, so the problem of choosing the optimal wavelet basis must be solved first. Cui et al. chose the complex Morlet wavelet with the right balance between time and frequency localization as the mother wavelet [7]. Here, we opt for wavelet packet analysis over wavelet analysis, converting Channel Impulse Response (CIR) into energy images via One-dimensional Wavelet Packet Analysis (ODWPA) [8]. Compared with wavelet analysis, wavelet packet analysis can divide the frequency band into multiple levels, further decompose the high frequency band which is not subdivided by wavelet analysis, and can select the appropriate frequency band adaptively according to the characteristics of the signal analyzed. It is worth mentioning that we found most of the CIRs have extremely low magnitude during the initial phase of signal reception after observing the CIRs in several experimental scenarios. Therefore, the raw CIR H(t) is subject to threshold operation:

$$h'(t) = H(t) - \rho \tag{14.1}$$

where ρ is the threshold related to the experimental scenario or experimental requirements. In this chapter, we have the following definition:

$$h'(t) \in L^2(\boldsymbol{R}) \tag{14.2}$$

where $L^2(\boldsymbol{R})$ is the signal space of $h'(t)$.

Next, we consider how to decompose this signal space $L^2(\boldsymbol{R})$ in detail. We use the "Haar" wavelet, which is defined as:

$$\Psi_H(t) = \begin{cases} 1, & 0 \le t \le \frac{1}{2} \\ -1, & \frac{1}{2} < t \le 1 \\ 0, & \text{others} \end{cases} \tag{14.3}$$

The scaling function of $\Psi_H(t)$ is given as:

$$\Phi(t) = \begin{cases} 1, & 0 \le t \le \\ 0, & \text{others} \end{cases} \tag{14.4}$$

In multi-resolution analysis, $L^2(\boldsymbol{R}) = \oplus W_j (j \in \boldsymbol{Z})$, i.e., the signal space $L^2(\boldsymbol{R})$ is decomposed into the orthogonal sum of all subspaces $W_j (j \in \boldsymbol{Z})$ according to different scale factors j, where $\oplus$ denotes the direct sum operation, $\{W_j\}$ are the wavelet subspaces of wavelet function $\Psi_H(t)$ and $\boldsymbol{Z}$ is the integer set. On this basis, we consider further subdivision of the wavelet subspace W_j. First, we define a new space $U_j^n (n \in \mathbf{Z}_+)$, where $\mathbf{Z}_+$ is the positive integer set and n refers to the ordinal number of each node of the decomposition tree:

$$\begin{cases} U_j^0 = V_j \\ U_j^1 = W_j \end{cases}, \quad j \in \mathbf{Z} \tag{14.5}$$

where V_j is the scaling space. According to the multiresolution analysis [9], the scaling space V_j is represented as:

$$V_j = V_{j-1} \oplus W_{j-1}; j \in Z \tag{14.6}$$

Thus, according to (14.5) and (14.6), U_{j+1}^0 can be expressed as:

$$U_{j+1}^0 = V_{j+1} = V_j \oplus W_j = U_j^0 \oplus U_j^1, \quad j \in Z \tag{14.7}$$

By excursion, (14.7) becomes

$$U_{j+1}^n = U_j^{2n} \oplus U_j^{2n+1}, \quad j \in Z, n \in \mathbf{Z}_+ \tag{14.8}$$

Therefore, according to (14.5) and (14.8), the subspace W_j can be subdivided as follows,

$$\begin{cases} W_j = U_j^1 = U_{j-1}^2 \oplus U_{j-1}^3, \\ = U_{j-2}^4 \oplus U_{j-2}^5 \oplus U_{j-2}^6 \oplus U_{j-2}^7, \\ \cdots \\ = U_{j-k}^{2^k} \oplus U_{j-k}^{2^k+1} \oplus \cdots \oplus U_{j-k}^{2^{k+1}-1}, \\ \cdots \\ = U_0^{2^j} \oplus U_0^{2^j+1} \oplus \cdots \oplus U_0^{2^{j+1}-1}, \end{cases} \tag{14.9}$$

where $k (k \in \mathbf{Z})$ is the position index, $j = 1, 2, \ldots, N (N \in \mathbf{Z}_+)$. Finally, the signal space $L^2(\boldsymbol{R})$ is decomposed as:

$$L^2(\boldsymbol{R}) = \oplus W_j = \cdots \oplus W_{-1} \oplus W_0 \oplus U_0^2 \oplus U_0^3 \oplus \cdots \tag{14.10}$$

As the scaling factor j increases, the spatial resolution of the corresponding wavelet basis function becomes higher. The division of the subspace W_j is represented by a binary tree as shown in Fig. 14.1, which is the wavelet packet decomposition tree. This case is the three-level decomposition and the number of levels is mainly determined by the specific signals and experimental requirements.

In this chapter, we utilize "Haar" wavelets to decompose $h'(t)$ up to the fifth layer. Figure 14.2 displays colored coefficient images for the root nodes of the wavelet

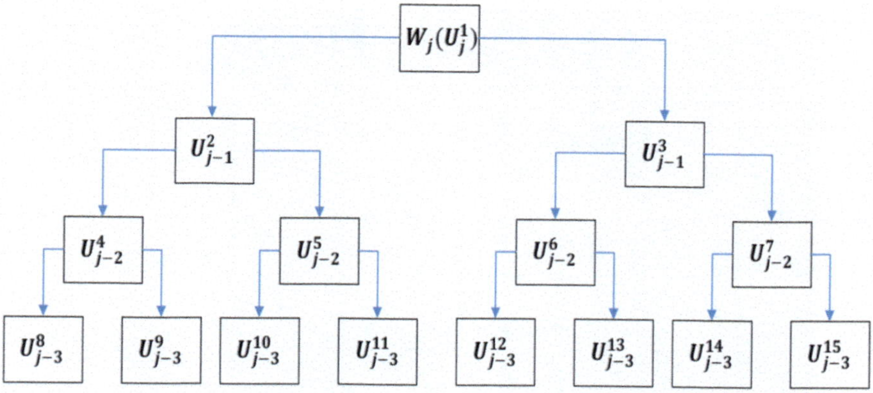

Fig. 14.1 The wavelet packet decomposition tree. Here W_j is decomposed into three layers as an example

decomposition tree for $h'(t)$ across three UWB signal propagation channels. The coefficient distributions at the root node vary among the propagation channels after one-dimensional wavelet packet decomposition. Larger color scales indicate higher coefficients at the root node, reflecting stronger signals at that layer. In the LOS propagation channel (Fig. 14.2a), the root node's coefficient distribution is highly concentrated, suggesting dominance of the first arrival path of the UWB signal. In the HNLOS propagation channel (Fig. 14.2c), the root node's coefficient distribution is more scattered, indicating significant influence from NLOS and pronounced multipath effects, resulting in signal distortion. The SNLOS propagation channel (Fig. 14.2b) exhibits intermediate characteristics between the LOS and HNLOS scenarios.

Convolutional neural network

In comparison to other machine learning techniques, convolutional neural networks (CNNs) excel in feature learning, especially with grayscale matrix inputs (e.g., images), offering stability and simplicity. A typical CNN architecture comprises convolutional layers, pooling layers, and fully connected layers, with adjustments to layer counts and hyperparameters based on specific tasks. Figure 14.3 illustrates the standard structure of a CNN, where feature map sizes decrease progressively across layers, culminating in outputs from the fully connected layer.

The initial step involves data input. Following the acquisition of colored coefficient images for root nodes from the wavelet packet decomposition tree across different signal propagation channels, we refrain from direct usage as input for the CNN model. Instead, we first engage in data preprocessing. Initially, a segment of the acquired image database is set aside as training data for the CNN. The remaining data serves as the validation and test sets for subsequent evaluation. All participating images are then converted into the *tfrecords* data format, a binary file type containing sequences

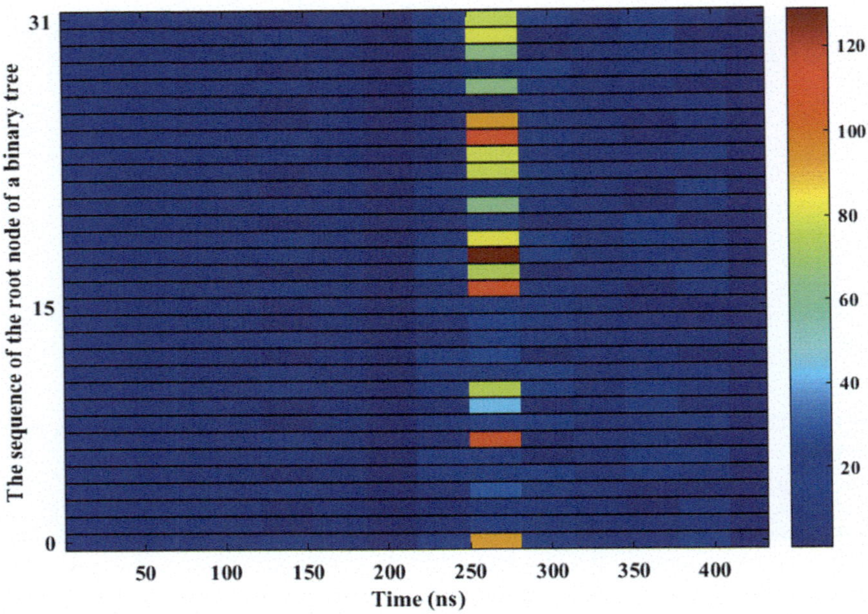

(a)

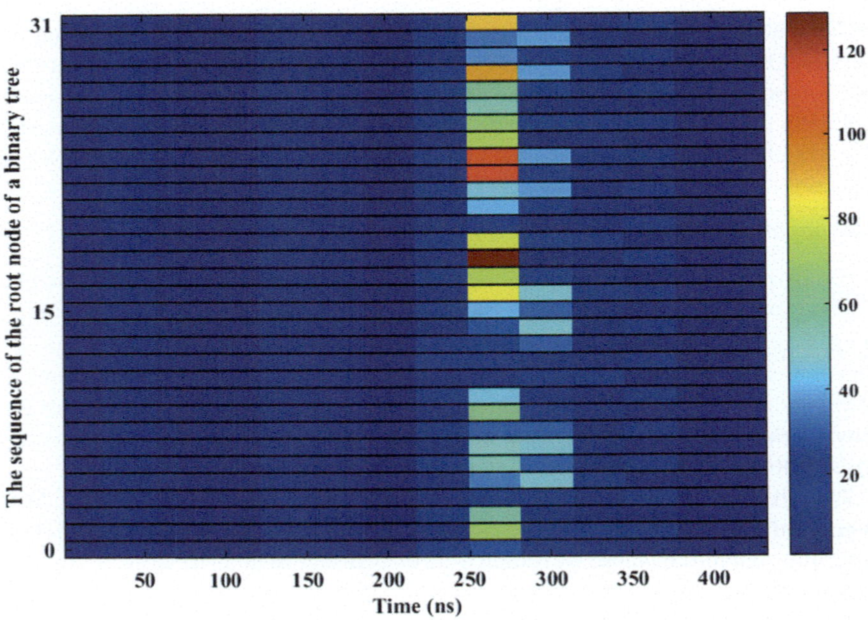

(b)

Fig. 14.2 **a**, **b**, and **c** are the colored coefficients images in LOS, SNLOS, and HNLOS propagation channels, respectively. The number of decomposition layers was set to 5

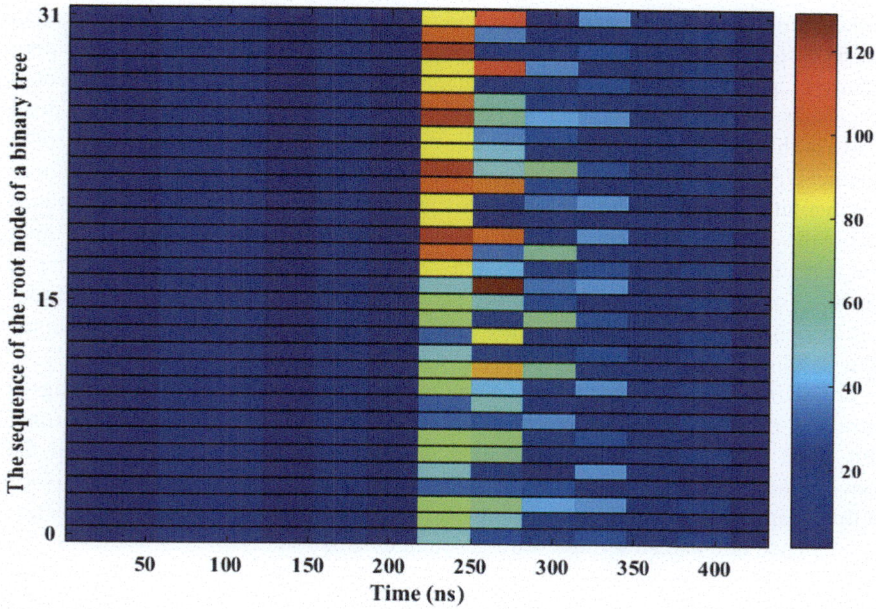

(c)

Fig. 14.2 (continued)

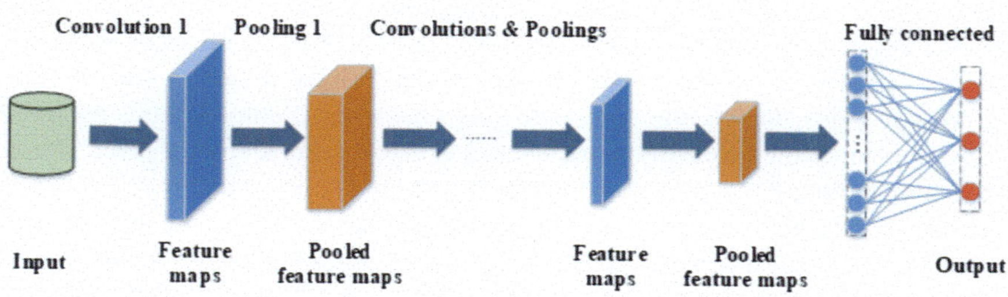

Fig. 14.3 The basic structure of a traditional CNN

of byte strings. These converted images are in RGB color mode with a pixel size of 128 × 128.

The next phase involves the convolution layer, which fundamentally entails a specialized matrix operation between input images and a convolution kernel. This operation computes by multiplying each weight (i.e., pixel value) in the convolution kernel f by the corresponding pixel value in the input image I, summing the results (considering only one channel in the calculation) as follows:

$$G(i,j) = (I * f)[i,j] = \sum_m \sum_n f(m,n) I(i-m, j-n) \tag{14.11}$$

In this layer, m and n represent the row and column indices of the convolution kernel matrix respectively, while i and j represent the row and column indices of the output feature map of the layer respectively. Three primary hyperparameters are essential here: filter size, strides, and padding. Convolutional kernels typically consist of pixel arrays with odd rows and columns. The filter size is determined by the number of input channels (3 in this chapter) and the number of output channels (equal to the number of convolutional kernels). Strides indicate the distance each convolution kernel slides during the operation. Larger strides theoretically lead to more significant loss of input image features. To mitigate feature loss after several convolution operations, we pad the edges of the feature map with empty pixels, known as padding. Padding includes two categories: "Same" (ensuring the size of the feature map remains unchanged after padding) and "Valid" (indicating no padding is applied).

Following the convolution operation, instead of directly applying pooling as shown in the basic CNN structure in Fig. 14.3, we introduce an activation function layer. This layer's primary function is to nonlinearly map the output of the convolutional layer, allowing the CNN model to fit complex functions effectively. In this chapter, the rectified linear unit (ReLU) function is employed due to its fast convergence rate and straightforward gradient calculation:

$$f\left(p_{i,j}\right) = max\left\{0, v_{i,j}\right\} \tag{14.12}$$

where $v_{i,j}$ is the pixel value of the feature map at position (i, j).

The subsequent pooling layer compresses the input feature images (termed feature downsampling) to simplify network computational complexity and extract key features. Pooling generally involves two types: max pooling and average pooling. Max pooling selects the maximum value within each local region, while average pooling computes the average value. In this chapter, we employ max pooling, which requires setting several hyperparameters akin to the convolutional layer. These include the pooling window size and the strides of the pooling window.

To enhance the CNN model's training speed and generalization capability, a batch normalization (BN) layer is incorporated. Unlike other normalization methods like local response normalization (LRN), BN normalizes each layer's input to maintain fixed mean and variance within a defined range. This normalization helps stabilize gradient descent, leading to improved training speed and generalization ability of the CNN model. A mini-batch $\mathbf{\Omega}$ of size m is denoted as $\mathbf{\Omega} = \left\{a_{1,}a_{2,\dots}a_m\right\}$, where $\{a_i\}$ denotes the input activation parameter. The essence of BN can be expressed as:

$$\begin{cases} \frac{1}{m}\sum_{i=1}^{m} a_i \rightarrow \mu_{\mathbf{\Omega}} \\ \frac{1}{m}\sum_{i=1}^{m} (a_i - \mu_{\mathbf{\Omega}})^2 \rightarrow {\sigma_{\mathbf{\Omega}}}^2 \\ \frac{a_i - \mu_{\mathbf{\Omega}}}{\sqrt{{\sigma_{\mathbf{\Omega}}}^2 + \varepsilon}} \rightarrow \hat{a}_i \\ BN_{\gamma,\beta}(a_i) \equiv \gamma\hat{a}_i + \beta \rightarrow y_i \end{cases} \tag{14.13}$$

where μ_{Ω} and ${\sigma_{\Omega}}^2$ are the mean and variance of $\boldsymbol{\Omega}$ respectively, $y_i(i = 1, 2, \ldots, m)$ is the output activation parameter, γ and β are the scale and shift parameters respectively, which are continuously updated during the normalization, and ε is a constant for numerical stability. More details about the BN can be seen in [10].

Then the last part of the CNN is the fully connected layer and the Softmax layer. The function of the fully connected layer is to map the obtained feature map to L real numbers on the range of $(+\infty, -\infty)$:

$$G = W^T x + b \tag{14.14}$$

where W^T is the weight matrix, x denotes the input of the fully connected layer, b is the bias, and G is the output column vector of the fully connected layer. The function of the Softmax layer is to map L real numbers on the range of $(+\infty, -\infty)$ to L real numbers within the range of (0, 1). Thus, the probability that a certain ODWPA-converted test image X belongs to category l is denoted as:

$$\begin{aligned} p(X, l) = softmax(g_l) = \frac{e^{g_l}}{\sum_L e^{g_l}}, \\ subject\ to \sum_L p(X, l) = 1 \end{aligned} \tag{14.15}$$

where g_l is the lth element in the column vector G and L is the total number of categories to be identified. Therefore, in the case of three channel categories, LOS, soft NLOS (SNLOS) and hard NLOS (HNLOS), the final classification result of the image can be denoted as:

$$P(X) = argmax\{p(X, l)\} = \begin{cases} 0, X \in SNLOS \\ 1, X \in HNLOS \\ 2, X \in LOS \end{cases} \tag{14.16}$$

In many instances, two channel categories—NLOS and LOS are commonly assumed for simplicity. Using the classification results mentioned above, we can determine the category of the signal propagation channel corresponding to the image, specifically the channel impulse response (CIR).

14.2.2 Experimental Analysis

Two experimental sites are chosen, representing typical indoor scenarios: a teaching building and an underground car park at China University of Mining and Technology. For each scenario, UWB devices are strategically positioned following guidelines from [11, 12]. Figure 14.4 illustrates the layout of UWB devices at these two experimental sites.

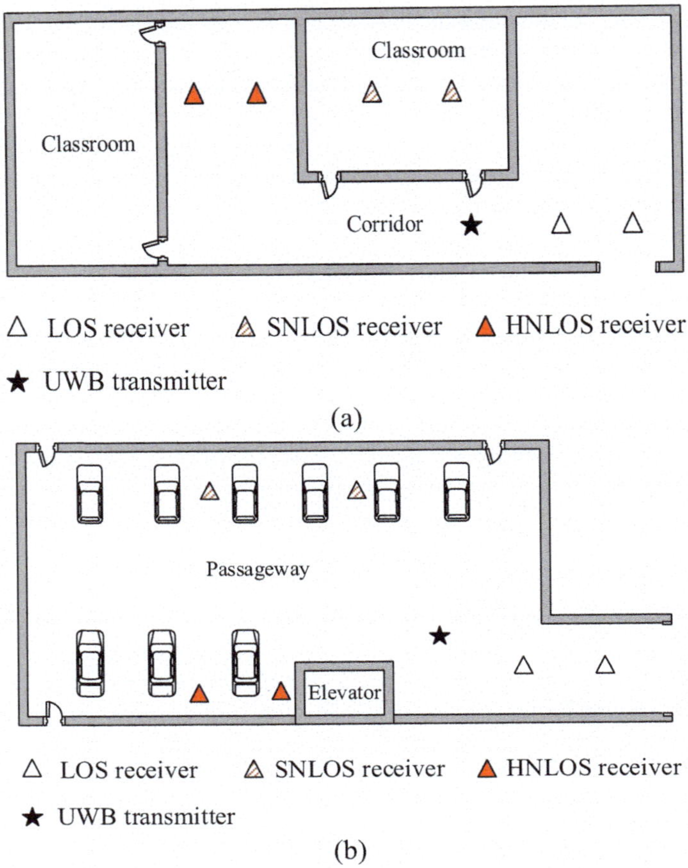

Fig. 14.4 **a** and **b** are the layouts of the experimental scenario #1 and scenario #2, respectively. The walls in both scenarios are concrete. The pentagram and triangle indicate the UWB signal transmitter and receiver respectively, and the triangles of the three styles (hollow, shaded and solid) respectively represent the UWB signal receivers in the three UWB signal propagation channels (LOS, SNLOS, and HNLOS)

Based on the previous section's description, we designed two self-built CNN models differing in complexity (including layer count, gradient descent rate, parameters per layer, and activation size). We then evaluated their performance, along with conventional CNN models, to select the optimal UWB signal propagation channel classifier. Tables 14.1 and 14.2 detail the structures and parameters of our two self-built CNNs (CNN_A and CNN_B) respectively. Further details on AlexNet, ResNet, and Inception_v3 can be found in [13, 14, 15] respectively. As shown in the tables, we transformed $h'(t)$ into 128×128 RGB energy images using ODWPA for CNN input.

Table 14.1 The structure and parameters of CNN_A

Layer	Hyper parameters	Activation shape	Activation size	Parameters of layer
INPUT		(128, 128, 3)	49, 152	0
CONV1	f = 3, s = 1, p = same, n = 64	(128, 128, 64)	1, 048, 576	640
POOL1	f = 3, s = 2, p = same	(64, 64, 64)	262, 144	0
CONV2	f = 3, s = 1, p = same, n = 16	(64, 64, 16)	65, 536	160
BN1	$\beta_0 = 0, \gamma_0 = 1, \varepsilon = 1e{-}3$	(64, 64, 16)	65, 536	0
POOL2	f = 3, s = 2, p = same	(32, 32, 16)	16, 384	0
FC3		(128, 1)	128	2, 097, 153
FC4		(128, 1)	128	16, 385
SM5		(3, 1)	3	385

Table 14.2 The structure and parameters of CNN_B

Layer	Hyper parameters	Activation shape	Activation size	Parameters of layer
INPUT		(128, 128, 3)	49, 152	0
CONV1	f = 3, s = 2, p = same, n = 32	(64, 64, 32)	131, 072	320
BN1	$\beta_0 = 0, \gamma_0 = 1, \varepsilon = 1e{-}3$	(64, 64, 32)	131, 072	0
POOL1	f = 3, s = 2, p = same	(32, 32, 32)	32, 768	0
FC2		(1024, 1)	1024	33, 554, 433
FC3		(256, 1)	256	262, 145
SM4		(3, 1)	3	769

*"f" denotes the size of the convolutional kernel, "s" denotes the strides, "p" denotes the padding, and "n" denotes the number of the convolutional, kernels. "β_0" and "γ_0" denote the initial values of scale and shift in batch normalization, respectively, and ε is a constant for numerical stability

During the training phase, 70% of the data collected from scenario #1 were allocated as the training dataset for the five CNNs. The learning rate regulates how quickly CNN parameters are updated during training, while batch size determines the number of samples processed per training iteration. Consistency between predicted classification labels and true labels is quantified using loss metrics. To ensure fair comparison, identical learning rates and batch sizes were set for all CNN models. The smaller the loss value, the closer the predicted label is to the real label. When it stabilizes around a certain value, it means that the training of the model has reached convergence. To monitor the loss variations during training of the five CNNs, we utilized Tensorboard for visualization, as depicted in Fig. 14.5. Clearly, the losses of the two self-built CNN models quickly approach zero and then stabilize, indicating rapid convergence. In comparison, AlexNet requires more epochs to converge. ResNet

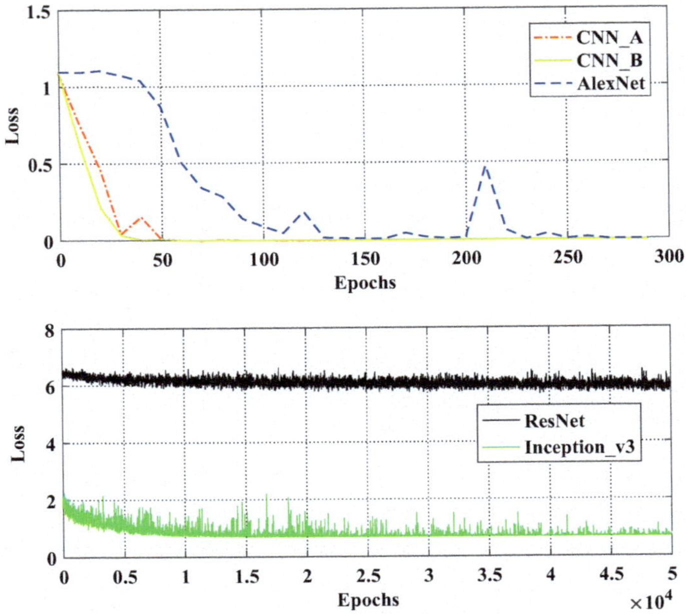

Fig. 14.5 The loss of five CNNs during training

and Inception_v3 exhibit prolonged training periods before reaching a stable state, with final steady-state losses around 5.9 and 0.7, respectively.

Next, the CNNs underwent testing. We saved the trained UWB signal propagation channel classifiers individually, and the remaining 30% of the dataset served as the test set to evaluate each classifier's performance. We assessed the classifiers' ability to identify three categories of UWB signal propagation channels using precision, recall, and F1-score metrics [16]. Table 14.3 compares the performance of CNN_A, CNN_B, AlexNet (at Epochs = 300), ResNet, and Inception_v3 (at Epochs = 50,000) after reaching convergence in training. Both self-built models demonstrated superior identification across all three propagation channels compared to the classical models. Notably, CNN_A achieved an average precision and recall of 100%. Furthermore, CNN_A exhibited 1.6, 17.4, and 2.8% higher average precision compared to AlexNet, ResNet, and Inception_v3, respectively. Additionally, performance across the three categories varied significantly, even within the same classifier. For instance, the identification recalls of AlexNet on HNLOS and LOS is 2.5% lower than that on SNLOS; the identification precision of ResNet on HNLOS is 8.4% lower than that on LOS; and the recall of Inception_v3 on LOS is 5.3% higher than that on HNLOS.

To clearly illustrate the identification performance of the five classifiers, Fig. 14.6 presents a box plot of their F1-scores. CNN_A demonstrates significantly superior identification performance across all three channel categories compared to the other classifiers. On average, CNN_A achieves F1-scores approximately 0.4%, 1.7,

Table 14.3 The precision and recall of five classifiers in scenario #1

	CNN_A		CNN_B		AlexNet		ResNet		Inception_v3	
	Precision	Recall	Precision	Recall	Precision	Recall	Precision	Recall	Precision	Recall
SNLOS	100.0	100.0	100.0	100.0	97.6	100.0	85.2	78.9	98.9	96.8
HNLOS	100.0	100.0	98.9	100.0	97.5	97.5	77.1	92.6	96.7	94.7
LOS	100.0	100.0	100.0	98.9	100.0	97.5	85.5	74.7	95.9	100.0
Mean	100.0	100.0	99.6	99.6	98.4	98.3	82.6	82.1	97.2	97.2

22.1, and 3.0% higher than CNN_B, AlexNet, ResNet, and Inception_v3, respectively. Despite its extensive training period (50,000 epochs), ResNet exhibits the poorest identification performance across the three channel categories. Additionally, we conducted experiments to verify if increasing the number of training samples impacts classifier performance. We collected an additional 3000 UWB signal waveforms under various signal propagation channels in scenario #1. However, the results show only marginal performance improvement, ranging from 0.1 to 0.6%. Therefore, we decided against adding more training samples in subsequent experiments. Table 14.4 details the time consumption of the five CNNs during training and testing phases. Notably, "training time consumption" refers to the duration from the start of training until convergence. CNN_A has the shortest total time consumption at 38.793 s. The self-built CNN models also exhibit shorter training times: 17.155 s for one and 8.947 s for the other, as observed in Fig. 14.5. Conversely, ResNet and Inception_v3 require longer training times to achieve convergence due to their larger number of convolutional kernels, neurons, and layers. While the training phase can be conducted offline, the time consumed during testing directly impacts the real-time performance of propagation channel identification for localization. Based on comprehensive evaluation and analysis, we selected the UWB signal propagation classifier based on CNN_A to proceed with the next phase of the experiment.

To further validate the superiority, robustness, and generalizability of the selected classifier, we combined UWB signal waveforms collected from scenario #1 and scenario #2 to create a new mixed scenario. This approach allows the classifier built in this mixed scenario to handle NLOS identification across multiple sub-scenarios, eliminating the need for repeated modeling efforts. The identification

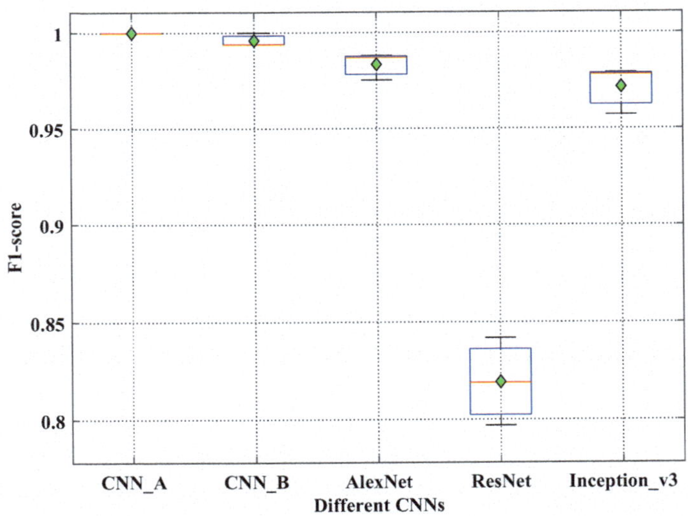

Fig. 14.6 The F1-score of five classifiers in scenario #1. The green diamonds indicate the mean, the red dashes indicate the median, and short black solid lines indicate the minimum or maximum

Table 14.4 The time of five CNNs for training and testing

Phase	CNN_A	CNN_B	AlexNet	ResNet	Inception_v3
Training	17.155 s	8.947 s	29.161 s	> 24 h	> 24 h
Testing	21.638 s	85.688 s	29.929 s	37.272 s	50.576 s
Total	38.793 s	94.635 s	59.090 s	> 24 h	> 24 h

performance of our proposed method across three experimental scenarios was also compared with several existing methods. In general, Support Vector Machine (SVM) excels in handling problems with a small number of samples, nonlinearity, and high-dimensional features, but it may lack model stability. K-Nearest Neighbors (KNN) offers higher accuracy but is computationally intensive. Random Forest (RF) can manage high-dimensional problems but may be prone to overfitting. It's important to note that the choice of feature sets used for training SVM, RF, and KNN can significantly impact their identification performance. We adopted specific feature sets based on previous research for each method to maintain consistency. For SVM, we utilized {maximum amplitude, mean excess delay} as described in [17]; for RF, {standard deviation, skewness, kurtosis} as detailed in [18]; and for KNN, {root-mean-square delay spread, mean excess delay, kurtosis, maximum amplitude, skewness, rise time} as outlined in [17, 18]. Additionally, to demonstrate the superiority of our proposed method, we evaluated and analyzed the performance of a CNN-based scheme without employing ODWPA. In this alternative approach, referred to as CNN_A', we utilized raw CIR data as input instead of the colored coefficients images processed by ODWPA. This strategy allows us to validate the effectiveness of our approach and compare it comprehensively against existing methods.

Table 14.5 presents the identification performance of six methods in scenario #1, highlighting precision and recall values above 90%. Both CNN-based methods (CNN_A and CNN_A') exhibit higher precision and recall compared to the other four methods. Importantly, our proposed method successfully identifies all three UWB signal propagation channels, a feat not achieved by previous approaches. SVM ("rbf"), SVM ("linear"), and RF exhibit poorer performance on certain UWB signal propagation channels, resulting in lower average precision and recall. For example, SVM ("linear") achieves only 62.3% precision for HNLOS, and RF achieves only 63.0% recall for HNLOS. This suggests these methods may struggle to effectively differentiate between the three UWB signal propagation channels using the chosen feature sets, leading to inadequate classifiers that lack sensitivity to specific propagation channel categories. Figure 14.7 displays the F1-score of the six methods in scenario #1, demonstrating that our proposed method achieves an average F1-score of 1.0, indicating significantly superior overall identification performance compared to the other methods.

Table 14.6 and Fig. 14.8 present the precision, recall, and corresponding F1-score of the six methods in scenario #2, an underground car park with complex conditions due to parked cars. In Table 14.6, RF achieves average precision and recall of 97.0 and 96.6% respectively in scenario #2, slightly lower than those of

Table 14.5 The Precision and Recall of Different Methods in Scenario #1

	CNN_A		CNN_A'		SVM ("rbf")		SVM ("linear")		RF		KNN	
	Precision	Recall	Precision	Recall	Precision	Recall	Precision	Recall	Precision	Recall	Precision	Recall
SNLOS	100.0	100.0	97.5	97.5	96.1	79.0	100.0	59.0	100.0	97.8	96.8	92.4
HNLOS	100.0	100.0	97.6	100.0	75.5	83.3	62.3	91.5	96.7	63.0	96.9	89.9
LOS	100.0	100.0	100.0	97.5	89.1	98.3	93..4	98.3	72.7	98.0	89.6	100.0
Mean	100.0	100.0	98.4	98.3	86.9	86.9	85.3	82.9	89.8	86.3	94.4	94.1

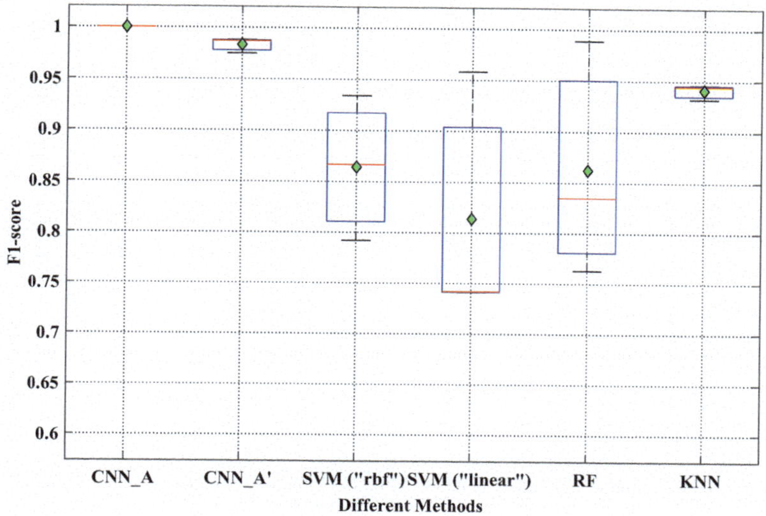

Fig. 14.7 The F1-score of different methods in scenario #1

the proposed method. However, overall, the average precision, recall, and F1-score of all methods in scenario #2 are lower compared to scenario #1, except for RF. Notably, the proposed method maintains the highest average precision and recall at 98.1%, while SVM ("rbf") and SVM ("linear") both achieve average precision and recall below 80%. Regarding the anomalous performance of RF in scenario #2, where it outperforms other methods, we speculate it may relate to the specific training feature set used ({standard deviation, skewness, kurtosis}). RF might be particularly adept at identifying UWB signal propagation channels in scenario #2 due to these features. Nonetheless, the proposed method, CNN_A', and RF exhibit significantly better overall identification performance compared to the other three methods, as depicted in Fig. 14.8.

Table 14.6 and Fig. 14.9 depict the precision, recall, and corresponding F1-score of the six methods in the mixed scenario, aimed at assessing the stability and adaptability of the classifier. In Table 14.6, only the proposed method achieves both average

Table 14.6 The Precision and Recall of Different Methods in Scenario #2

	CNN_A		CNN_A'		SVM ("rbf")		SVM ("linear")		RF		KNN	
	Pre	Rec	Pre	Rec	Pre	Rec	Pre	Rec	Pre	Rec	Pre	Rec
SNLOS	95.7	95.7	100.0	87.5	54.7	77.4	50.6	75.5	100.0	98.0	70.1	100.0
HNLOS	100.0	98.8	82.6	95.0	82.6	76.3	89.3	84.7	94.3	98.0	100.0	100.0
LOS	98.8	100.0	94.9	92.5	87.6	78.0	90.8	75.4	96.8	93.8	100.0	60.8
Mean	98.1	98.1	92.5	91.7	75.0	77.2	76.9	78.5	97.0	96.6	90.0	86.9

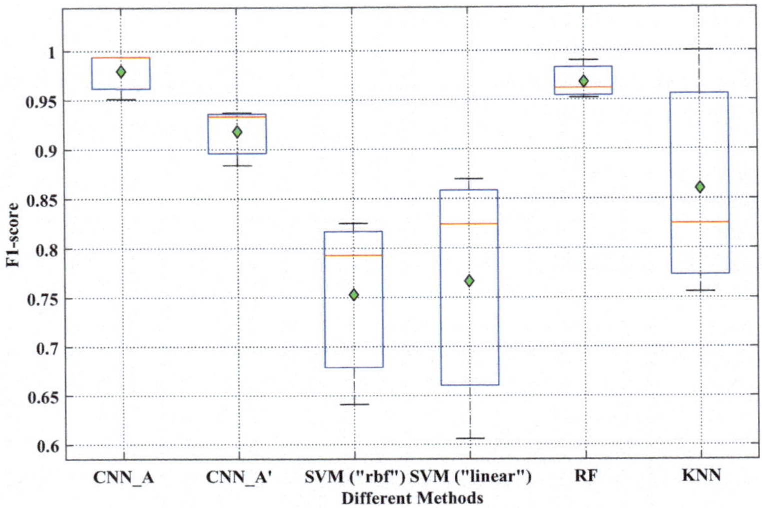

Fig. 14.8 The F1-score of different methods in scenario #2

precision and recall above 90%. SVM ("linear") exhibits the lowest average precision and recall at 69.6 and 64.1%, respectively. Figure 14.9 illustrates that the proposed method consistently achieves a significantly higher average F1-score across the three UWB signal propagation channels compared to the other methods. This highlights the robustness and effectiveness of the proposed method in handling mixed scenarios and its superior performance in accurately identifying UWB signal propagation channels.

Figure 14.10 illustrates the identification accuracy of various methods across three scenarios, evaluated using accuracy as the metric, which measures the ratio of correctly classified samples to the total number of samples. Both CNN-based methods (CNN_A and CNN_A') consistently achieve the highest accuracy across all scenarios, except for CNN_A' being slightly lower than RF in scenario #2, highlighting the effectiveness of CNN and ODWPA. Overall, the accuracy of nearly all methods declines from scenario #1 to the mixed scenario, with SVM ("linear") experiencing the largest decrease of approximately 36%. In contrast, the proposed method maintains an overall accuracy consistently above 90%, with a decrease of about 7% across scenarios. This underscores the robustness and autonomous learning capability of the proposed method compared to other approaches. It's important to note that each method identifies a single sample within milliseconds, making them suitable for real-time identification and localization tasks. However, the proposed method has only been tested in two individual scenarios and a simple hybrid scenario. For more challenging indoor environments (e.g., large shopping malls) and complex hybrid scenarios involving multiple distinct scenarios, a more sophisticated NLOS identification scheme may be necessary.

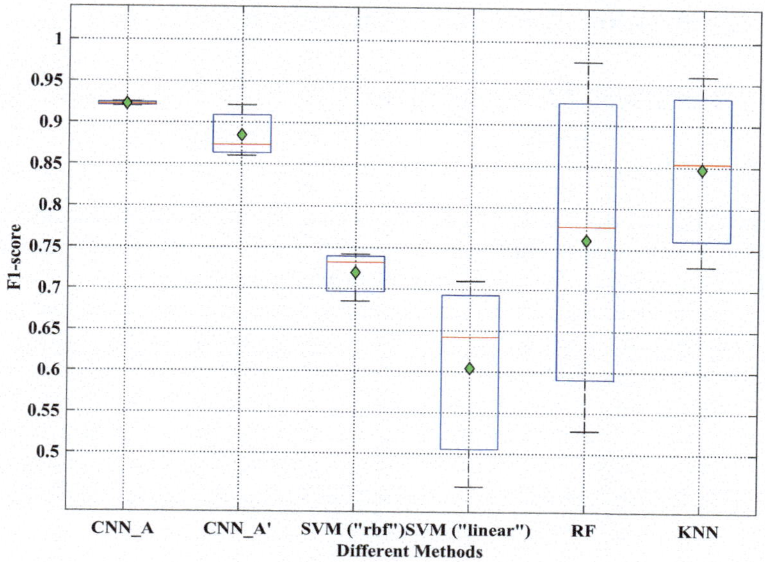

Fig. 14.9 The F1-score of different methods in scenario mixed

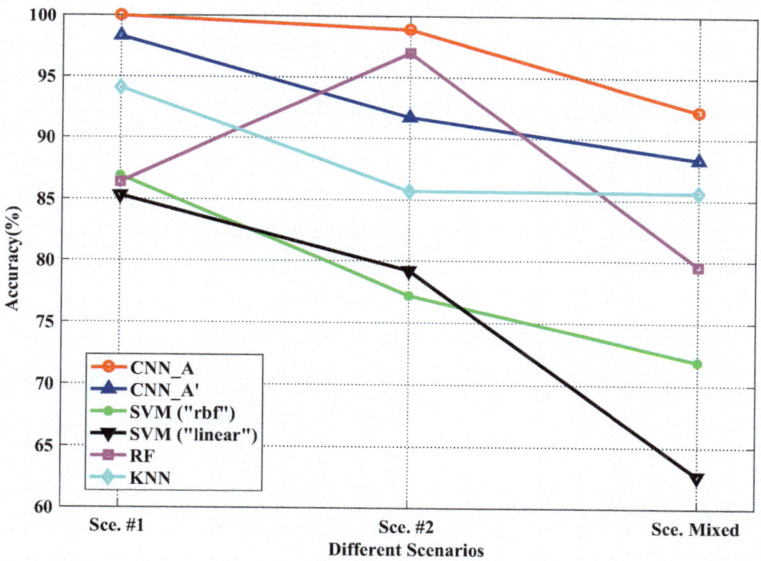

Fig. 14.10 The overall accuracy of different methods in different scenarios

14.3 Localization Based on Distance Error Mitigation

14.3.1 Modeling of Ranging Error

Method Description

Since the ranging errors in LOS and NLOS scenarios have different sources and properties, in order to make the ranging values closer to the real values, we consider modeling the ranging errors in these two environments separately.

Newton's method, gradient descent method and least squares method are usually used for curve fitting. In particular, the least squares method is widely used because it has the advantages of low complexity, less resource consumption, and globally optimal solution, so we consider using the least squares method for curve fitting. Assuming that there is a set of sample data points to be fitted $\{(x_i, y_i), i = 1,2, \ldots, n\}$, and the fitting function is $y = \varphi(x)$; then the sum of the squares of the deviations of this curve from each data point is minimized, and the mathematical expression is as follows:

$$min \sum_{i=1}^{n} \delta_i^2 = min \sum_{i=1}^{n} (\varphi(x_i) - y_i)^2 \tag{14.17}$$

where x_i is the ranging value in the LOS/NLOS environment, and y_i is the absolute or relative error corresponding to x_i. In addition, when choosing a function model, the distribution of the data to be fitted should be analyzed first, and then the interpretability and rationality of the function model should also be considered. The common nonlinear functions for fitting include polynomial function, Gaussian function and exponential function. Based on the data characteristics and previous research results, high-order polynomial and exponential function are used to model the relative error of ranging in LOS/NLOS environments, respectively.

The high-order polynomial function model is described as:

$$y = a_0 + a_1 x + \cdots + a_k x^k \tag{14.18}$$

where $\{a_i\}$ are the unknown polynomial coefficients to be determined. By substituting the n data points $\{(x_i, y_i), i = 1,2, \ldots, n\}$ into Eq. (14.18) generates:

$$Y = XA \tag{14.19}$$

where $Y = [y_1, \ldots, y_k]^T$, $X = \begin{bmatrix} 1x_1 \ldots {x_1}^k \\ 1x_2 \ldots {x_2}^k \\ \ldots \\ 1x_n \ldots {x_n}^k \end{bmatrix}$ and $A = [a_0, a_1, \ldots, a_k]^T$ is the coefficient vector. The least squares solution to (14.19) is given by:

$$A = \left(X^T X\right)^{-1} X^T Y \tag{14.20}$$

The exponential function model is expressed as:

$$y = ae^{bx} \tag{14.21}$$

where $\{a, b\}$ are the two model parameters to be determined. Taking logarithms on both sides of Eq. (14.21) yields:

$$\overline{y} = bx + C \tag{14.22}$$

where $\overline{y} = \ln y$ and $C = \ln a$. Substituting $\{(x_i, y_i), i = 1,2, \ldots, n\}$ into Eq. (14.21) yields a linear equation of compact form:

$$L = KB \tag{14.23}$$

where

$L = \left[\hat{y}_1, \hat{y}_2, \ldots, \hat{y}_n\right]^T, K = \begin{bmatrix} 1x_1 \\ 1x_2 \\ \cdots \\ 1x_n \end{bmatrix}, B = [c, b]^T$. The least squares solution to (14.23) is given by

$$B = \left(K^T K\right)^{-1} K^T L \tag{14.24}$$

In addition, regarding the degree of goodness of fitting, we consider using the coefficient of determination (R-square) as an evaluation index. A larger R-square indicates better fitting performance of the model. R-square is defined as follows:

$$R\_square = \frac{SSR}{SST} = \frac{SST - SSE}{SST} = 1 - \frac{SSE}{SST} \tag{14.25}$$

where SSE is the sum of squares of the difference between the fitted value $\hat{y}_i = \varphi(x_i)$ and the corresponding original value y_i. SST is the sum of squares of the difference between the original value and its mean. That is:

$$SSE = \sum_{i=1}^{n} \left(\hat{y_i} - y_i\right)^2 \tag{14.26}$$

$$SST = \sum_{i=1}^{n} \left(y_i - \frac{1}{n}\sum_{i=1}^{n} y_i\right)^2 \tag{14.27}$$

Experimental Analysis

This experiment is chosen to be carried out in the underground car park of the fifth teaching building in Nanhu Campus of China University of Mining and Technology (CUMT). The experimental equipment is Time Domain P450 module.

In both LOS and NLOS environments (obstacles with wooden boards, iron sheets or pedestrians), ranging data were collected at 28 locations separated by 50 cm. At each location, collection was repeated 500 times, to form a group of data. Among the 28 groups of data, 24 groups were used for modeling and 4 groups were used for model validation. The roughness of each group of ranging data was firstly eliminated, and then its root mean square value was computed and the corresponding absolute and relative errors were calculated.

First, the ranging error in the LOS environment is modeled and validated for analysis. According to the results, the error model based on the third-order polynomial function in the LOS environment can be established, as shown in Fig. 14.11.

The polynomial function can fit the trend of the relative error curve better. By observing this relative error curve and according to the characteristics of the exponential function, the transformed exponential function model is considered:

$$y = ae^{bx} + ce^{dx} \tag{14.28}$$

The fitting principle is similar to those applied to Eqs. (14.20) and (14.21). Thus, the functional expressions for the two models are thus obtained as:

$$\varphi_1(x) = -1.062e - 08 * x^3 + 3.445e - 05 * x^2 - 0.03547 * x + 10.79 \tag{14.29}$$

$$\varphi_2(x) = 55.3 * \exp(-0.00348 * x) + (1.107e + 10) * \exp(-0.4214 * x) \tag{14.30}$$

The evaluation metrics of model fitting errors are shown in Table 14.7.

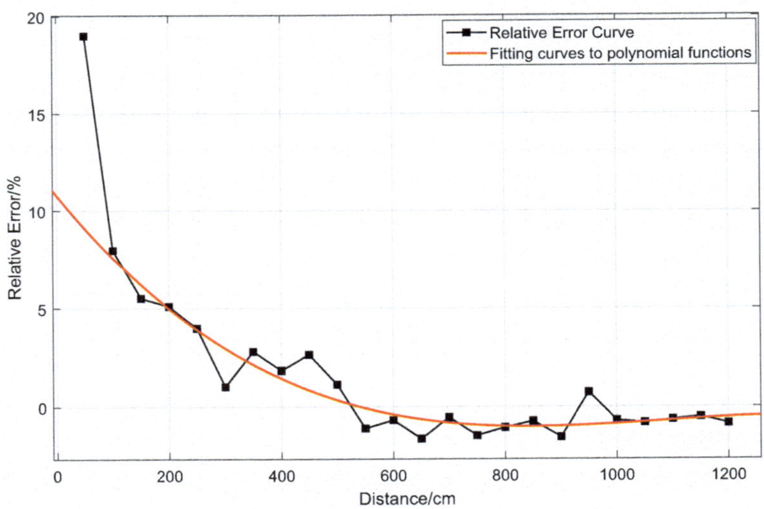

Fig. 14.11 Error model based on third-order polynomial function in LOS

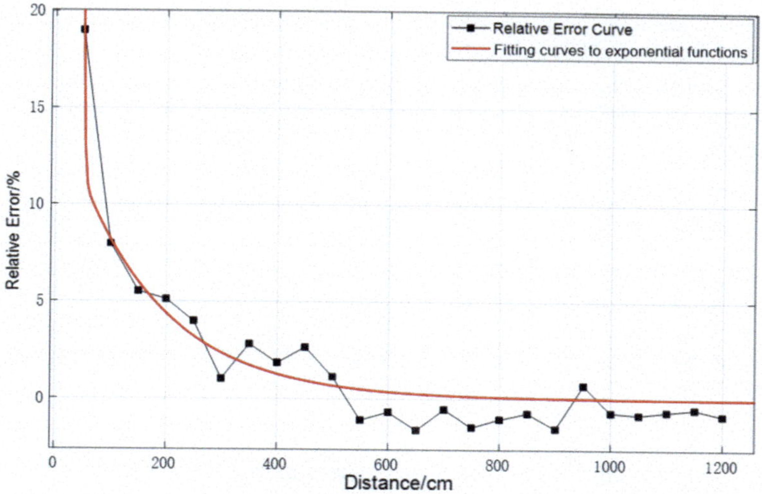

Fig. 14.12 Error model based on exponential function in LOS

Table 14.7 Evaluation metrics of the model for fitting the ranging error in LOS

Error model	$\varphi_1(x)$	$\varphi_2(x)$
SSE	35.2	45.79
R-square	0.9245	0.9147

It can be seen from Table 14.7 that the polynomial function-based model $\varphi_1(x)$ has SSE and higher R-square than the exponential function based model $\varphi_2(x)$, so the former is considered as the ranging error compensation model in the LOS environment. Next, the $\varphi_1(x)$ model is validated using the remaining ranging data, and the validation results are shown in Table 14.8.

From Table 14.8, after the relative error estimation by the $\varphi_1(x)$ model and then the correction of the ranging values, the ranging accuracy is significantly improved, by 38% on average. Thus, the polynomial function-based error model $\varphi_1(x)$ can significantly improve the ranging accuracy in the LOS environment.

Table 14.8 The validation results of model $\varphi_1(x)$

The real value/cm	Ranging value/cm	Ranging error/cm	Relative error estimate $\varphi_1(x_i)$/%	Corrected ranging value/cm	Corrected ranging error/cm
1250	1237.0	−13.0	−0.462	1244.2	−5.8
1300	1291.5	−8.5	−0.433	1294.4	−5.6
1350	1337.4	−12.6	−0.439	1344.1	−5.9
1400	1392.4	−7.6	−0.487	1393.2	−6.8

Similarly, based on the third-order polynomial function as well as the exponential function, the error models in the NLOS environment are established respectively. It is observed that the error model based on exponential function achieves better performance with SSE of 81.18 and R-square of 0.7451. The mathematical expression of the model is:

$$f_2(x) = 32.6 * \exp(-0.01638 * x) + 0.675 * \exp(-0.0001901 * x) \tag{14.31}$$

In summary, the polynomial function-based error model $\varphi_1(x)$ and the exponential function-based error model $f_2(x)$ are selected for error compensation of ranging in LOS and NLOS environments, respectively. They will be used in the localization experiments to improve the localization accuracy as discussed in the next subsection.

14.3.2 Improved Chan-Kalman Localization Algorithm

Method Description

Due to the good characteristics of Kalman filter in solving the localization and tracking problems, this chapter integr ates Chan algorithm, a traditional UWB localization algorithm, with Kalman filter. Before proceeding further, the original observation information needs to be processed to reduce ranging error, in order to improve the accuracy of localization. The proposed algorithm mainly consists of three stages.

(1) Stage I: Acquire the original ranging information and then utilize the UWB signal propagation classifier for NLOS identification.
(2) Stage II: According to the recognition results in (1), the proposed error model is utilized to compensate the errors of LOS/NLOS ranging values respectively, so as to obtain the corrected ranging information.
(3) Stage III: Based on the corrected ranging information in (2), the Chan algorithm is used to obtain the initial position estimate of the target tag, and then the Kalman filter is used to obtain the final position estimate of the target tag. Stages I and II have been elaborated and experimentally analyzed earlier, and the specific implementation details of Chan-Kalman localization algorithm in stage III will be presented next. Assuming that the position of the target tag determined by Chan algorithm at the moment k is $(x_{t,k}, y_{t,k})$, the position of the ith (i = 1, 2,…,m) base station is (x_i, y_i), and the estimated position of the target tag at the moment k is $(\hat{x}_{t,k}, \hat{y}_{t,k})$, i.e., the state vector at the moment k is $X(k) = [\hat{x}_{t,k}, \hat{y}_{t,k}]^T$. Then, according to Kalman filter, the state equation of the system is:

$$X_k = A_0 X_{k-1} + W_{k-1} \tag{14.32}$$

where $A_0 = [1, 0; 0, 1]$ is the state transfer matrix, the system control matrix is 0, and W_{k-1} is the process noise vector whose covariance matrix is a diagonal matrix denoted by Q. The measurement equation of the system is:

$$Z_k = H_0 X_k + V_k \tag{14.33}$$

where Z_k is the tag position vector $[x_{t,k}, y_{t,k}]^T$ determined by the Chan algorithm at the kth moment, and V_k is the measurement noise vector whose covariance matrix is also a diagonal matrix denoted by R. The observation matrix is $H_0 =$ [1,0;0,1]. The implementation of the Kalman filter is realized as follows:

$$\hat{X}_k^- = A_0 X_{k-1} \tag{14.34}$$

$$P_k^- = A_0 P_{k-1} A_0^T + Q \tag{14.35}$$

$$K_k = P_k^- H_0^T \left(H_0 P_k^- H_0^T + R\right)^{-1} \tag{14.36}$$

$$\hat{X}_k = X_k^- + K_k (Z_k - H_0)\hat{X}_k^- \tag{14.37}$$

$$P_k = (I - K_k H_0) P_k^- \tag{14.38}$$

With the above three stages, the estimated position of the tag can finally be made as accurate and reliable as possible.

Experimental Analysis

The site used in this localization experiment is LOS/NLOS hybrid environment. The mobile tag is Tag_{104}, the positions of four base stations BS_{100}, BS_{101}, BS_{102} and BS_{103} are (0, 0), (1500 cm, 0), (1500 cm, 700 cm), (0, 700 cm), and rectangle ABCD is the preset trajectory of the mobile tag, in which the positions of A, B, C, D are (100 cm, 100 cm), (1100 cm, 100 cm), (1100 cm, 600 cm), and (100 cm, 600 cm), respectively, as shown in Fig. 14.13.

First, static localization experiments are conducted. The mobile tag Tag_{104} is placed on four known points A, B, C, and D successively, and 1000 times of location solving is performed on each point. Therefore, the RMSE of Chan algorithm, Chan-Kalman algorithm and the proposed algorithm (improved Chan-Kalman algorithm based on NLOS recognition) for location solving at different points can be obtained, as shown in Table 14.9.

From the above experimental results, regardless of which point the mobile tag is at, the localization accuracy of Chan-Kalman algorithm is always higher than Chan algorithm, with an average improvement of about 10.7%. Among them, when the mobile tag is at point B, Chan-Kalman algorithm improves the localization accuracy

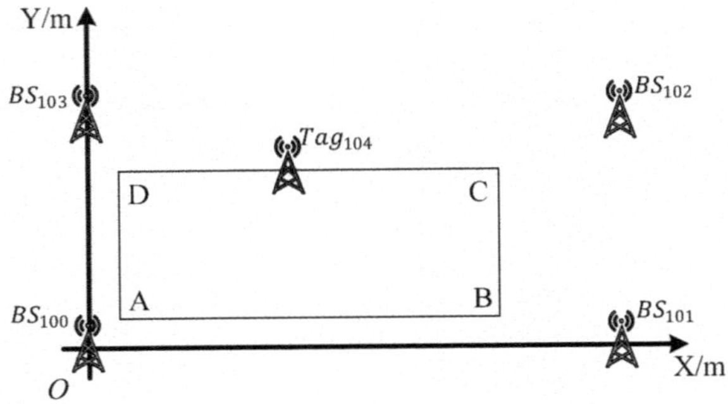

Fig. 14.13 Deployment of base stations and preset trajectory of mobile tag

Table 14.9 RMSE (cm) of different algorithms for position solving at four points A, B, C and D

	A	B	C	D
Chan algorithm	30.1	9.4	14.9	28.9
Chan-Kalman algorithm	28.2	6.9	13.7	28.4
The proposed algorithm	18.1	4.8	5.7	21.9

of Chan algorithm by about 26.6% at most. And since both Chan algorithm and Chan-Kalman algorithm utilize the raw ranging information to complete the localization solution, this proves the gain effect of Kalman filter in the localization algorithm. The average localization accuracy of the proposed algorithm at the four points is 12.6 cm, which is about 32.8% better than the other two algorithms on average, indicating that the proposed algorithm receives a good gain by utilizing the ranging information after the error compensation.

Next, the dynamic localization experiment is carried out. The mobile tag Tag_{104} is moved slowly along the preset trajectory ABCDA. Therefore, the tag trajectory obtained by Chan algorithm, Chan-Kalman algorithm and the proposed algorithm can be obtained, as shown in Fig. 14.14.

As can be seen from the figure, the trajectory estimated by the proposed algorithm is the closest to the real one, and the estimated trajectory is very smooth, almost unaffected by the ranging error. The average error in the X-coordinate and Y-coordinate is about 10.7 and 8.9 cm, respectively, which are much lower than the other two algorithms.

14.4 Conclusions

Aiming at the problem that UWB indoor localization technology is greatly affected by NLOS propagation in complex environments, this chapter carries out a detailed study on UWB non-line-of-sight propagation identification and localization. First, an NLOS identification method based on ODWPA and CNN is proposed for improving

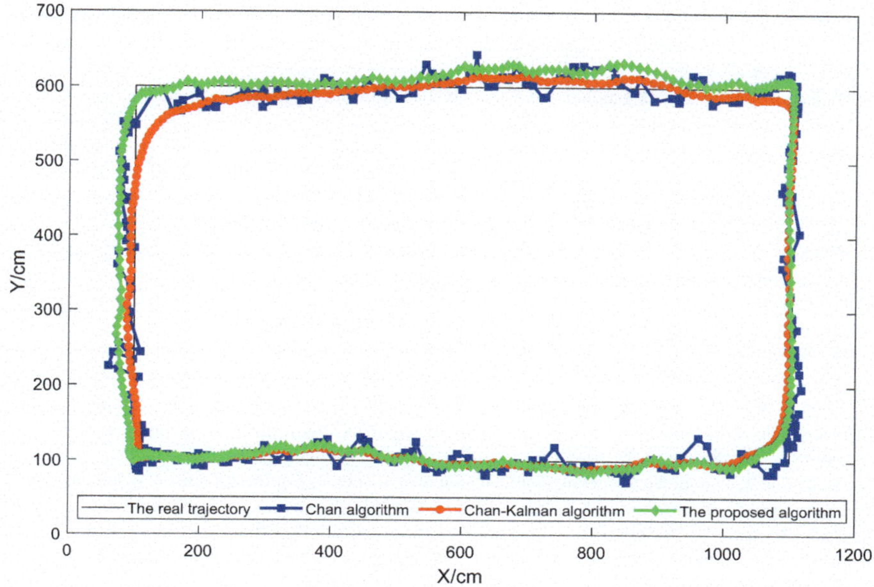

Fig. 14.14 Trajectories estimated by different algorithms

the LOS/NLOS identification accuracy in complex indoor environments. Secondly, the error models in LOS and NLOS environments are established for suppressing UWB ranging errors, respectively. Finally, an improved Chan-Kalman localization algorithm based on NLOS identification is proposed to improve the localization accuracy in static and dynamic scenarios.

References

1. Shao H, Zhang X, Wang Z, Wang Z (2007) An efficient closed-form algorithm for aoa based node localization using auxiliary variables. IEEE Wirel Commun 14(4):90–96
2. Alameda-Pineda X, Horaud R (2014) A geometric approach to sound source localization from time-delay estimates. IEEE/ACM Trans Audio, Speech, Lang Process 22(6):1082–1095
3. Gezici S (2008) A survey on wireless position estimation. Wireless Pers Commun 44:263–282
4. Foy WH (1976) Position-location solutions by Taylor-series estimation. IEEE Trans Aerosp Electron Syst 2:187–194
5. Fang BT (1990) Simple solutions for hyperbolic and related position fixes. IEEE Trans Aerosp Electron Syst 26(5):748–753
6. Chan YT, Ho KC (1994) A simple and efficient estimator for hyperbolic location. IEEE Trans Signal Process 42(8):1905–1915
7. Cui Z, Gao Y, Hu J, Tian S, Cheng J (2020) LOS/NLOS identification for indoor UWB positioning based on Morlet wavelet transform and convolutional neural networks. IEEE Commun Lett 25(3):879–882

8. Wang J, Yu K, Bu J, Lin Y, Han S (2022) Multi-classification of UWB signal propagation channel based on one-dimensional wavelet packet analysis and CNN. IEEE Trans Veh Technol 71(8):8534–8547
9. Baussard A, Nicolier F, Truchetet F (2004) Rational multiresolution analysis and fast wavelet transform: application to wavelet shrinkage denoising. Signal Process 84(10):1735–1747
10. Ioffe S, Szegedy C (2015) Batch normalization: Accelerating deep network training by reducing internal covariate shift. In International conference on machine learning. Pmlr, pp 448–456
11. Venkatesh S, Buehrer RM (2007) Non-line-of-sight identification in ultra-wideband systems based on received signal statistics. IET Microwaves Antennas Propag 1(6):1120–1130
12. Barral V, Escudero CJ, García-Naya JA, Maneiro-Catoira R (2019) NLOS identification and mitigation using low-cost UWB devices. Sensors 19(16):3464
13. Krizhevsky A, Sutskever I, Hinton GE (2012) Imagenet classification with deep convolutional neural networks. Adv Neural Inf Process Syst 60(6):84–90
14. He K, Zhang X, Ren S, Sun J (2016) Deep residual learning for image recognition. In Proceedings of the IEEE conference on computer vision and pattern recognition, pp 770–778
15. Szegedy C, Vanhoucke V, Ioffe S, Shlens J, Wojna Z (2016) Rethinking the inception architecture for computer vision. In Proceedings of the IEEE conference on computer vision and pattern recognition, pp 2818–2826
16. Cui Z, Liu T, Tian S, Xu R, Cheng J (2020) Non-line-of-sight identification for UWB positioning using capsule networks. IEEE Commun Lett 24(10):2187–2190
17. Yu K, Wen K, Li Y, Zhang S, Zhan K (2018) A novel NLOS mitigation algorithm for UWB localization in harsh indoor environments. IEEE Trans Veh Technol 68(1):686–699
18. Ramadan M, Sark V, Gutierrez J, Grass E (2018) NLOS identification for indoor localization using random forest algorithm. In: WSA 2018; 22nd international ITG workshop on smart antennas. VDE, pp 1–5

The manufacturer's authorised representative in the EU is Springer
Nature Customer Service Centre GmbH, Europaplatz 3, 69115 Heidelberg,
Germany. If you have any concerns regarding our products, please
contact ProductSafety@springernature.com

Printed and bound by CPI Group (UK) Ltd, Croydon, CR0 4YY
30/09/2026
02225253-0013